AF851148

Frequency Synthesis

Theory, Design & Applications

Frequency Synthesis

Theory, Design & Applications

Věnceslav F. Kroupa, Ph.D.
of the Institute of Radio Engineering and Electronics
Czechoslovak Academy of Sciences

UNIV. COLL. N.W.
BANGOR
LIBRARY

GRIFFIN 1820 LONDON

734122

Charles Griffin & Company Limited
42 Drury Lane, London WC2B 5RX

Copyright © 1973

All rights reserved

No part of this publication may be reproduced, stored in a retrieval system, or transmitted, in any form or by any means, electronic, mechanical, photocopying, recording or otherwise, without the prior permission of Charles Griffin & Company Limited.

First published 1973

234 × 156 mm, xxi + 295 pages
183 illustrations
ISBN 0 85264 190 7

Set by Wm Clowes & Sons Ltd, London, Colchester and Beccles
Printed in Great Britain by Butler & Tanner Ltd, Frome

Preface

The present book is the first systematic introduction to the new and rapidly developing field of frequency synthesis, which deals with methods of generating any arbitrary frequency from a given standard frequency.

In the last 15 years I have been confronted with many frequency-synthesis problems. The need to replace the empirical methods, generally used, led finally to the development of a general theory built on sound mathematical concepts.

In the arrangement of the material of this book three groups of readers have been considered: students specializing in communications and electronic measuring engineering; systems designers; and the equipment users who operate frequency synthesizers and produce technical specifications of requirements.

The book has been written to be useful to practising electronic engineers and many of the practical frequency synthesis problems can be understood without a specialist knowledge of theory. However, the book also covers the rapidly developing mathematical theory of frequency synthesis and for this some familiarity with number theory, Fourier and Laplace transforms, feedback network theory, and noise and spectral analysis is necessary.

The book contains much entirely new material, but it must of course draw on the results of other research workers throughout the world. References to the sources of such information are made in the text and I here wish to thank all concerned for permission to reproduce illustrations and text as appropriate.

My particular thanks are due to the Institute of Radio Engineering and Electronics of the Czechoslovak Academy of Sciences, and especially to the Institute's director, Dr. V. Zima, for his generous support of the present work.

October 1972 Věnceslav F. Kroupa

DEDICATED TO MY PARENTS
ON THEIR GOLDEN WEDDING

Contents

6 PHASELOCK LOOP

GLOSSARY

I Symbols

Capitals in general, are used to designate peak values, amplitudes of sinusoidal quantities, complex amplitudes in Fourier series expansions, and Fourier and Laplace transforms, whereas lower-case letters, generally, indicate instantaneous values or the time domain representation. An asterisk designates complex conjugate quantities; a bar above a symbol indicates the mean value.

A	amplitude; operational amplifier gain
a.c.	alternating current
ADM	additive decimal module
AGC	automatic gain control
A_k	numerator of the kth continued fraction approximation
a_m	amplitude modulation index
AM	amplitude modulation; amplitude modulated
APC	automatic phase control
$A(t)$	amplitude, slowly varying function of time
AVC	automatic (volume) control
$B_1 < B < B_2$	normalized pass band of the mixer output filter
BCD	binary coded decimal
BDC	binary decimal code
B_i	input phase-noise bandwidth
B_k	denominator of kth continued fraction approximation
B_L	'Loop noise bandwidth' of PLL
BPF, BPF 1	band pass filter
BW	bandwidth
C	capacity
CCIR	Committée consultative international de radiodiffusion (International Radio Consultative Committee)
C_d	minority-charge capacity
CRT	cathode ray tube
$C(v)$, C_v	voltage dependent capacity
d.c.	direct current
D_1, D_2	diodes
E, e	voltage
E_B, E_0	biasing voltage
$e_{d,b} = [\overline{e^2_{d,b}(t)}]^{\frac{1}{2}}$	effective beat-note voltage

F	PLL loop filter
f	frequency
FET	field-effect transistor
f_c	cut-off frequency of the low pass filter
f_i	input frequency; intermodulation frequency
FM	frequency modulation
F_m	mixer output filter (PLL feedback filter)
$F_M(s)$, $\mid F_M(\omega) \mid$	modulation transfer function of the IF filter; attenuation of the IF filter
$f_{p1} < f_p < f_{p2}$	output-filter passband
FR.DET.	frequency detector
$F(s)$	transfer function of the PLL loop filter
f_s	standard frequency; synthesizer output frequency
f_1, f_2	mixer input frequencies ($f_1 < f_2$)
G, g	conductance
G_d	equivalent forward-bias diode conductance
HF	high frequency
HG	harmonic generator
$H(s), H'(s)$	transfer function of a PLL
I, i	current
IC	integrated circuit
$\mathrm{I}_{c,0}, \mathrm{I}_{c,00}$	d.c. component of the collector current
I_{DC}	d.c. component of the diode current
I_{ES}	emitter saturation current
IF	intermediate frequency
IM, IM'	power level of intermodulation signals
$I_m(y), \bar{I}_m(y)$	modified Bessel functions
i_p	peak current
I_s	diode reverse current
j	imaginary operator $\sqrt{-1}$
$J_m(z)$	Bessel functions of the first kind
K	detuning function; gain of the PLL
K'	reduced or enhanced gain of the PLL
K_d	phase detector gain
$K_0, K_0(v)$	VCO gain
K_v	d.c. gain in the PLL
L	inductance; conversion loss; division factor
LC	inductance capacitance circuit
LF	low frequency
L.O.	local oscillator
lg	Napierian logarithm
LPF	low pass filter
m_a	amplitude modulation index

m_p, m_{p0}	phase modulation index; output phase modulation index
MOSFET	metal-oxide-surface field-effect transistor
M.S., m.s.	mean square
N	multiplication or division factor; the overall multiplication factor in a frequency synthesizer; order of the intermodulation frequency.
n	constant in switching mixers; multiplication or division factor; arbitrary integer
$n_c(t)$	noise voltage (the cosine component)
N_i	input spectral density
$n_s(t)$	noise voltage (the sine component)
$n(t)$	noise voltage
P	power
PD, P.DET	phase detector
PLL	phase locked loop
PLDM	phase locked decimal module
P_N	noise power
P_{NA}	power of the amplitude noise
P_{norm}	power coefficient in varactor frequency multipliers
P_{NP}	power of the phase noise
Q	figure of merit of resonant circuit; figure of merit of varactor (Q = reactance/resistance)
Q, q	charge
q	electron charge; damping constant of the differentiated rectangular waveform; mixing ratio, $f_2/f_1, f_2 > f_1$
q_d	total charge in the bulk diode material; lower bound of the mixer output frequency normalized to the lower of the input frequencies
q_h	upper bound of the mixer output frequency normalized to the lower of the input frequencies
R	resistance; residue of the continued fraction approximation
r	the rth multiple of the lower mixer input frequency, f_1
R_b, r_b	equivalent reverse-bias diode resistance
$r_{bb'}$	lumped base resistance
RC	resistance-capacitance network
R_d, r_f	equivalent forward-bias diode resistance
RF	radio frequency
R.M.S., r.m.s.	root mean square
R_r, R_s	remainder of a series
$R_s, R_s(v)$	bulk resistance of the semiconductor diode; signal generator resistance
$R_\varphi(\tau), R_{\dot\varphi}(\tau)$	autocorrelation of the random phase, frequency departure
R_0	series resistance of the LC circuit at resonance

S, S_1, S_2	pulse area
s	Laplace transform operator; product of the coupling coefficient and circuit Q; the sth multiple of the higher mixer input frequency, f_2
$\frac{S}{N}$, SNR	signal-to-noise ratio
$(SNR)_L$	'loop signal-to-noise ratio'
$SN_\varphi R$	signal-to-phase-noise ratio
$(SN_\varphi R)_{SSB}$	single sideband $SN_\varphi R$
SPD	sampling phase detector
SRD	step-recovery diode
SSB	single-side-band
S/S_pSB	signal-to-spurious-side-band ratio
$s(t), S(v)$	elastance, time function; voltage function
$S_\varphi(\omega), S_\varphi(f)$	two-sided spectral density of phase
$S'_\varphi(f)$	one-sided spectral density of phase
$S_{\varphi_{an}}(\omega)$	spectral density of auxiliary-oscillator phase-noise
$S_{\varphi_{in}}(\omega)$	input phase noise spectral density
$S_{\varphi_s}(\omega)$	additive phase noise spectral density
$S_{\varphi_{xn}}(\omega)$	output phase noise spectral density
T	absolute temperature; period duration; time delay; time constant of the PLL filter
t	time, variable
T_{L1}, T_{L2}	lock-in time in the first order, second order, loop
T_p	pull-in time
T_1, T_2	time constants of the PLL filter; transistors
u	integer-numerator of the rational mixing ratio q
UHF	ultra high frequency
$u_i(t)$	interfering signal
u_m	integer
$u(t)$	input waveform; periodically discontinuous function
V, v	voltage
v	integer-denominator of the rational mixing ratio q
V_B	biasing voltage; junction breakdown voltage
V_C	amplitude of the switching signal
V_{CC}	d.c. collector supply voltage
$V_{CE\ sat}$	collector-emitter saturation voltage
VCO	voltage controlled oscillator
VCXO	voltage controlled crystal oscillator
VHF	very high frequency
$v(i)$	voltage, current function
VLF	very low frequency
V_P	biasing voltage in varactor frequency multipliers

V_0	biasing voltage
$V_0, v_0(t)$	VCO output voltage
$v_2, V_2(s)$	VCO control voltage; – – –, Laplace transformed
$v_{1,m}, v_{2,m}$	bounds of $V_{I,m}$
$v_{3,m}, v_{4,m}$	bounds of $V_{II,m}$
$V_{I,m}, V_{II,m}$	intervals on rational number line for which frequency synthesis equation has solution
x_{ai}	normalized frequency of the ith auxiliary oscillator
X_0	normalized frequency of the auxiliary branch in frequency synthesizers
$x_0, x_0^{(i)}$	normalized frequency of the transporting carrier; normalized frequency of the transporting carrier in the ith decimal module
x_0', x_0''	components of the transporting carrier
x_{0i}	integers ($i = 0, 1, \ldots, 9$)
X_1	integer, numerator of the normalized frequency, f_x/f_s
$X_1, X_2, \ldots$	products of small and large prime numbers
$X_1', X_2', \ldots$	products of large prime numbers, only
$x_1, x_2, \ldots$	products of small prime numbers only
$y(t), Y(s)$	dependent variable, time function: – –, Laplace transformed
Y_1	integer, denominator of the normalized frequency, f_x/f_s
$y_1, y_2, \ldots y_r \ldots$	partial division factors
$Y_2, Y_3, \ldots$	partial division factors
Z	impedance; matching impedance in mixers
$z_1, z_2, \ldots z_k$	product of small prime numbers
α	diode constant ($\alpha = n\lambda$)
α_F	forward short-circuit current gain (common base connexion)
α_n	normalized amplitude (A_n/A_1)
α_R	reverse short-circuit current gain (common base connexion)
α_T	temperature coefficient of the resonant circuit
$\alpha(t)$	normalized amplitude
α_v	voltage coefficient of the resonant circuit
α_+, α_-	normalized amplitude
β_F, β_R	forward, reverse, short-circuit current gain (common emitter connexion)
γ	doping profile constant
γ_n	phase angle of the nth harmonic component
$\gamma(v)$	voltage-dependent doping profile characteristic
Δ	designates small but finite change, amplitude of a disturbance, etc.
δ	damping factor
$\Delta f_{P\max}, \Delta f_{P\min}$	maximum, minimum, pull-in frequency

Δf_0	peak value of the frequency instability of the free-running VCO; frequency offset introduced in the multiple-period technique
$\Delta\omega$	difference between input (reference) frequency and the free-running frequency of the VCO
$\Delta\omega_H$, $\Delta\omega'_H$	hold-in range
$\Delta\omega_L$, $\Delta\omega'_L$	lock-in range
$\Delta\omega_P$, $\Delta\omega'_P$	pull-in range
$\Delta\omega_{PO}$	pull-out frequency
$\Delta\omega_0$	peak value of the frequency instability of the free-running VCO
$\in$	symbol designating a member of a set
ϵ	dielectric constant; arbitrarily small quantity
ζ	PLL damping factor
η	efficiency
Θ, Θ_H	current flow angle, – – –, at half pulse amplitude
λ	diode and transistor characteristic constant
ν	frequency (modulating –, etc.); off-resonance detuning
π	3·14159
Π	symbolic representation of multiple multiplication
Σ	symbolic representation of summation
$\sigma^2[\Delta_\tau\varphi]$	variance of accumulated phase, etc.
τ	differential time delay; time interval
τ_d	minority carrier lifetime
τ_H	pulse duration at half pulse amplitude
τ_1, τ_2	pulse duration; propagation delays
Φ	contact potential
φ	phase angle
φ_a	'acceleration error' or 'dynamic tracking error'
φ_a, φ_b	phase shifts in a regenerative frequency divider
$\varphi_{in}(t)$	input phase noise
φ_n, $\Phi_n(t)$	phase shift of the nth sinusoidal component
$\varphi_{ni}(t)$, $\varphi_{no}(t)$	input, output phase jitter (noise)
$\varphi_s(t)$	system phase noise
$\dot{\varphi}(t) = \frac{d\varphi(t)}{dt}$	frequency
$\langle\varphi\rangle_{t,\tau}$	average phase in the time interval τ, etc.
φ_v	'static phase error' or 'velocity error'
$\varphi_{xn}(t)$	output phase noise
$\Phi_{0,sp}(s)$	disturbance of the PLL output phase, Laplace transformed, caused by the spurious voltages $V_{sp}(s)$ added to v_d
$\Phi_0(\Omega)$	amplitude of the disturbance of the PLL output phase caused by the spurious voltage V_{sp} added to the input voltage $v_i(t)$

ψ, ψ_ν	phase angle; – – of the signal with frequency ν
Ω	a low frequency (2πHz); difference of two higher frequencies
ω	frequency (2πHz)
$\omega_{\alpha F}$	forward current gain cut-off frequency (common base connexion)
ω_B	3-dB bandwidth of LC resonant circuit
ω_c	varactor cut-off frequency; frequency of the free-running VCO
ω_n	natural frequency of PLL; the nth frequency in a multi-frequency signal
ω_r	series resonance frequency of a quartz resonator; reference frequency
ω_0	resonance frequency of LC circuit; PLL output frequency

II Inferiors

b	base
c	collector
d	detector
e	emitter; error; equivalent
g, G	generator
H, HF	high frequency; – – limit
i, in	input; ith decimal module; ith loop, etc.
if	intermediate frequency
k	summation index; kth triggering
L	load; amplitude of a limited waveform
L, LF	low frequency; – – limit
LO	local oscillator
m	summation index; amplitude of the input sinusoidal waveform; mth integer
n	summation index; nth harmonic component; nth approximation
nm	side-band in transistor mixer
0, out	output
p, q, r, s	summation index
r	reference
rs	intermodulation side-band
S	signal
s	stray
sp	spurious
x, $x0$	output
φ	phase
$\dot{\varphi}$	frequency

1 Generation of Harmonic Frequencies

Circuit elements—passive or active—with nonlinear voltage-current characteristics may generate signals the frequencies of which are exact multiples (sometimes sub-multiples) of the fundamental input frequency. In this chapter we shall recall from the theory of nonlinear circuits only principles and results that are necessary for the understanding of frequency synthesis problems, in general, and of harmonic series generation, in particular.

1.1 GENERAL NONLINEAR ELEMENT

We begin with the inspection of a simple nonlinear element connected in a closed circuit as in Fig. 1.1. If v_g is the input voltage and $v(i)$ the voltage drop on the nonlinear element then the circulating current is given by

$$i = \frac{v_g - v(i)}{Z_G + Z_L} \tag{1.1}$$

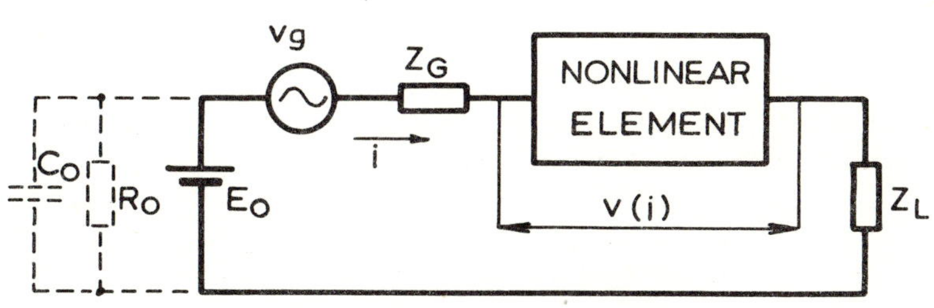

Fig. 1.1 Circuit with a nonlinear element.

In the small-signal case, $v(i)$ may be approximated by Taylor series expansion:

$$v(i) = (v)_{i_0} + \left(\frac{\partial v}{\partial i}\right)_{i_0} (i - i_0) + \frac{1}{2!}\left(\frac{\partial^2 v}{\partial i^2}\right)_{i_0} (i - i_0)^2 \cdots \tag{1.2}$$

Without any loss of generality we may put $i_0 = 0$.

$$v(i) = v_0 + \sum_{n=1}^{\infty} \frac{1}{n!}\frac{\partial^n v}{\partial i^n} i^n \tag{1.3}$$

Assuming further $v_g = V_{g1} \cos \omega t$ and

$$V_{g1} \gg |v(i)| \tag{1.4}$$

i.e. the current-driven case, then the current, i, must have a large component $i_1 \cos (\omega t + \varphi)$ and

$$i \approx i_1 \cos(\omega t + \varphi) = \frac{i_1}{2}\{\exp[j(\omega t + \varphi)] + \exp[-j(\omega t + \varphi)]\} \tag{1.5}$$

After substituting in (1.3) we get

$$v(i) = v_0 + \sum_{n=1}^{\infty} \frac{1}{n!}\left(\frac{i_1}{2}\right)^n \frac{\partial^n v}{\partial i^n} \{\exp[j(\omega t + \varphi)] + \exp[-j(\omega t + \varphi)]\}^n \tag{1.6}$$

Applying the Binomial Theorem and arranging equation (1.6) according to the multiples of exp $(jk\omega t)$ we have

$$v(i) = \sum_{k=-\infty}^{\infty} v'_k \exp(jk\omega t) \tag{1.7}$$

It may easily be proved that v'_{-k} and v'_k are complex conjugates of each other, which means that the term on the right-hand side of eq. (1.7) is merely the complex Fourier expansion of $v(i)$.

Substituting in (1.1) reveals that the current i must also have components of higher harmonic frequencies and consequently that its expansion into a similar complex Fourier series is justified.

$$i = \sum_{k=-\infty}^{\infty} i'_k \exp(jk\omega t)^{\dagger} \tag{1.8}$$

To arrive at the complete solution, the current given by (1.8) must be substituted in (1.3) instead of the first-order approximation used.

---

† As stated, in equations (1.7) and (1.8) $v'_{-k} = v_k'^*$ and $i'_{-k} = i_k'^*$ because the actual voltages and currents are real physical quantities. Further, it should be remembered that if one of the variables were to be put in the trigonometric form $v = v_n \cos(n\omega t + \lambda)$, then

$$|v_n| = |v'_{-n}| + |v'_n| = 2|v'_n|.$$

Employing the theory of the general intermodulation product, detailed in Appendix 4.1, amplitudes of the individual harmonics may be found. Generally the process is prohibitively complicated, but there are some simplifications which enable an approximate solution to be obtained in almost all cases of practical interest.

1.2 IDEAL SWITCH

In this case the voltage $v(i)$ on the nonlinear element is either zero, or equal to $v_g(t)$ if both impedances Z_G and Z_L have no reactive components.

For the input voltage

$$v_g(t) = E_m \cos \omega t - E_0 \tag{1.9}$$

the switching begins when

$$E_0 = E_m \cos \frac{\theta}{2}; \tag{1.10}$$

thus the current flows only in the interval $(-\theta/2, +\theta/2)$ (Fig. 1.2). The respective Fourier expansion is

$$i(t) = \frac{1}{\pi} \frac{i_p}{1 - \cos(\theta/2)} \left[\sin\frac{\theta}{2} - \frac{\theta}{2}\cos\frac{\theta}{2} + \left(\frac{\theta}{2} - \sin\frac{\theta}{2}\cos\frac{\theta}{2}\right)\cos\omega t + \sum_{n=2}^{\infty} \frac{2}{n^2-1}\left(\frac{1}{n}\sin\frac{n\theta}{2}\cos\frac{\theta}{2} - \cos\frac{n\theta}{2}\sin\frac{\theta}{2}\right)\cos n\omega t \right] \tag{1.11}$$

where the peak current i_p is given by

$$i_p = i_m\left(1 - \cos\frac{\theta}{2}\right) = \frac{E_m}{R_G + R_L}\left(1 - \cos\frac{\theta}{2}\right) \tag{1.12}$$

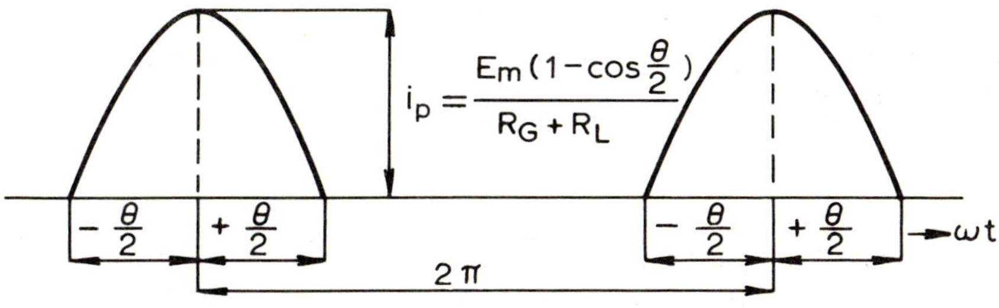

Fig. 1.2 Sine-wave pulses with current flow angle θ.

The relative amplitudes of the first seven harmonic components in the current $i(t)$ of eq. (1.11) are shown as functions of the current flow angle θ in Fig. 1.3(a) and 1.3(b).

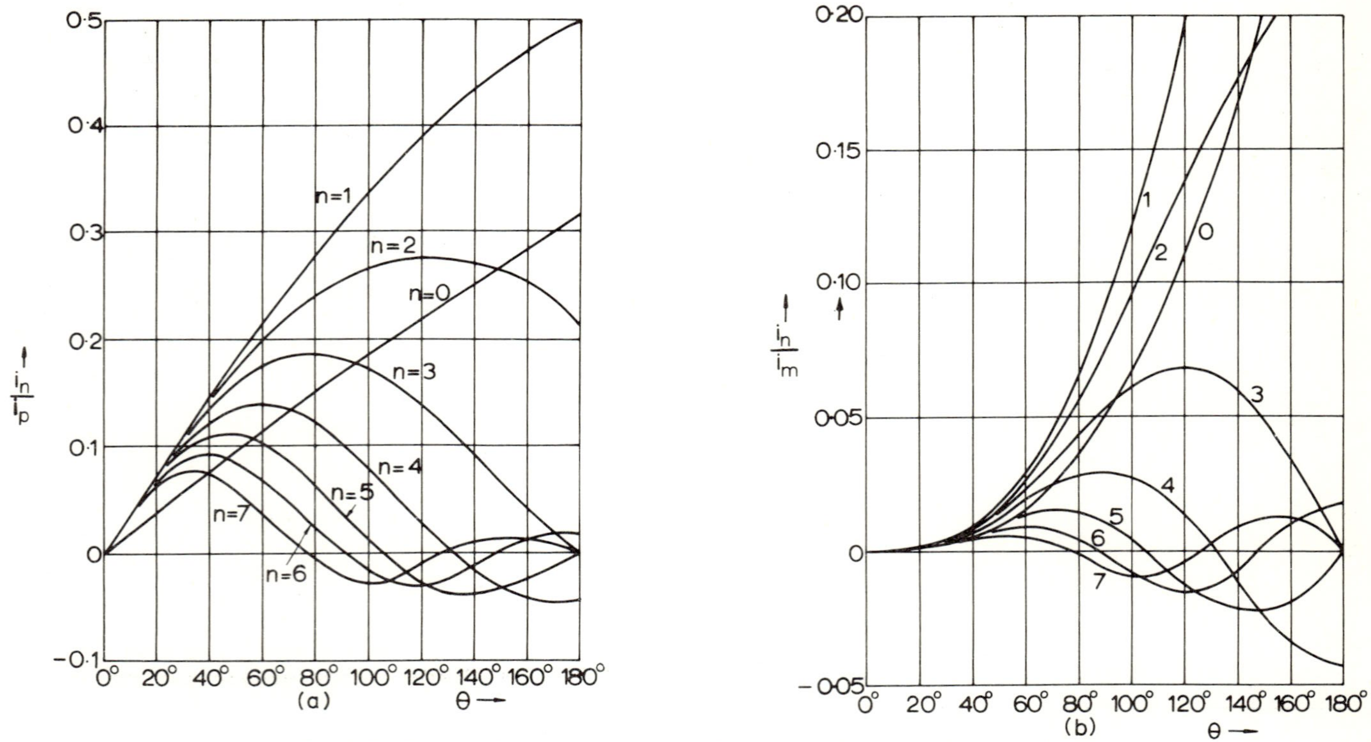

Fig. 1.3 Harmonic components of a sine-wave pulse train generated by the 'ideal-switch' action: (a) harmonic amplitudes i_n normalized to the peak current $i_p = E_m[1 - \cos(\theta/2)]/(R_G + R_L)$ vs. current flow angle θ; (b) harmonic amplitudes i_n normalized to the input amplitude $i_m = E_m/(R_G + R_L)$ vs. current flow angle θ.

Using eq. (1.11) the harmonic content of several important waveforms may be readily found.

(A) Half-Wave Rectified Sinusoid

$$\frac{\theta}{2}=\frac{\pi}{2}; \quad i_p=i_m=\frac{E_m}{R_G+R_L};$$

$$i(t)=\frac{2}{\pi}\,i_m\left[\frac{1}{2}+\frac{\pi}{4}\cos\omega t+\frac{1}{3}\cos 2\omega t-\frac{1}{15}\cos 4\omega t+\frac{1}{35}\cos 6\omega t\mp\ldots\right] \tag{1.13}$$

(B) Push-Push Connexion of Two Ideal Switches

produces the waveform shown in Fig. 1.4 which is merely addition of another Fourier expansion (1.11) delayed by π.

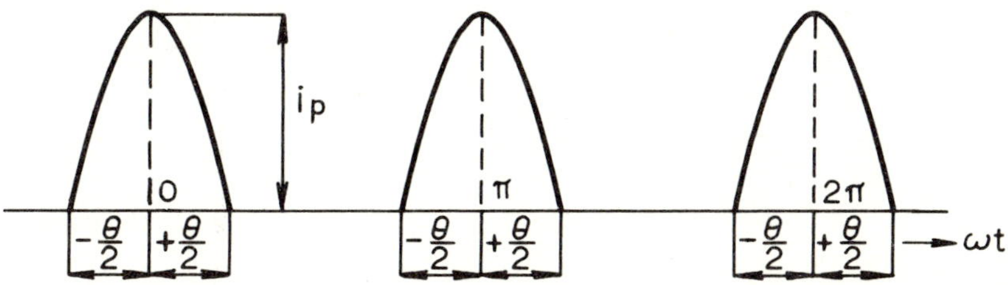

Fig. 1.4 The push-push addition of sine-wave pulses with current flow angle θ.

Through the interference the odd harmonics are cancelled and the even harmonics doubled:

$$i(t)=\frac{2}{\pi}\frac{i_p}{1-\cos(\theta/2)}\left[\sin\frac{\theta}{2}-\frac{\theta}{2}\cos\frac{\theta}{2}\right.$$
$$\left.+\sum_{n=2,4}^{\infty}\frac{2}{n^2-1}\left(\frac{1}{n}\sin\frac{n\theta}{2}\cos\frac{\theta}{2}-\cos\frac{n\theta}{2}\sin\frac{\theta}{2}\right)\cos n\omega t\right] \tag{1.14}$$

(C) Full-Wave Rectified Sinusoid

$$\frac{\theta}{2}=\frac{\pi}{2}, \quad i_p=i_m$$

$$i(t)=\frac{4}{\pi}\,i_m\left[\frac{1}{2}+\frac{1}{3}\cos 2\omega t-\frac{1}{15}\cos 4\omega t+\frac{1}{35}\cos 6\omega t\mp\cdots\right] \tag{1.15}$$

(D) Push-Pull Connexion of Two Ideal Switches

produces the waveform shown in Fig. 1.5. In contrast to **(B)** subtraction of the π-delayed Fourier expansions must be performed.

$$i(t) = \frac{2}{\pi}\frac{i_p}{1-\cos(\theta/2)}\left[\left(\frac{\theta}{2} - \sin\frac{\theta}{2}\cos\frac{\theta}{2}\right)\cos\omega t + \sum_{n=3,5}^{\infty}\frac{2}{n^2-1}\left(\frac{1}{n}\sin\frac{n\theta}{2}\cos\frac{\theta}{2} - \cos\frac{n\theta}{2}\sin\frac{\theta}{2}\right)\cos n\omega t\right] \tag{1.16}$$

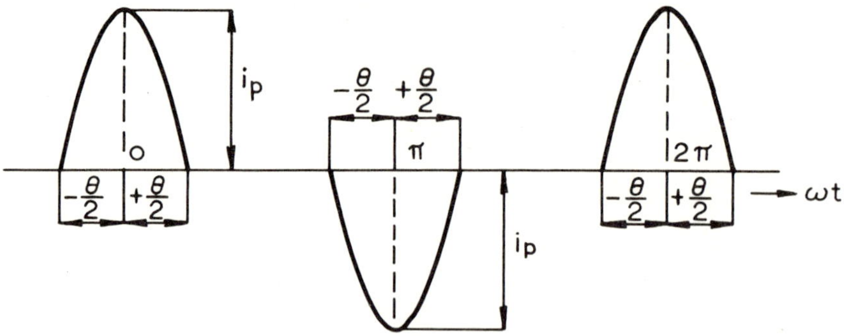

Fig. 1.5 The push-pull addition of sine-wave pulses with current flow angle θ.

1.3 SEMICONDUCTOR DIODES

The diffusion and the diode theories [1.1] predict (see Section 2.2A) a current-voltage characteristic of the form

$$i = A[\exp(\alpha v) - 1] \tag{1.17}$$

where v is the voltage across the rectifying barrier and A and α are constants. From the theory

$$\alpha_{\text{theor}} = \lambda = \frac{q}{kT} \tag{1.18}$$

q is the electron charge = $1{\cdot}6 \times 10^{-19}$ (coulomb)
k is the Boltzmann's constant = $1{\cdot}38 \times 10^{-23}$ (joule/°K)
T is absolute temperature (°K);

at room temperature (25°C)

$$\lambda \approx 40 \quad (\text{volts})^{-1} \tag{1.19}$$

Only modern epitaxial Schottky barrier diodes have $i - v$ characteristics exponential over a range of several decades of current with the slope approaching λ. But generally $\alpha = n\lambda$ where n lies between 1 and 2. It can be seen from Fig. 1.6

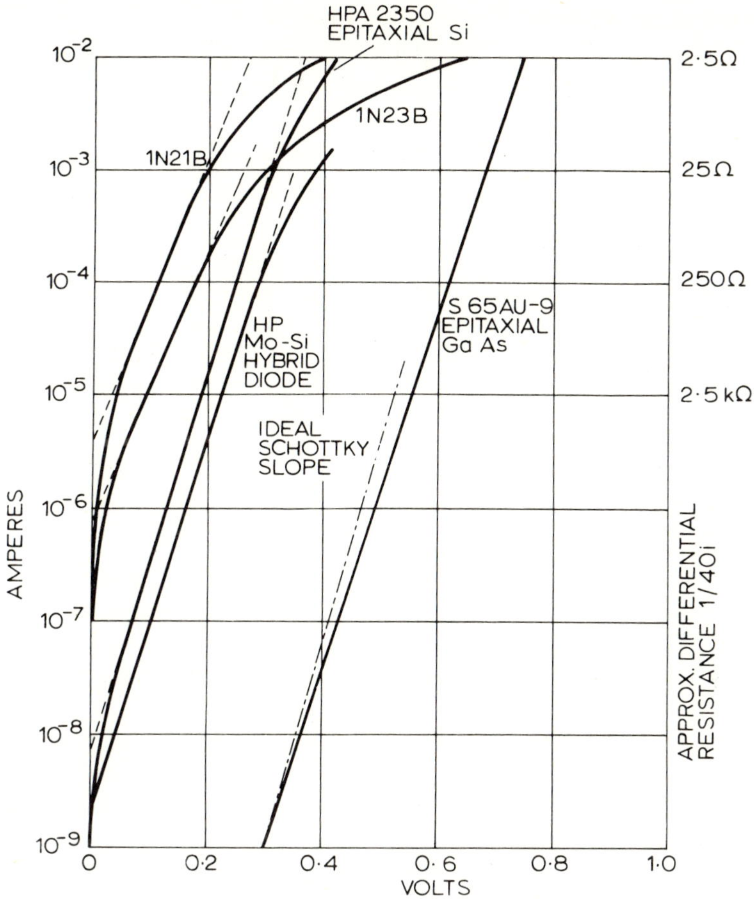

Fig. 1.6 Voltage–current characteristics of different semiconductor diodes. (From Barber [4.3].)

that (log i) follows a straight line (of slope α but smaller than the theoretical 40 volts^{-1}) up to 0·2 or 0·3 volt, and then curves. Subtracting the voltage drop on the *bulk resistance* R_s of the semiconductor the curve may be made to continue with the slope α to somewhat higher voltages.

$$i = A\{\exp[\alpha(v - iR_s)] - 1\} \qquad (1.20)$$

However, there is another difficulty because the constants α and A are not valid for the negative part of the diode characteristic. A partial remedy is the addition of a small parallel conductance $1/R_B$ which accounts for the finite slope of the reverse characteristic.

The advantage of eq. (1.20) is that it gives a very close approximation to the actual diode characteristic and that the compact analytical form may easily be expanded into power series, see (1.2) and (1.3), as well as into Fourier series. According to Appendix 1.1, for $v = E_0 + E_m \cos \omega t$ we get

$$i = A\{\exp[\alpha(E_0 + E_m \cos \omega t)] - 1\} = A\{\exp(\alpha E_0)[I_0(\alpha E_m)$$

$$+ 2I_1(\alpha E_m)\cos \omega t + 2I_2(\alpha E_m)\cos 2\omega t + \cdots] - 1\} \tag{1.21}$$

where $I_0(\alpha E_m)$, $I_1(\alpha E_m)$, etc. are modified Bessel functions of the argument (αE_m).

The important feature of (1.21) is that all harmonic frequencies with steadily decaying amplitudes are present. But the useful range of the expansion (1.21) is limited only to those cases where the positive peak voltage $E_0 + E_m$ is below 0·3 V and the sum of source and load resistances is small compared with the equivalent diode resistance R_d defined by (2.11).

For larger voltages, or resistance $R_G + R_L$, the use of eq. (1.20) would be necessary.

The application of the Taylor expansion leads to the power series,

$$i = a_0 + a_1 v + a_2 v^2 + \cdots \tag{1.22}$$

The coefficients a_k have been evaluated, for example, by Mills [1.2].

When only the lower-order harmonics are of interest the piecewise linearization of the diode characteristic (see Section 2.2 and Fig. 2.3) gives sufficiently accurate results. Thus the actual working conditions are simplified to those of an ideal switch (1.11) in series with $R_G + R_L + r_d$,† which is particularly justified for large values of $(R_G + R_L)$. The advantage is that the approximate solution may be readily found in diagrams (see Fig. 1.3(a) and (b)).

A better approximation of the conducting part of the diode characteristic gives the power-law approximation. The advantage is the simple analytical expression which enables the amplitudes of harmonic currents i_n to be expressed in particularly compact form [1.3]. The relative amplitudes i_n/i_p ($n = 0, 1, \ldots 7$) for the square-law characteristic are drawn as functions of the current flow angle θ in Fig. 1.7.

1.4 LIMITERS

Earlier, in Section 1.2, we showed that by the switching action the voltage on the nonlinear element (of the 'ideal switch' type) is a one-sided clipped-sinusoid. A symmetrically clipped or limited sinusoid is created by connecting another ideal switch with opposite biasing polarity, in parallel, see the simple limiter in Fig. 1.8.

The harmonic content of the limited sinusoidal voltage (see Fig. 1.9) is found when subtracting from the fundamental voltage, $E_m \cos \omega t$, the clipped peaks using eq. (1.11)

---

† According to Fig. 2.3, $r_d = R_s + R_d$.

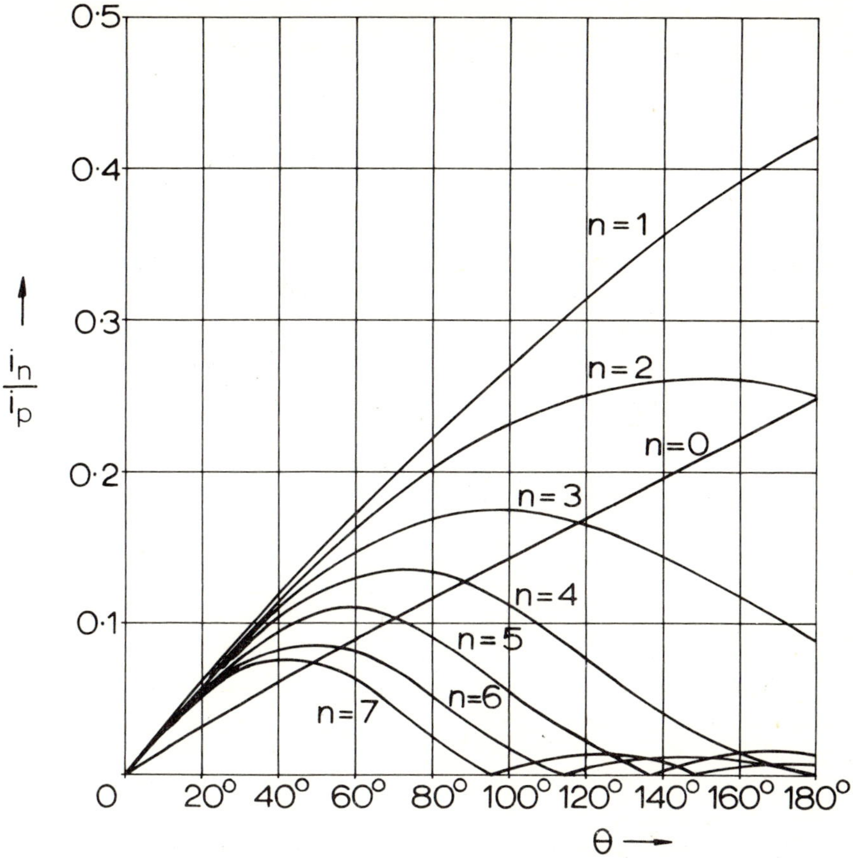

Fig. 1.7 Amplitudes of harmonic currents normalized to the peak current i_p as functions of the current flow angle θ in the case of the square-law diode characteristic.

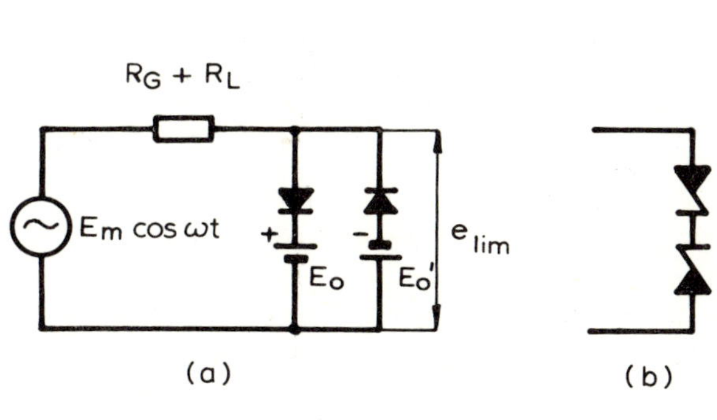

Fig. 1.8 A simple large-signal limiter: (a) using two biased diodes in parallel; (b) gaining the biasing voltage through the Zener breakdown.

$$e_{\lim} = E_m \left\{ \frac{1}{\pi} \left[-\left(\sin\frac{\theta_1}{2} - \frac{\theta_1}{2}\cos\frac{\theta_1}{2} \right) + \left(\sin\frac{\theta_2}{2} - \frac{\theta_2}{2}\cos\frac{\theta_2}{2} \right) \right] \right.$$

$$+ \left[1 - \frac{1}{2\pi}(\theta_1 + \theta_2 - \sin\theta_1 - \sin\theta_2) \right] \cos\omega t$$

$$- \frac{1}{\pi} \sum_{n=2,3,\ldots}^{\infty} \frac{2}{n^2-1} \left[\left(\frac{1}{n}\sin\frac{n\theta_1}{2}\cos\frac{\theta_1}{2} - \cos\frac{n\theta_1}{2}\sin\frac{\theta_1}{2} \right) \right.$$

$$\left. \left. - (-1)^n \left(\frac{1}{n}\sin\frac{n\theta_2}{2}\cos\frac{\theta_2}{2} - \cos\frac{n\theta_2}{2}\sin\frac{\theta_2}{2} \right) \right] \cos n\omega t \right\} \quad (1.23)$$

$$\frac{\theta_1}{2} = \arccos\frac{E_{L1}}{E_m}; \qquad \frac{\theta_2}{2} = \arccos\frac{E_{L2}}{E_m} \quad (1.24)$$

For symmetrical limitation $E_{L1} = E_{L2} = E_L$, consequently $\theta_1/2 = \theta_2/2 = \theta/2$, and the even harmonics are suppressed.

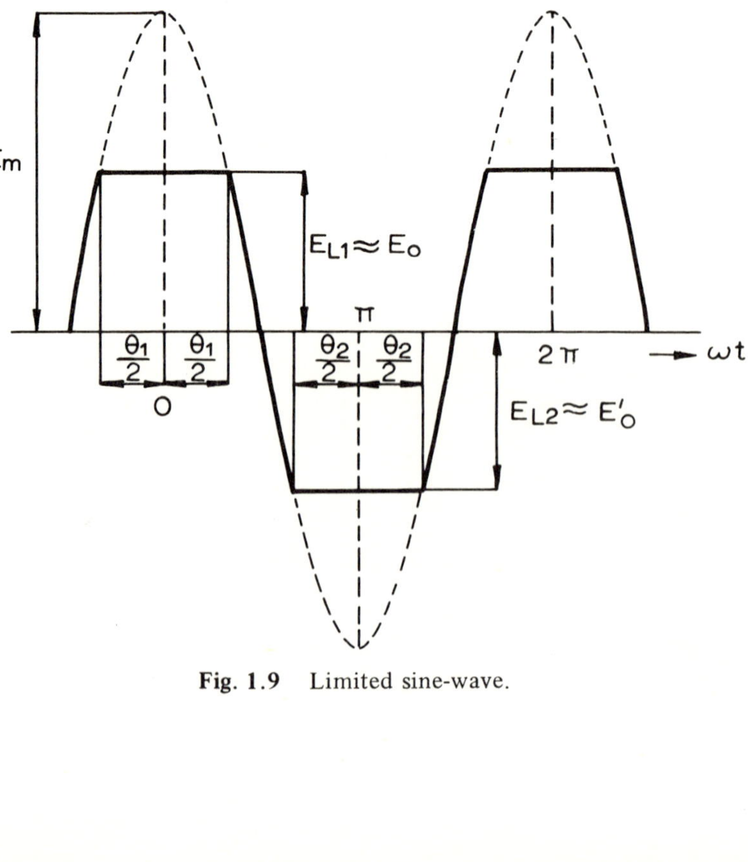

Fig. 1.9 Limited sine-wave.

$$e_{\lim} = E_m \left\{ \left[1 - \frac{1}{\pi}(\theta - \sin\theta) \right] \cos\omega t \right.$$

$$\left. - \frac{2}{\pi} \sum_{n=3,5,\ldots}^{\infty} \frac{2}{n^2 - 1} \left(\frac{1}{n} \sin\frac{n\theta}{2} \cos\frac{\theta}{2} - \cos\frac{n\theta}{2} \sin\frac{\theta}{2} \right) \cos n\omega t \right\} \quad (1.25)$$

where

$$\frac{E_L}{E_m} = \cos\frac{\theta}{2}$$

The diagram in Fig. 1.10 shows the change of the normalized amplitude e_1/E_L of the first harmonic as a function of the normalized input amplitude E_m/E_L.

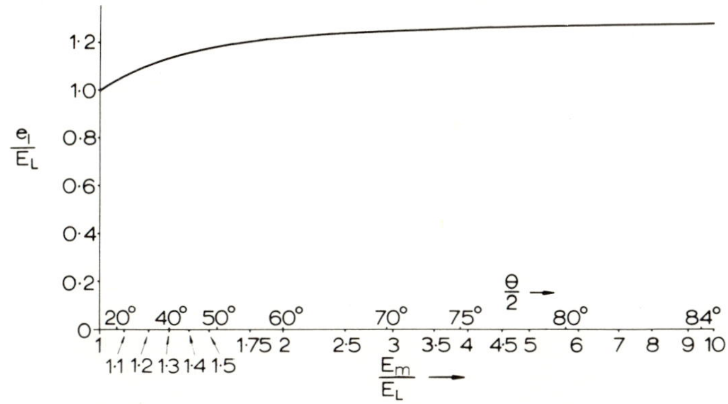

Fig. 1.10 Normalized amplitude e_1/E_L of the first harmonic in the limited sine-wave as a function of the normalized input amplitude E_m/E_L.

When two or three limiters with power gain (as shown in the functional form in Fig. 1.11) are connected in series the final relative limiting level is very low and

$$\frac{\theta}{2} \rightarrow \frac{\pi}{2}$$

In this case the expansion (1.25) simplifies to

$$e_{\lim} \approx \frac{4}{\pi} E_L \left[\cos\omega t - \frac{1}{3} \cos 3\omega t + \frac{1}{5} \cos 5\omega t - \cdots \right.$$

$$\left. + (-1)^{(n-1)/2} \frac{1}{n} \frac{\sin n[(\pi - \theta)/2]}{n(\pi - \theta)/2} \cos n\omega t + \cdots \right] \quad (1.26)$$

The most interesting feature of the above expansion is that, in the first approximation, the amplitude of the fundamental harmonic is independent of the current flow angle θ and consequently of the input amplitude E_m (see Fig. 1.10).

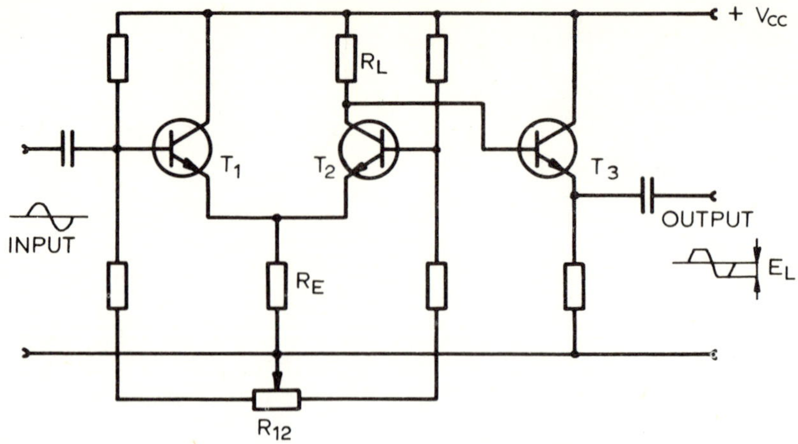

Fig. 1.11 Limiter with power gain – the emitter coupled clipper. Symmetry is adjusted by the balancing control R_{12}. Limited voltage level $|E_L| \approx |V_{CC}/2| R_L/(R_E + R_L)$; T_3 serves as an emitter-follower isolation device as well as driver for the next stage. (From Maxwell [8.3].)

1.5 REGENERATIVE SWITCHING

In the preceding sections we have considered the waveforms which do not differ substantially from the basic sinusoid or its part. We have found, generally, a rapid decline of higher harmonic amplitudes, and often a complete disappearance of some harmonics. These shortcomings may be removed by introducing sharp transients, coherent with the fundamental frequency. Miscellaneous circuits, multivibrators, Schmitt triggers, flip-flops, saturated magnetic cores, step recovery diodes, etc., fulfil the preceding requirement. The generated waveforms may be of different shape, nevertheless some parts will be found that are common to almost all of them: transients with different time constants. Because of the superposition possibilities, knowledge of the Fourier expansion of a simple waveform containing only one transient is of the utmost importance.

(A) Differentiated rectangular waveform

$$f(t) = \begin{cases} A\exp(-q\omega t) & (0 < \omega t < \theta) \\ B & (\omega t = \theta) \qquad \text{(see Fig. 1.12)} \\ 0 & (\theta < \omega t < 2\pi) \end{cases} \tag{1.27}$$

$$f(t) = \frac{A-B}{2\pi q} + \frac{1}{\pi}\sum_{n=1}^{\infty} \frac{1}{q^2+n^2}\{[Aq - B(q\cos n\theta - n\sin n\theta)]\cos n\omega t$$

$$+ [An - B(q\sin n\theta + n\cos n\theta)]\sin n\omega t\} \tag{1.28}$$

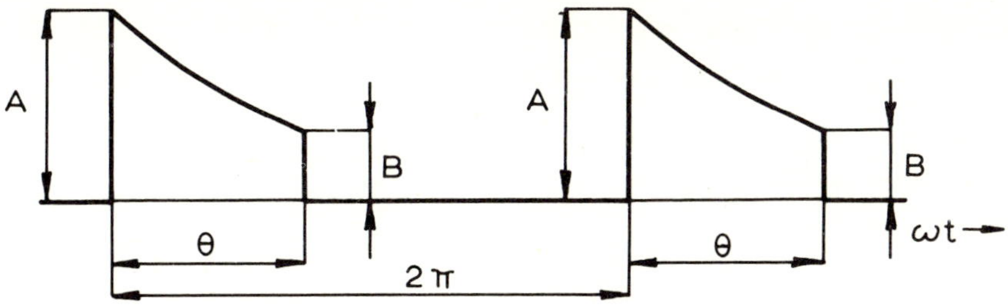

Fig. 1.12 Differentiated rectangular waveform.

where $q = (1/2\pi)\cdot(T/\tau) = (1/\theta)\lg(A/B)$, τ is the time constant determining the rate of change of the initial amplitude A, and T is the duration of the fundamental period. Calculating the absolute value of the nth harmonic amplitude from (1.28) we get

$$c_n = \frac{1}{\pi}\sqrt{\frac{A^2 + B^2 - 2AB\cos n\theta}{q^2 + n^2}} \tag{1.29}$$

It is evident that no harmonic disappears as long as $A > B$.

Fourier expansions of some other important waveforms may be readily found as limiting cases of (1.28).

(B) Rectangular Pulses

$A = B$, $q = 0$, and advancing the origin about $\theta/2$ (see Fig. 1.13) gives

$$f(t) = 2\frac{A}{\pi}\left[\frac{\theta}{4} + \sum_{n=1}^{\infty} \frac{1}{n} \sin\frac{n\theta}{2} \cos n\omega t\right] \tag{1.30}$$

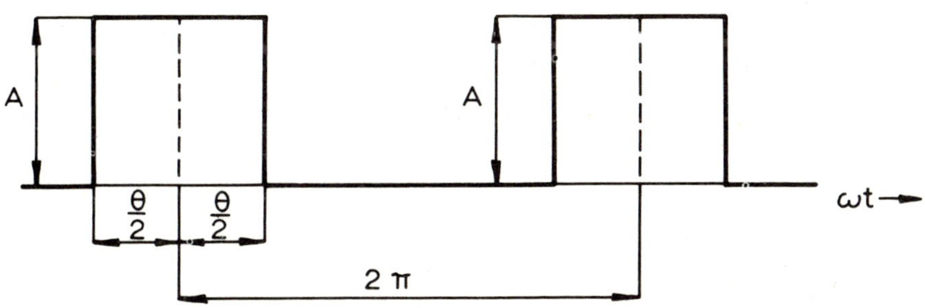

Fig. 1.13 Rectangular pulses.

In the expansion there are zero amplitudes whenever

$$\sin\frac{n\theta}{2} = 0 \quad \text{or} \quad \frac{n\theta}{2} = k\pi \tag{1.31}$$

(C) Differentiated Pulses (Fig. 1.14)

$$B \to 0 \text{ (before } \omega t \text{ reaches } 2\pi)$$

$$f(t) = \frac{A}{\pi}\left[\frac{1}{2q} + \sum_{n=1}^{\infty} \frac{1}{\sqrt{q^2+n^2}} \sin(n\omega t + \gamma_n)\right] \tag{1.32}$$

$$\tan\gamma_n = \frac{q}{n} \tag{1.33}$$

Amplitudes of all harmonics $n < q$ are practically uniform with a slow decline to -3 dB for $q = n$.

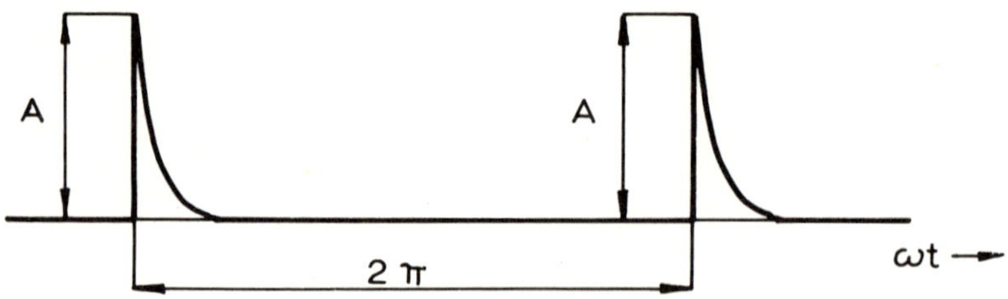

Fig. 1.14 Differentiated pulses.

(D) Arbitrary Narrow Pulses

The common feature is that the effective current-flow angle θ is very small compared with 2π.

Evidently, the δ-function representation is appropriate. However, in this case the pulse area is more important than the actual pulse shape.

The respective Fourier expansion is

$$f(t) \approx \frac{2}{T} S\left[\frac{1}{2} + \sum_{n=1}^{n<r} \cos n\omega t\right] + R_r \tag{1.34}$$

where S is the area of a pulse in the $(a; t)$ plane (see Fig. 1.15), and R_r is the sum of all remaining terms of the series.

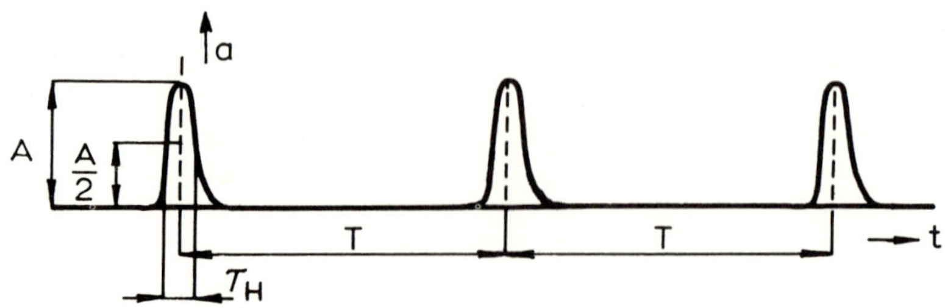

Fig. 1.15 Arbitrary narrow pulses.

Because of the arbitrary shape of pulses the definition of the current flow angle θ at zero level might sometimes be vague. Therefore the angle θ_H or the duration $\tau_H = T(\theta_H/2\pi)$ between the half-pulse amplitudes $A/2$ is introduced. The advantage of this parameter is that it may easily be identified experimentally; and from a knowledge of this parameter one is in a position to estimate r in (1.34). When performing an approximate integration one finds that

$$S \approx \tau_H A \tag{1.35}$$

Further, a rough estimation of the power spectrum (based on samples† with amplitudes $A/2, A, A/2$ at times $-\tau_H/2, 0, +\tau_H/2$) shows that the first minimum should be expected for

$$n \approx \frac{T}{\tau_H} \quad \text{or} \quad n \approx \frac{2\pi}{\theta_H}\text{‡} \tag{1.36}$$

Evidently, each circuit changing the sinusoidal input signal into a sequence of sufficiently narrow pulses may be used as a wide-band harmonic generator:

(i) At lower frequencies (below 1 MHz approximately) any type of regenerative switch may be employed, e.g., synchronized multivibrators or Schmitt triggers followed by differentiating circuits and if necessary gates suppressing pulses of the undesired polarity. Another switching phenomenon used in these frequency ranges is associated with highly nonlinear inductances [1.5, 1.6].

(ii) At higher frequencies (up to 20 GHz with contemporary techniques) the interruption of the current through a charge-controlled switch (a highly nonlinear capacitor–step recovery or charge storage diode, see also Section 2.2 and Fig. 2.3) is used for the generation of the extremely narrow pulses (tens of picoseconds) [1.7, 1.8]. Dense harmonic series, far beyond 5 GHz, may be realized in this way [1.9].

1.6 MODULATION OF HARMONIC SERIES

Very dense spectra are often generated by the amplitude, or phase modulation of the original pulse train. The advantage is that some spectral lines, generally those of the fundamental spectrum, may be made more prominent than the newly added *side-bands,* which helps in orientation. On the other hand when the modulating frequency is varied, then the side-bands sweep the intervals between the original spectral lines making the interpolation problem easier.

---

† A deeper analysis of the problem may be found in a surveying article [1.4].

‡ Departures from this rule may be expected particularly for highly unsymmetrical pulses; cf. the Fourier expansion (1.32) of the train of differentiated pulses.

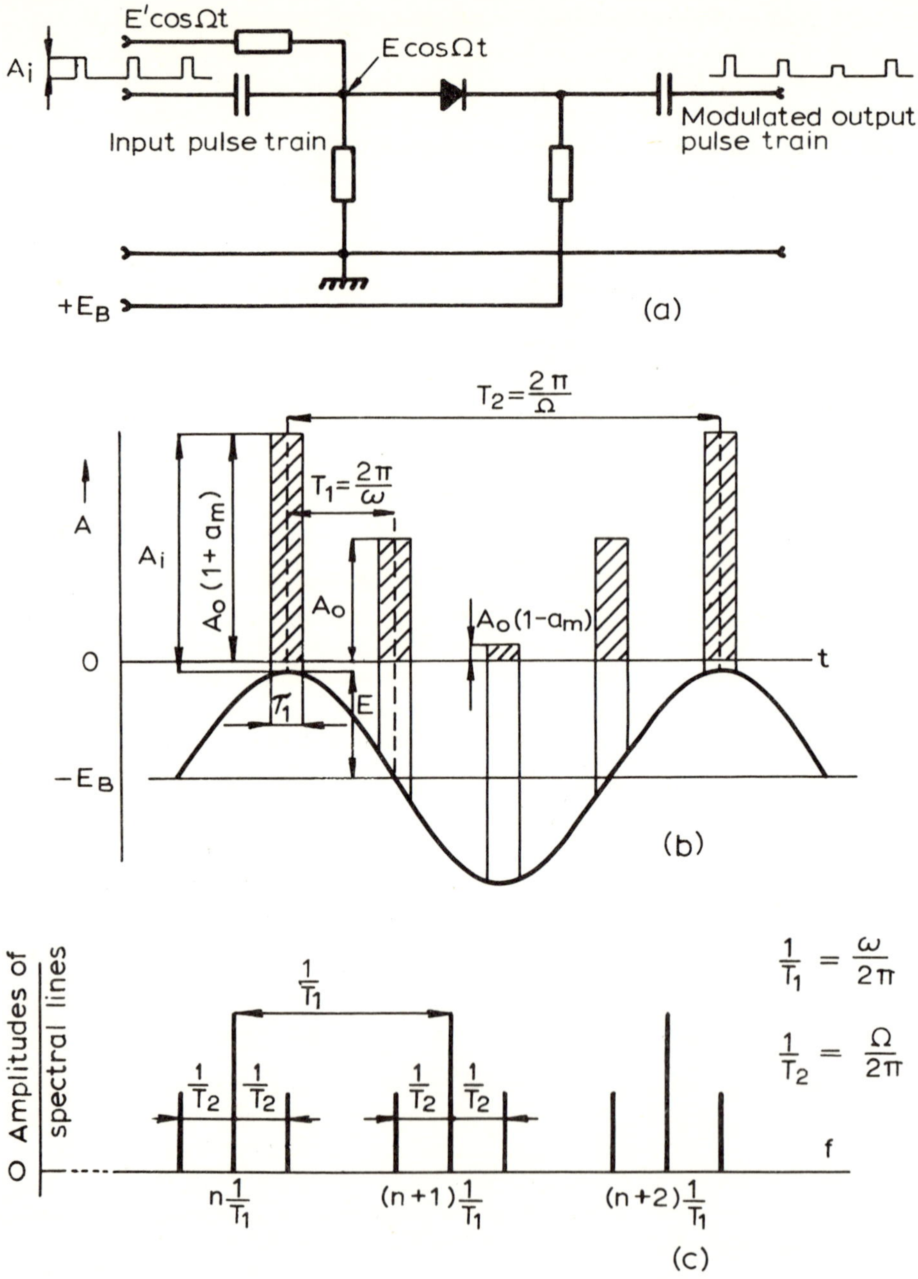

Fig. 1.16 Amplitude modulation of a pulse train: (a) functional form of a modulator; (b) time domain representation of the modulated output pulse train; (c) frequency domain representation of the modulated output pulse train.

(A) Amplitude Modulation of a Pulse Train

The area S of a pulse in eq. (1.34) is generally a nonlinear function of its amplitude A. Applying the Taylor expansion we get

$$S = (S)_{A_0} + \left(\frac{\partial S}{\partial A}\right)_{A_0} (A - A_0) + \frac{1}{2}\left(\frac{\partial^2 S}{\partial A^2}\right)_{A_0} (A - A_0)^2 + \cdots$$

$$= S_0 + S_1(A - A_0) + S_2(A - A_0)^2 + \cdots \tag{1.37}$$

If the amplitude A is periodically varied with the frequency Ω then

$$A = A_0(1 + a_m \cos \Omega t) \quad 0 < a_m \leqslant 1 \tag{1.38}$$

Substituting (1.37) and (1.38) in (1.34) leads to the frequency domain representation of the amplitude-modulated train of narrow pulses.

$$f(t) \approx \frac{2}{T} [S_0 + S_1\, a_m\, A_0 \cos \Omega t + S_2\, a_m^2\, A_0^2 \cos^2 \Omega t + \cdots]$$

$$\times \left(\tfrac{1}{2} + \sum_{n=1,2,\ldots} \cos n\omega t\right) \tag{1.39}$$

Expansion of products $\cos(\Omega t) \cdot \cos n\omega t = \frac{1}{2}[\cos(\Omega + n\omega)t + \cos(n\omega - \Omega)t]$ reveals that each spectral line in the frequency domain has two *side-bands* at intervals $\pm\Omega$ (see Fig. 1.16(c)).

The nonlinearity between S and A, i.e., the presence of higher terms $S_2(A - A_0)^2$, etc., in (1.37) is eventually responsible for higher order side-bands at intervals $\pm 2\Omega$, $\pm 3\Omega$, etc.

The amplitude modulation of a pulse train may be realized using a simple diode gate. The functional form of such a modulator is shown in Fig. 1.16(a).

(B) Impulse Modulation of a Pulse Train

Modulation of a pulse train $(\tau_1; T_1)$ with a rectangular wave, having the period $T_2 = kT_1$ (k being a positive integer), the same time origin, and the pulse duration τ_2 such that (see Fig. 1.17(a))

$$\tau_1 < \tau_2 < 2T_1 - \tau_1, \tag{1.40}$$

is of considerable importance. The Fourier expansion is found as the sum of two series

$$f(t) = 2\frac{S_1}{T_1}\left[\frac{1}{2} + \sum_{n=1}^{n<r} \cos n\frac{2\pi}{T_1}t\right] + R_r$$

$$+ 2\frac{S_2}{T_2}\left[\frac{1}{2} + \sum_{m=1}^{m<s} \cos m\frac{2\pi}{T_2}t\right] + R_s \tag{1.41}$$

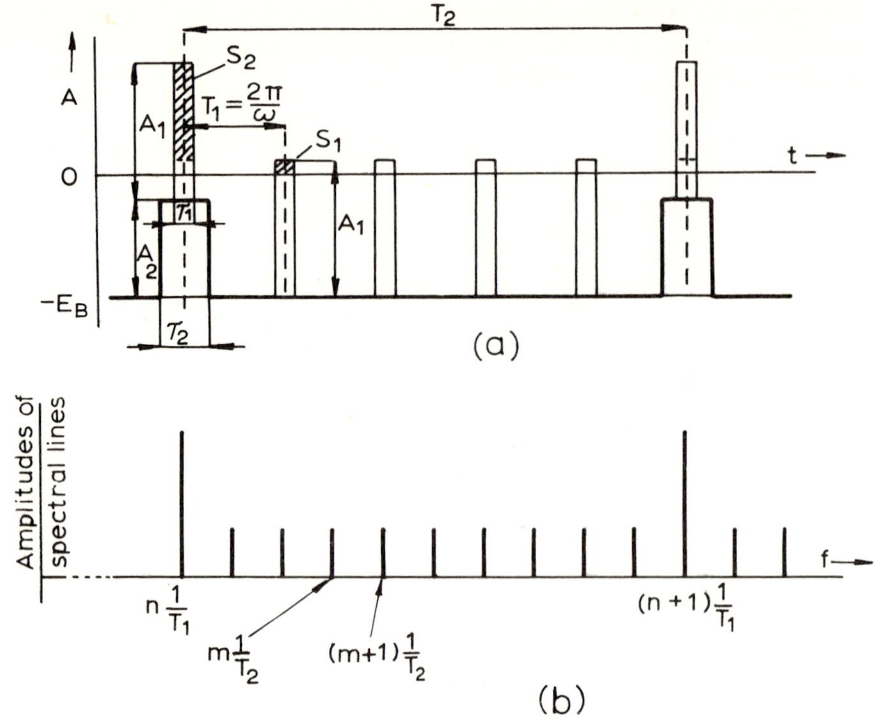

Fig. 1.17 Impulse modulation of a pulse train: (a) time domain representation of the modulated pulse train; (b) the respective frequency domain representation.

The distinguishing feature of the frequency domain representation is the dominating role of the $n\ (1/T_1)$ frequency lines (see Fig. 1.17(b)). This is easily proved by adding on the right-hand side of (1.41) terms with equal frequencies. Decimal spectral scales are often made in this way, particularly at lower frequency bands.

The case of the 100% modulation, when $S_1 = 0$, deserves attention. The impulse modulator, e.g. of the type shown in Fig. 1.16(a), changes to a generator of very narrow pulses; the other advantage is that the absolute time jitter of individual pulses (caused by noise, uncertainty of switching levels in pulse generators, or other spurious signals) remains the same as that of the original denser pulse train. Consequently the phase jitter is lowered in the proportion T_1/T_2. When phase-jitter reduction is the only aim then the 100%-impulse modulator is called a *pulse selector* and finds extensive use in frequency-dividing circuits [1.10].

(C) Phase or Position Modulation of a Pulse Train

The triggering effect which is responsible for the generation of a pulse train is started when the input waveform reaches a predetermined level. When superimposing another signal the crossings of the switching threshold are either delayed or advanced and consequently the generated pulse train is position (phase) modulated (see Fig. 1.18).

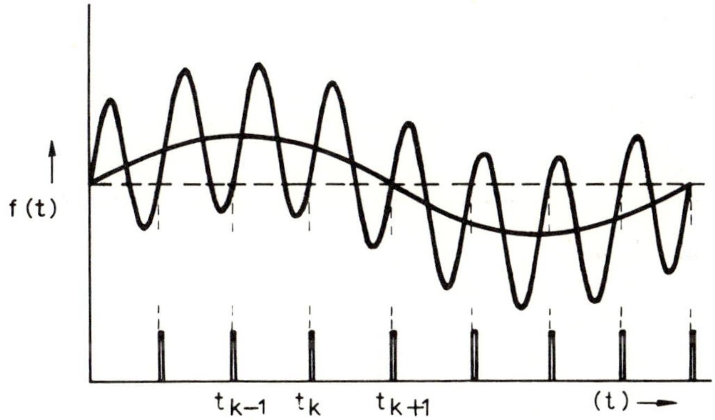

Fig. 1.18 Advance and delay of zero-crossings in the presence of a small superimposed sinusoidal signal.

The respective time-function may be written as

$$f(t) = \sum_{k=-\infty}^{\infty} p(t - t_k) \tag{1.42}$$

where $p(t)$ is the basic pulse shape.

In the case of the sinusoidal modulation (uniform sampling)

$$t_k = kT - \Delta T \sin \Omega kT \qquad \left(T = \frac{2\pi}{\omega} \right) \tag{1.43}$$

Using [1.11] the Fourier transform of (1.42) is easily found to be

$$f(t) = \frac{1}{T} \sum_{m=-\infty}^{\infty} \sum_{n=-\infty}^{\infty} P(m\omega + n\Omega) J_n(\Delta T[m\omega + n\Omega]) \times \exp[j(m\omega + n\Omega)t] \tag{1.44}$$

where $P(\nu)$ is the Fourier transform of $p(t)$.

Considering a train of very narrow pulses (1.34)

$$f(t) \approx \frac{S}{T} \left\{ 1 + 2 \sum_{m=1,2,\ldots} \sum_{n=0,\pm 1,\pm 2,\ldots} J_n(\Delta T[m\omega + n\Omega]) \cos[m\omega + n\Omega]t \right\}^{\dagger} \tag{1.45}$$

† The low frequency components ($m = 0$, $n \neq 0$) as well as part of the high frequency components have been neglected.

UNIV. COLL. N.W. BANGOR LIBRARY

It is clear from the above expansion that each harmonic component has a family of side-bands. Their amplitudes depend on the actual modulation index

$$2\pi\frac{\Delta T}{T}\left[m+n\frac{\Omega}{\omega}\right]\approx 2\pi m\frac{\Delta T}{T} \tag{1.46}$$

which is approximately proportional to the harmonic multiple considered.

As long as the modulation index is small, in practice if

$$2\pi\frac{\Delta T}{T}\left[m+n\frac{\Omega}{\omega}\right]<0{\cdot}2, \tag{1.47}$$

then only the first side-bands ($m\omega\pm\Omega$) are present. But with larger modulation indices one must be careful not to lose orientation in the frequency domain line-spectrum representation because the amplitude of the 'carrier' may even be zero. However, this shortcoming is compensated by the simplicity of the methods by which the small index phase modulation of a pulse train may be performed. Because the operation takes place before the actual pulse shaping, obviously the final form of the output pulses is not degraded by the phase-modulation process, which need not necessarily be the case with amplitude or impulse modulation discussed above. The phase modulation of a pulse sequence is therefore preferred when dense spectra at very high frequencies must be generated [1.9].

1.7 MODULATION OF HIGH-FREQUENCY CARRIERS

Very often the solving of frequency-synthesis or frequency-measurement problems requires a dense spectrum, but in a limited frequency band only. In such a case substantially larger power may be transferred to the individual harmonic components if they are generated as modulation products of a suitable midband-frequency carrier, and not derived as direct harmonic multiples according to the methods discussed in previous sections.

(A) Linear Amplitude Modulation

Linear amplitude modulation of a high-frequency carrier $A_H \sin(\omega t+\varphi)$ with a low-frequency signal $f_L(t)$ having only positive values and peak to peak amplitude A_L (see Fig. 1.19(a) and (b)) is described by the equation

$$f_H(t)=\left\{1+a_m\left[2\frac{f_L(t)}{A_L}-1\right]\right\}A_H\sin(\omega t+\varphi) \tag{1.48}$$

where the amplitude modulation index a_m has a value

$$0<a_m<1(a_m=A_L/2A_H) \tag{1.49}$$

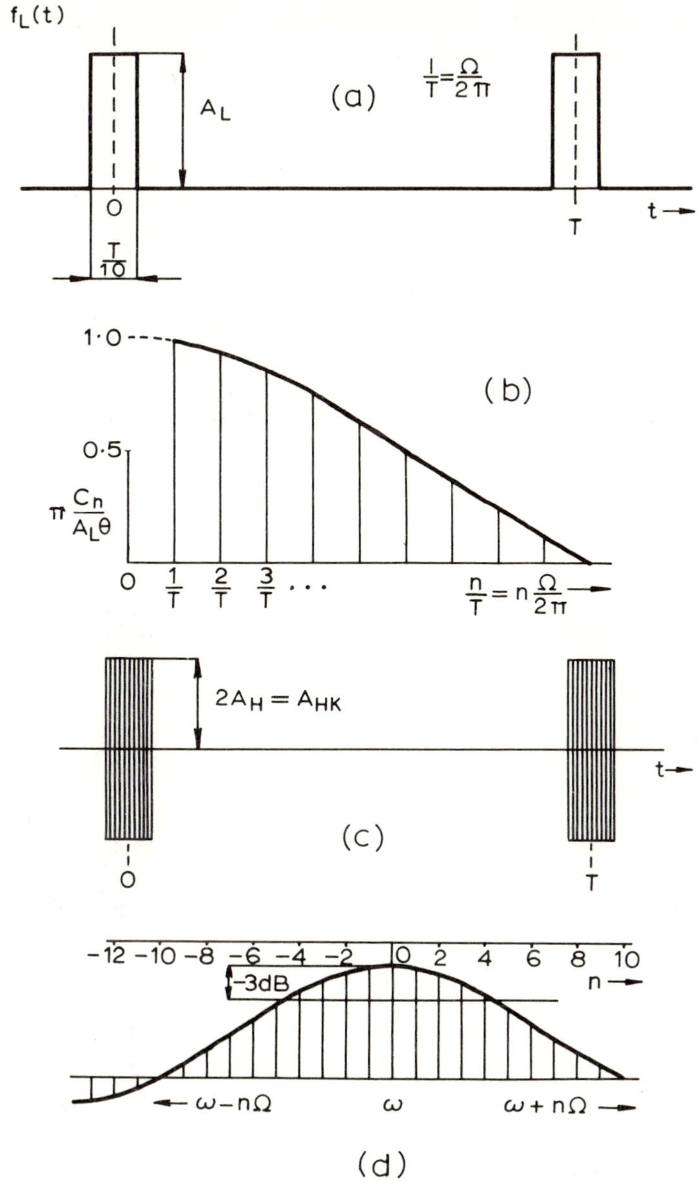

Fig. 1.19 Modulation of a high-frequency carrier $A_H \sin(\omega t + \varphi)$ with a low-frequency signal $f_L(t)$: (a) time domain representation of a rectangular waveform $f_L(t)$, ($\theta = \pi/5$), (b) the respective frequency domain representation; (c) time domain representation of the modulated carrier $f_H(t)$, (d) the respective frequency domain representation.

If $f_L(t)$ is a waveform rich in higher harmonic components (e.g. any of the waveforms discussed in the preceding sections), eq. (1.48) may be written as

$$f_H(t) = \left[1 - a_m + 2\frac{a_m}{A_L}\sum_{n=0}^{\infty} c_n \cos(n\Omega t + \psi_n)\right] A_H \sin(\omega t + \varphi)$$

$$= (1 - a_m)\, A_H \sin(\omega t + \varphi) + \frac{1}{2}\sum_{n=0} c_n [\sin(\omega t + n\Omega t + \varphi + \psi_n)$$

$$+ \sin(\omega t - n\Omega t + \varphi - \psi_n)] \qquad (1.50)$$

Evidently around the carrier there is a family of sidebands, the amplitudes of which decay according to the same law as the harmonic-component amplitudes of the original low-frequency signal (see Fig. 1.19(b) and (d)). Special attention must be paid to the carrier because it forms part of the spectrum envelope only when the first term in (1.50) is zero, i.e.

$$a_m = 1 \qquad (1.51)$$

and when

$$\lim_{n>0} c_n = 2c_0 \qquad (1.52)$$

The second condition is not fulfilled, for example, if $f_L(t)$ is a rectangular wave with a large current flow angle θ (i.e. $2\pi - \theta \ll 2\pi$). In this case the carrier amplitude passes, eventually, many times over the envelope.

(B) On-Off Keying

On-off keying of a high-frequency signal A_{HK} sin $(\omega t + \varphi)$ is equivalent to 100% modulation with a low-frequency rectangular waveform. If this low-frequency signal is a periodic one then the amplitudes of individual spectral components are readily found using (1.50) and (1.30). Note. $A_L = 2A_H = A_{HK}$!

$$f_{HK}(t) = A_{HK}\frac{\theta}{2\pi}\left\{\sin(\omega t + \varphi) + \sum_{n=1}^{\infty}\frac{\sin(n\theta/2)}{n\theta/2}[\sin(\omega t + n\Omega t + \varphi)\right.$$

$$\left. + \sin(\omega t - n\Omega t + \varphi)]\right\} \qquad (1.53)$$

When sidebands of very high orders are not desirable in the output signal, rounding of abrupt edges of keying pulses will help.†

(C) Coherent Modulation

For $\omega = k\Omega$ both (1.50) and (1.53) change into a simple Fourier series of the fundamental frequency Ω. The amplitudes of the individual harmonics may be

† Compare the removing of key clicks in the transmitting technique [1.12, Section 24.3].

derived by the following procedure. The two sums on the right-hand side combine easily into one when extending the summation to the negative n as well. Introducing a new summation rule, viz. $s = k - n$, and further assuming the validity of (1.51) and (1.52) we get a compact form of the time function $f_{HC}(t)$.

$$f_{HC}(t) = \sum_{s=-\infty}^{\infty} c_{k-s} \sin(s\Omega t + \varphi - \psi_{k-s}) \tag{1.54}$$

For the negative s and sufficiently large k the coefficients c_{k+s} are very small and consequently, without any appreciable error, summation only over positive s may be retained

$$f_{HC}(t) \approx \sum_{s=0}^{\infty} c_{k-s} \sin(s\Omega t + \varphi - \psi_{k-s}) \tag{1.55}$$

Three different cases encountered in practice can be evaluated using (1.55):

(i) Actual Modulation between Two Coherent Signals

Both the carrier (ω) and the modulating signals (Ω) are generated separately but from the same source. The low-frequency waveform must be properly shaped since it determines the width of the HF spectrum. Because of its practical importance the problem is more closely dealt with in the following example.

Example 1.1

In Section 5.3 we shall see that the variable frequency synthesizer requires, according to eq. (5.61), generation of the following part of a harmonic series

$$f_0;\ f_0 + f_r;\ f_0 + 2f_r;\ \ldots;\ f_0 + 9f_r$$

where

$$\frac{f_0}{f_r} = x_0 \gg 1 \quad (x_0 \text{ is a large integer})$$

Further requirements are: a constant amplitude of all useful harmonic components and 'zero' amplitudes of all signals beyond the pass-band ($f_0; f_0 + 9f_r$).

A 'carrier' – either $f_0 + 4f_r$ or $f_0 + 5f_r$ – will be generated and then keyed in the rate of f_r. The angle of the current flow θ must be chosen according to the permitted drop of amplitudes in the frequency-band ($f_0; f_0 + 9f_r$). For the –4 dB drop the angle of the current flow θ should be

$$\theta \approx \frac{\pi}{5}$$

The actual relative amplitudes for $\theta = \pi/5$ are shown in Fig. 1.19(d).
For the –1 dB drop one may find

$$\theta \approx \frac{\pi}{10}$$

The suppression of the harmonic components outside the pass-band ($f_0; f_0 + 9f_r$) will be increased with an appropriate low-frequency keying filter.

(ii) Damped Waveform in the Shock-excited Resonant Circuit

Generation of the 'carrier' in separate circuits may be omitted if a train of sufficiently narrow pulses excites a resonant circuit. The damped waveform (see Fig. 1.20) is to a very high degree expressed by (1.55) and (1.28). The damping factor is

$$\delta = \frac{R}{2L} = \frac{\omega}{2Q} = q\Omega \tag{1.56}$$

(Q being the circuit Q).

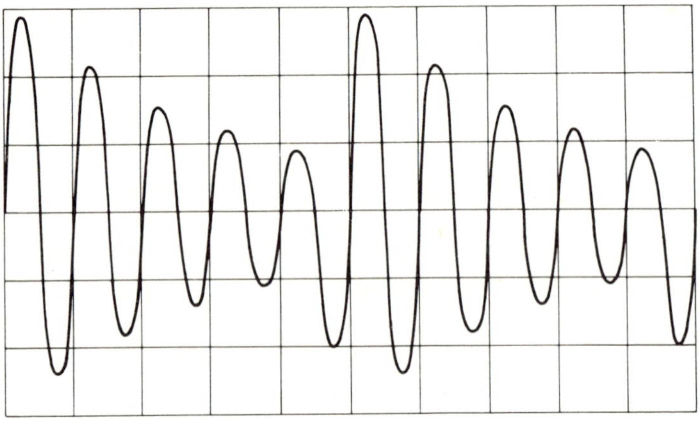

Fig. 1.20 Damped waveform in shock-excited resonant circuit.

In an actual device the circuit resonant frequency will not be exactly a kth multiple of the exciting frequency Ω.

The phase coherence can be maintained but through a spurious phase modulation, i.e. $\varphi = \varphi_0 + \varphi(t)$

$$|\varphi(t)| < |\omega - k\Omega| T_\Omega \leqq \frac{\Omega}{2} T_\Omega \left(\frac{\Omega}{2} T_\Omega = \pi \right) \tag{1.57}$$

In this case a distortion of the spectral envelope with respect to (1.55) must be expected.

Harmonic generators of this type were used in carrier telephony for a considerable time [1.5]. Whereas broadband-high-order frequency multipliers in the range of tens of MHz to tens of GHz, based on step-recovery diodes, are much more recent [1.8].

(iii) Superregenerative Harmonic Generator

A superregenerative harmonic generator, like a superregenerative detector [1.12, Section 16.9], is essentially an ordinary oscillator which is alternated between an oscillating and a non-oscillating condition at a low frequency.

However, for proper operation the oscillations are forced to start always with the same phase, and must decay to an amplitude less than the circuit noise, before the oscillating condition is again restored (Fig. 1.21). In other words, irrespective of the instantaneous frequencies, there exists a phase coherence between the high-frequency oscillations and the low-frequency quench oscillations. The phenomenon is fully described by equations (1.55) and (1.57).

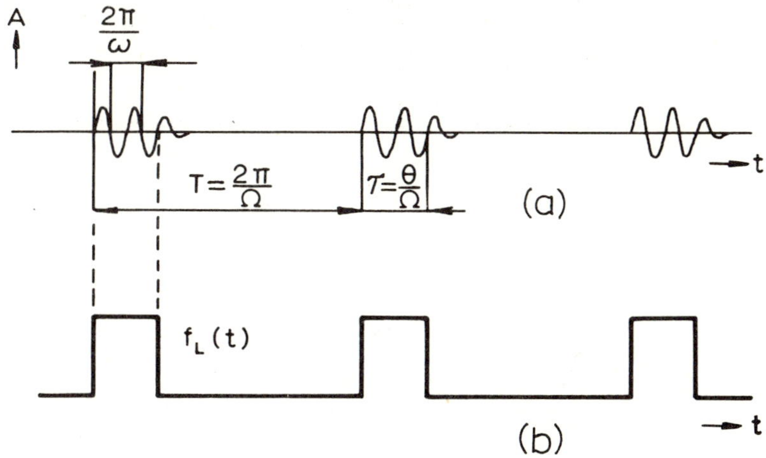

Fig. 1.21 (a) Idealized quenched oscillations of a superregenerative harmonic generator; (b) the low-frequency quenching waveform.

The maximum of the spectral envelope is at ω. However, no signal with frequency ω can be detected at the output as long as $\omega \neq k\Omega$. The amplitudes of individual spectral lines decay according to the $(\sin x)/x$ law for the idealized rectangular quenching envelope.

The advantages of the superregenerative harmonic generators are the simple circuitry (see Fig. 1.22) and the high output power. There are, however, disadvantages also. The device cannot be considered as fail-safe, because some exceptional circuit changes might prevent the oscillations from decaying below the noise level in which case the output spectral lines would be side-bands around ω rather than multiples of Ω.

Superregenerative harmonic generators may be realized in different ways depending on the means used for changing the oscillating condition to a non-oscillating one. Only one example is given, in Fig. 1.22; for a more profound discussion of individual features readers should refer to the literature [1.13, 1.14].

1.8 CONCLUSIONS

Harmonic generators may be divided into three main groups:

(i) where only low-frequency multiples are of concern,

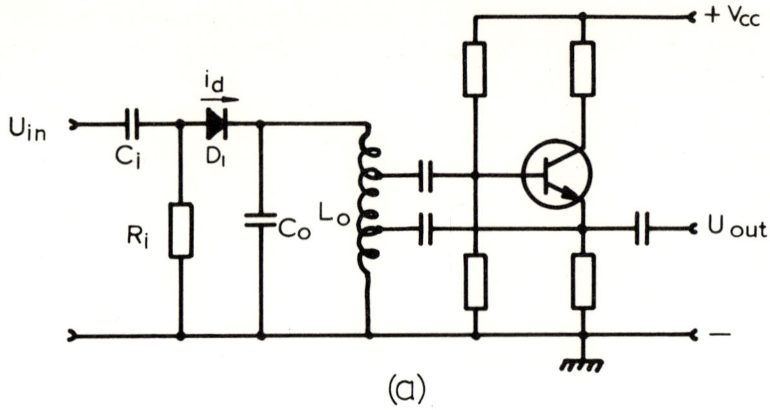

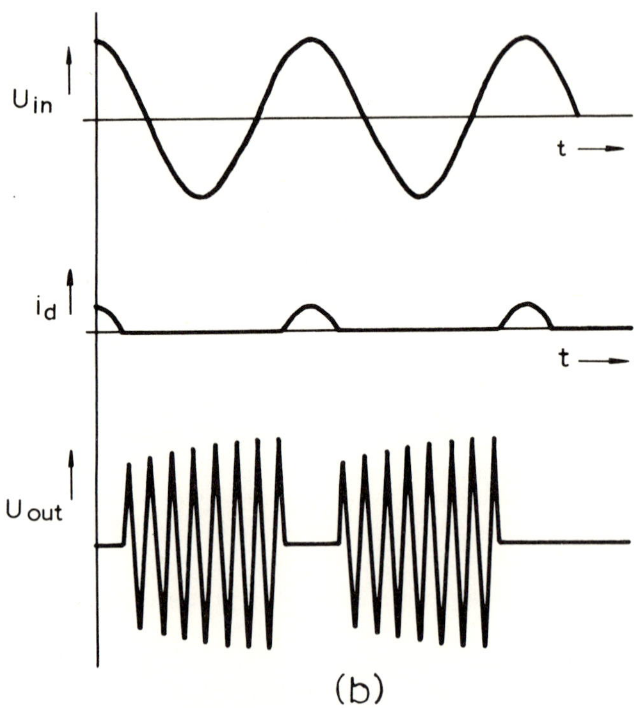

Fig. 1.22 Superregenerative harmonic generator (frequency multiplier): (a) a fundamental circuit; when diode D_1 is conducting (at positive peaks of U_{in}) then oscillating circuit L_0C_0 is heavily damped and oscillations are suppressed; (b) the respective time diagrams.

(ii) where a few high-frequency multiples in a limited range are required,
(iii) where dense and marked harmonic spectra in extended ranges have to be generated.

Simple nonlinear elements (diodes, transistors) are pertinent to the first group whereas switching circuits generating fast transients (Schmitt triggers, multivibrators, coils with rectangular magnetic characteristics at lower frequencies, and step recovery diodes at higher frequencies) are inevitable for the latter. In the third group modulation of the pulse trains is generally performed to increase the number of spectral lines.

Constant amplitude, wide range, harmonic generators, also called 'harmonic comb generators', may be considered as the first generation of multi-frequency synthesizers, but, this does not mean that they are no longer used. On the contrary these devices still retain their former importance at the highest frequency bands and find applications as calibrators, mark generators, etc. Their simplicity and low price is appreciated.

In today's single- or variable-frequency synthesizers both types, (i) and (ii), of harmonic generator may be encountered. Because the desired output is always only one single frequency, all but one harmonic multiple must be suppressed. Knowledge of the spectral composition of the waveforms generated in nonlinear circuits helps with their proper choice and with the design of the pertinent filters.

APPENDIX 1.1

(A) Expansions of $\exp(\pm jz\cos\omega t)$ and $\exp(\pm jz\sin\omega t)$

These expansions are easily performed by using the Jacobi–Anger formula [1.15, Section 21.8–4].

$$\exp(\pm jz\cos\omega t) = \sum_{m=-\infty}^{\infty} J_m(\pm z)\left[\exp\left(j\frac{\pi}{2}\right)\cdot\exp(j\omega t)\right]^m$$

$$= \sum_{m=-\infty}^{\infty} (\pm j)^m J_m(z)\exp(jm\omega t)$$

$$= J_0(z) + 2\sum_{m=1}^{\infty} (\pm j)^m J_m(z)\cos m\omega t \tag{1.58}$$

$$\exp(\pm jz\sin\omega t) = \sum_{m=-\infty}^{\infty} J_m(\pm z)\exp(jm\omega t)$$

$$= J_0(z) + 2\sum_{m=1}^{\infty} [(\pm j)J_{2m-1}(z)\sin(2m-1)\omega t$$

$$+ J_{2m}(z)\cos 2m\omega t] \tag{1.59}$$

where the coefficients are the Bessel functions of the first kind. Some of their properties are:

$$J_{-m}(z) = (-1)^m J_m(z) \qquad (1.60)$$

$$J_m(-z) = (-1)^m J_m(z) \qquad m = 0, 1, 2, \ldots \qquad (1.61)$$

$$J_{m+1}(z) = 2\frac{m}{z} J_m(z) - J_{m-1}(z) = J_{m-1}(z) - 2\frac{d}{dz} J_m(z) \qquad (1.62)$$

$$J_m(z) = \sum_{k=0}^{\infty} \frac{(-1)^k (z/2)^{2k+m}}{k!(k+m)!} \qquad (1.63)$$

This reduces for small values of z to:

$$\begin{aligned} J_0(z) &\approx 1 \\ J_1(z) &\approx \frac{z}{2} \\ J_2(z) &\approx \frac{z^2}{8} \\ &\vdots \\ J_m(z) &\approx \left(\frac{z}{2}\right)^m \frac{1}{m!} \end{aligned} \qquad (1.64)$$

Inserting an imaginary argument $z = jy$ we have

$$J_m(jy) = j^m I_m(y) \qquad (1.65)$$

where $I_m(y)$ are modified Bessel functions. Some of their properties are readily obtained from (1.60) to (1.63) with the help of (1.65)

$$I_{-m}(y) = I_m(y) \qquad (1.66)$$

$$I_m(-y) = (-1)^m I_m(y) \qquad m = 0, 1, 2, \ldots \qquad (1.67)$$

$$I_{m+1}(y) = I_{m-1}(y) - \frac{2m}{y} I_m(y) = 2\frac{d}{dy} I_m(y) - I_{m-1}(y) \qquad (1.68)$$

$$I_m(y) = \sum_{k=0}^{\infty} \frac{(y/2)^{2k+m}}{k!(k+m)!} \qquad (1.69)$$

This reduces for small values of y to

$$
\begin{aligned}
I_0(y) &\approx 1 \\
I_1(y) &\approx \frac{y}{2} \\
I_2(y) &\approx \frac{y^2}{8} \\
&\vdots \\
I_m(y) &\approx \left(\frac{y}{2}\right)^m \frac{1}{m!}
\end{aligned}
\tag{1.70}
$$

(B) Expansions of exp ($\pm y$ cos ωt) **and** exp ($\pm y$ sin ωt)

These expansions are readily obtained from (1.58) and (1.59) by substituting $z = -jy$ and using (1.65)

$$\exp(\pm y \cos \omega t) = I_0(y) + 2 \sum_{m=1}^{\infty} (\pm 1)^m I_m(y) \cos m\omega t \tag{1.71}$$

$$\exp(\pm y \sin \omega t) = I_0(y) + 2 \sum_{m=1}^{\infty} (-1)^m [(\mp 1) I_{2m-1}(y) \sin (2m-1)\omega t + I_{2m}(y) \cos 2m\omega t] \tag{1.72}$$

where the coefficients are the modified Bessel functions (see above).

REFERENCES

[1.1] SZE, S. M. *Physics of Semiconductor Devices.* New York, London, Sydney, Toronto. Wiley, 1969.

[1.2] MILLS, H. D. On the equation $i = i_0 [\exp (v - Ri) - 1]$. *IBM Journal,* **11,** 553-4, Sept. 1967.

[1.3] BENNETT, W. R. The biased ideal rectifier. *Bell Sys. Tech. J.,* **26,** 139-69, Jan. 1947.

[1.4] RICHARDS, P. I. Computing reliable power spectra. *I.E.E.E. Spectrum,* **4,** 83-90, 1967.

[1.5] PETERSON, E., MANLEY, J. M. and WRATHALL, L. R. Magnetic generation of a group of harmonics. *Bell Sys. Tech. J.,* **16,** 437-55, Oct. 1937.

[1.6] HUSSEY, L. W. Nonlinear coil generators of short pulses. *Proc. I.R.E.,* **38,** 40-4, 1950.

[1.7] HAMILTON, S. and MOLL, R. Shunt-mode harmonic generation using step recovery diodes. *Microwave Journal,* **10,** 69-78, May 1967.

[1.8] Harmonic generation using step recovery diodes and SRD modules. *Hewlett-Packard, Application Note* 920.

[1.9] CARLSON, R. A frequency comb generator with a range from 1 MHz to beyond 5 GHz. *Hewlett-Packard J.,* **18,** 15-20, Sept. 1966.

[1.10] BECKER, G. Ausschaltung von Laufzeit und Phasenschwankungen bei Frequenzteilung mit Zählschaltungen. *Zeitschrift f. Instrumentenkunde,* **74,** 358-9, 1966.

[1.11] KROUPA, V. Spectra of position modulated pulse trains. *Proc. I.E.E.E.,* **57,** 708-9, 1969.

[1.12] TERMAN, F. E. *et al., Electronic and Radio Engineering,* 4th ed. New York, Toronto, London: McGraw-Hill, 1955.

[1.13] GOTTHARDT, G. Über ein Verfahren zur Erzeugung von Oberwellen höherer Ordnung. *Bul. SEV,* **55**, 675-80, 1964.

[1.14] BECKER, G. Frequenzverfielfachung mit grossen Faktoren mittels Anschwingverfahren. *Frequenz,* **14**, 412-16, 1960.

[1.15] KORN, G. A. and KORN, T. M. *Mathematical Handbook.* New York, Toronto, London: McGraw-Hill, 1961.

2 Frequency Multiplication

In principle, a frequency multiplier is a modified harmonic generator emphasizing only one desired multiple of the fundamental frequency. All other signals are suppressed or, more exactly, must be kept below a prescribed value. This is generally achieved when an ordinary harmonic generator is followed with filters or tuned amplifiers having suitable selectivity. Further, the power-level difference between signals with useful and spurious harmonic frequencies may be enhanced through an appropriate design of the harmonic generator itself. Particularly, the push-pull connexion of equal nonlinear circuits eliminates the even harmonics and the push-push connexion the odd harmonics (see Fig. 2.1(b)).

Today all requirements of frequency synthesis may be fulfilled with solid-state devices and we shall therefore limit our investigations in this chapter mainly to the diode and transistor multipliers.

2.1 IDEAL FREQUENCY MULTIPLIER

An ideal frequency multiplier is an electronic device changing the input signal $A_i \cos \omega t$ directly into the desired harmonic multiple $A_0 \cos n\omega t$; it is assumed that no other signals are present at the output. No filters are therefore needed, and the advantage of such a multiplier is that it may be used in wide frequency bands without any tuning.

The transfer characteristic of the suitable nonlinear element or device must fulfil the following parametric equations:

$$\begin{aligned} x &= A_i \cos \omega t \equiv X = \cos \omega t \\ y &= A_0 \cos n\omega t \equiv Y = \cos n\omega t \end{aligned} \qquad (2.1)$$

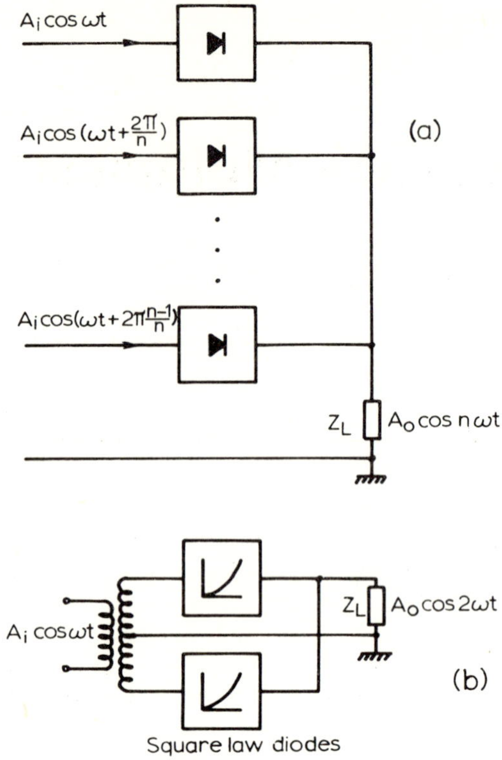

Fig. 2.1 Ideal frequency multiplier: (a) approximation of an ideal frequency multiplier with a system of simple nonlinear elements which are successively activated by a polyphase input signal; (b) an almost-ideal frequency doubler.

Elimination of the parameter t is easy by introducing Chebyshev polynomials [2.1]

$$Y = \cos(n \arccos X) = T_n(X)^{\dagger} \tag{2.2}$$

The graphical solution of (2.1) or (2.2) leads to the well-known single-valued (for Y) Lissajous figures.

Practical realization of frequency multiplying elements with an ideal transfer characteristic is a difficult one.

† Chebyshev polynomials are expanded as follows, see [1.15 Section 21.7-4].

$$T_1(X) = X \tag{2.3}$$
$$T_2(X) = 2X^2 - 1 \tag{2.4}$$
$$T_3(X) = 4X^3 - 3X \tag{2.5}$$
$$T_{n+1}(X) = 2XT_n(X) - T_{n-1}(X) \tag{2.6}$$

One approach uses a system of successively activated switches, saturated reactors, etc. (see Fig. 2.1(a)). Masked cathode ray tubes have been tried [2.2]: recently the use of tunnel diodes has been suggested [2.1].

The case, when $n = 2$, deserves special attention. According to eq. (2.4) the ideal multiplying characteristic is a parabola which may be easily synthesized through a push-push connexion of two square-law nonlinear elements (Fig. 2.1(b)). In this way aperiodic frequency doublers for ranges 0·5 to 500 MHz have been realized having at the output the spurious level of the 1st and 3rd harmonic below −30 dB (up to 50 MHz) [2.3].

2.2 DIODE FREQUENCY MULTIPLIERS

Semiconductor diodes have made it possible to obtain useful harmonic power with solid-state devices in an extremely wide frequency band, effectively from zero to some 2500 GHz. Both the nonlinear resistance and the nonlinear capacitance behaviours are exploited in practical frequency multipliers.

(A) Equivalent Diode Circuit

The mathematical model of the diode based on continuity and transport equations for semiconductor materials is the one of the transmission line with distributed analogue r-g-c parameters. The first-order approximation involving only two lumped values leads to the equivalent circuit in Fig. 2.2, according

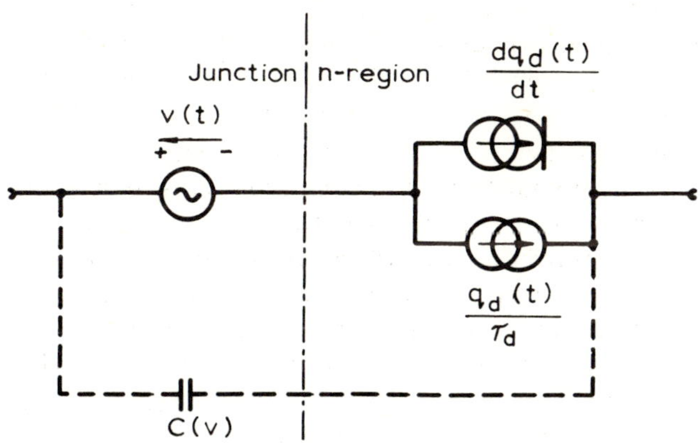

Fig. 2.2 The first-order charge-control equivalent circuit of the semiconductor diode. (Koehler [2.4]. Copyright, 1967, American Telephone & Telegraph Co., reprinted by permission.)

to ref. [2.4]. There is an ideal voltage generator in series with two current generators (exhibiting no voltage drop) shunted through the depletion-layer voltage-dependent capacitor. The diode current is given by the well-known charge-control equation

$$i_d(t) = \frac{dq_d(t)}{dt} + \frac{q_d(t)}{\tau_d} \tag{2.7}$$

q_d is the total charge in the bulk diode material and is a function of the junction voltage $v(t)$

$$q_d(t) = I_s \tau_d \{\exp [\lambda v(t)] - 1\} \tag{2.8}$$

I_s is the reverse current when $v \to -\infty$,
τ_d is the minority carrier lifetime, and
λ has been defined by (1.18)

The charge stored in the depletion layer capacitor is a function of the voltage $v(t)$ and of the doping profile in the region near the p-n junction [1.1, 2.5].

$$q_c + q_\Phi = \frac{v + \Phi}{(1-\gamma) S(v)} = \frac{(V_B + \Phi)^\gamma}{S_{\max}} (v + \Phi)^{1-\gamma} \frac{1}{1-\gamma} \dagger \tag{2.9}$$

where Φ is the contact potential,
q_Φ the respective charge,

$$S(v) = \frac{1}{C(v)} = \frac{dv}{dq} \tag{2.10}$$

is the incremental voltage-dependent elastance‡ with its maximum attained just before the junction break-down voltage V_B, and
γ is the doping profile constant; $\gamma = 0{\cdot}5$ for the abrupt-junction varactors, 0·33 for the graded-junction varactors, and may be as low as 0·1 for true charge-storage or step-recovery diodes, whereas for the hyper-abrupt junction varactors $\gamma > 0{\cdot}5$ (see Table 7.2).

The nonlinear equations (2.7), (2.8), and (2.9) cannot be solved without imposing further approximations. In the small-signal case few terms of the Taylor series expansion may be sufficient. On the other hand the piecewise linearization is generally best suited when large signals must be dealt with. The equivalent diode circuit, changed accordingly, is shown in Fig. 2.3.

When the forward conduction starts, switch S_1 closes and the junction capacitance is shunted with the parallel combination of an equivalent resistance R_d and a capacitance C_d (several orders larger than the junction capacitance $C(v)$). In long-hand calculations R_d may be assumed to be

$$R_d = \frac{\tau_d}{\lambda q_{\text{average}}} = \frac{1}{\lambda I_{\text{av}}} \approx \frac{2}{\lambda I_{\max}} \tag{2.11}$$

† In conformity with established practice the reverse-region diode voltages and charges are considered positive in varactor (voltage variable capacitor) applications.

‡ It is convenient to introduce the junction elastance instead of junction capacitance $C(v)$ which approaches ∞ when $v \to \Phi$, whereas $S(\Phi) \to 0$.

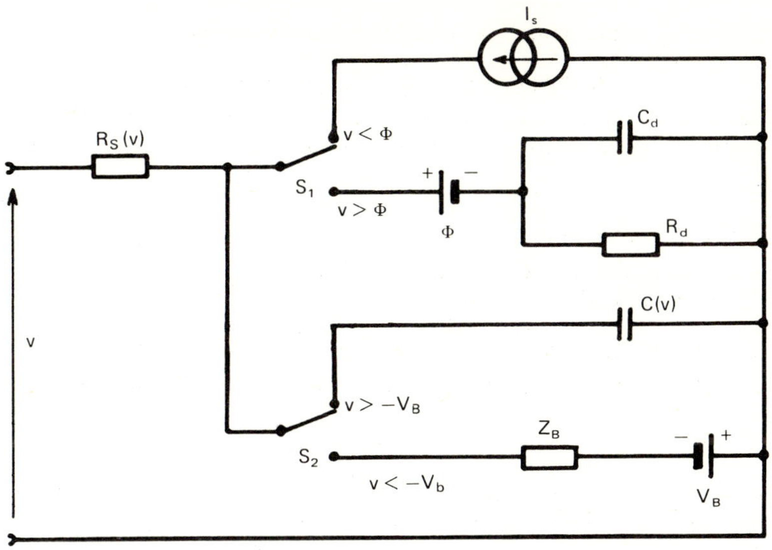

Fig. 2.3 Large-signal piecewise-linear equivalent circuit of the semiconductor diode.

(B) Nonlinear Resistor Operation

Because the diode resistance in the reverse bias region surpasses R_d by many orders it may be assumed to be infinite; it is therefore more convenient to introduce the conductance, which is effectively zero in these circumstances, into the general nonlinear equation.

$$i(t) = g(t)\, v(t) \tag{2.12}$$

In the frequency-multiplier operation the periodicity of the input signal causes all variables in (2.12) to be periodic functions of time with the same fundamental frequency. This suggests that the solution should be looked for through the frequency domain approach. Introducing the respective Fourier expansions

$$i(t) = \sum_{r=-\infty}^{\infty} I_r \exp(jr\omega t) \tag{2.13}$$

$$v(t) = \sum_{p=-\infty}^{\infty} V_p \exp(jp\omega t) \tag{2.14}$$

$$g(t) = \sum_{q=-\infty}^{\infty} G_q \exp(jq\omega t) \tag{2.15}$$

and comparing components with equal frequencies, one gets

$$I_r \exp(jr\omega t) = \sum_{p=-\infty}^{\infty} G_{r-p} V_p \exp(jr\omega t) \tag{2.16}$$

This equation system becomes particularly clear when written in the matrix form†

$$\begin{vmatrix} \vdots \\ I_1 \\ I_0 \\ I_{-1} \\ \vdots \end{vmatrix} = \begin{vmatrix} & \vdots & \vdots & \vdots & \vdots & \vdots & \\ \cdots & G_{-1} & G_0 & G_1 & G_2 & G_3 & \cdots \\ \cdots & G_{-2} & G_{-1} & G_0 & G_1 & G_2 & \cdots \\ \cdots & G_{-3} & G_{-2} & G_{-1} & G_0 & G_1 & \cdots \\ & \vdots & \vdots & \vdots & \vdots & \vdots & \end{vmatrix} \cdot \begin{vmatrix} \vdots \\ V_1 \\ V_0 \\ V_{-1} \\ \vdots \end{vmatrix} \tag{2.17}$$

The limited number of voltages in the actual multiplier circuit together with the piecewise linearization are the constraints making possible the solution of the above nonlinear matrix equation. Referring to Fig. 2.4, the input and output circuits present a short-circuit for all currents except the fundamental and the desired harmonic. As a result the effective diode voltage is

$$\begin{aligned} v(t) &= v_1 \cos(\omega t + \varphi_1) + v_n \cos(n\omega t + \varphi_n) - V_0 \\ &= \sum_{p=0,\pm 1,\pm n} V_p \exp(jp\omega t) \end{aligned} \tag{2.18}$$

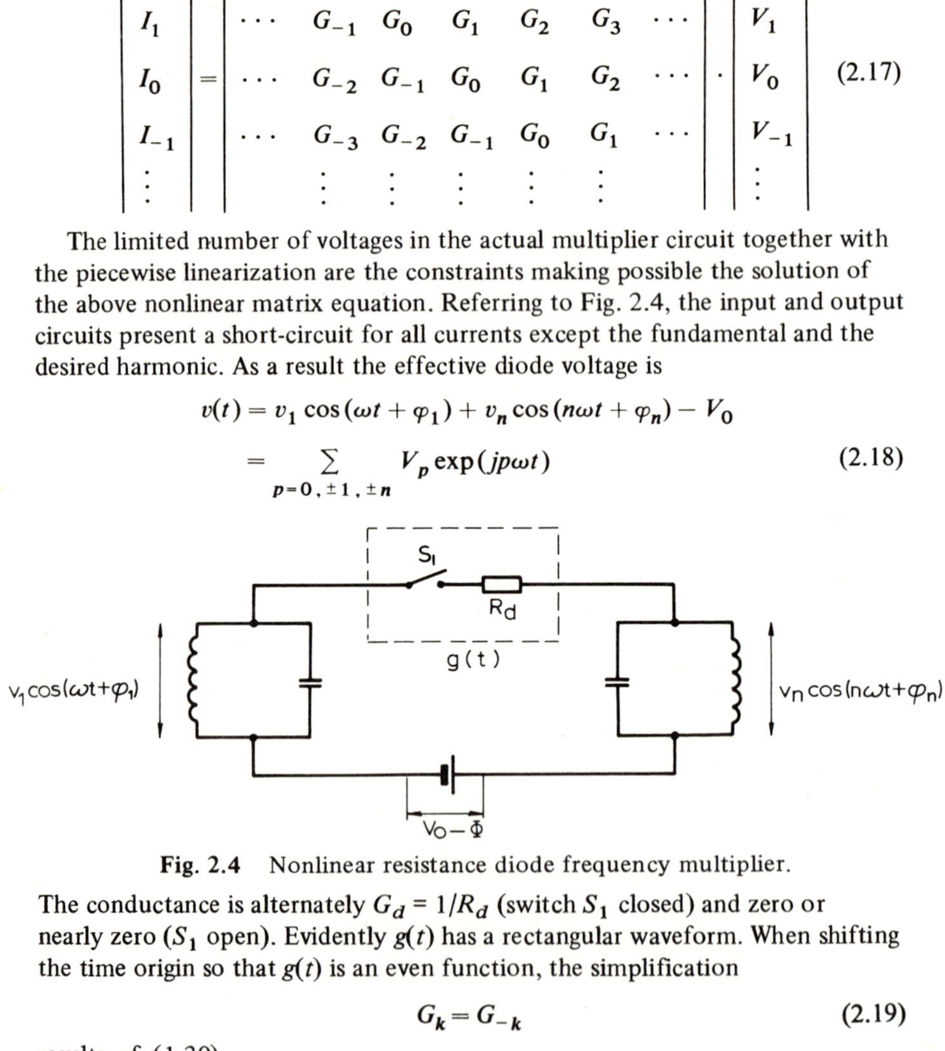

Fig. 2.4 Nonlinear resistance diode frequency multiplier.

The conductance is alternately $G_d = 1/R_d$ (switch S_1 closed) and zero or nearly zero (S_1 open). Evidently $g(t)$ has a rectangular waveform. When shifting the time origin so that $g(t)$ is an even function, the simplification

$$G_k = G_{-k} \tag{2.19}$$

results, cf. (1.30).

† The most important rules of matrix algebra are summarized in ref. [1.15, Section 13.2].

Now we are able to calculate the currents I_0, I_1, I_{-1}, I_n, and I_{-n} together with the input and output powers

$$P_{\text{in}} = V_1 I_{-1} + V_{-1} I_1 \tag{2.20}$$

$$-P_{\text{out}} = V_n I_{-n} + V_{-n} I_n \tag{2.21}$$

However, to keep the size of the book to reasonable limits we shall follow further only the leading ideas of the theory of nonlinear resistor frequency multipliers [2.15].

When substituting into (2.21) I_n and I_{-n} calculated from (2.17) the output power delivered into the load is

$$\begin{aligned} -P_{\text{out}} = G_{n-1}(V_1 V_{-n} + V_{-1} V_n) \\ + G_{n+1}(V_1 V_n + V_{-1} V_{-n}) + G_n V_0(V_n + V_{-n}) \\ + 2G_0 V_n V_{-n} + G_{2n}(V_n V_n + V_{-n} V_{-n}) \end{aligned} \tag{2.22}$$

Terms proportional to $|V_1| \cdot |V_n|$ are responsible for the power transfer and should be maximized; this leads to the condition $\varphi_1 \overset{+}{(-)} \varphi n = s\pi (s = 0, \pm 1, \ldots)$ and finally, taking into account the assumption (2.19), to

$$\varphi_1 = 0 \quad \text{i.e. } V_1 = V_{-1} \tag{2.23}$$

and

$$\varphi_n = 0, \text{ or } \pi \quad \text{i.e. } V_n = V_{-n} \tag{2.24}$$

By substituting for G_k using (1.30) and normalizing with respect to V_1 we get

$$\begin{aligned} P_{\text{out}} = \frac{2}{\pi} G_d V_1^2 \left\{ \frac{1}{n} \left| \frac{V_n}{V_1} \right| \left[\frac{\sin(n-1)\theta/2}{n-1} - \frac{\sin(n+1)\theta/2}{n+1} \right] \right| \\ - \left(\frac{V_n}{V_1} \right)^2 \left(\frac{\theta}{2} - \frac{\sin n\theta}{2n} \right) \bigg\} = P_{1,\text{norm}} \left\{ \frac{V_n}{V_1} a - \left(\frac{V_n}{V_1} \right)^2 b \right\} \end{aligned} \tag{2.25}$$

The biasing voltage V_0 has been removed by the condition for the current flow angle θ

$$2V_1 \cos \frac{\theta}{2} \overset{+}{(-)} 2V_n \cos n \frac{\theta}{2} - V_0 = 0 \tag{2.26}$$

Maximization of (2.25) in respect to $|V_n/V_1|$ and θ leads to

$$\frac{\theta}{2} = \frac{m}{n} \pi \quad (m = 1, 2, \ldots, < n) \tag{2.27}$$

and to

$$\frac{V_n}{V_1} = \frac{a}{2b} \tag{2.28}$$

Some results which are of interest to the designer are summarized in Fig. 2.5, where diagrams showing the optimized power loss (efficiency) and respective normalized output voltage $|V_n/V_1|$, biasing voltage $|V_0/2V_1|$, input conductance $G_g/G_d = R_d/R_g$, and output conductance $G_L/G_d = R_d/R_L$ are drawn as functions of the multiplication factor n for $m = 1$.

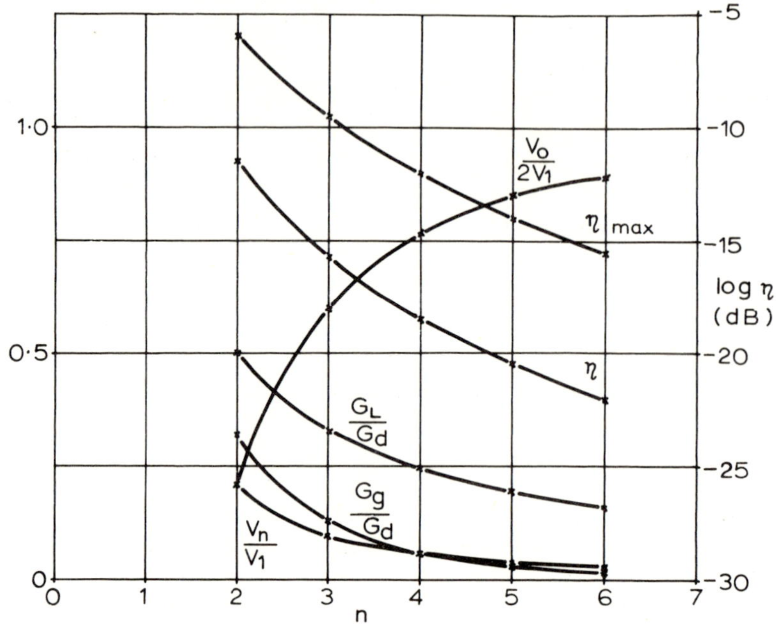

Fig. 2.5 Nonlinear resistance diode frequency multiplier: optimized power loss (efficiency), together with output voltage $|V_n/V_1|$, biasing voltage $|V_0/2V_1|$, and input and output conductances $G_g/G_d = R_d/R_g$ and $G_L/G_d = R_d/R_L$ respectively, as functions of the multiplication factor n (the operation mode chosen being $m = 1$, i.e. $\theta/2 = \pi/n$. For comparison $\eta_{max} = 1/n^2$ is indicated).

Maximization of the efficiency, $\eta = P_{out}/P_{in}$, instead of the output power, leads to different results for small current-flow angles because the first maximum appears for $m = 0$ and not for $m = 1$; the remaining maxima are approximately the same as in (2.27). Returning to the case of $m = 0$ one finds that the efficiency attains its maximum value $\eta_{max} = 1/n^2$. This is the value which is so often referred to as the optimum theoretical efficiency, but without indicating that besides matching difficulties ($G_g, G_L \to 0$) the output power tends to zero; this follows immediately from (2.25) when introducing $\theta/2 \to 0$. On the other hand the larger m is, the larger is the output power but the smaller the efficiency. A reasonable compromise, often used in practice, is to choose $\theta/2 = \pi/2$. In this case $\eta \approx (2/\pi)^2/n^4$ (n large and even), $G_g = G_L = G_d/2$, and $V_n/V_1 \approx a/2b$ as in (2.28).

(C) Nonlinear Capacitor (Varactor) Operation

For true varactor applications only the nonlinearity of the junction capacitance $C(v)$ in the voltage range from $+V_B$ to $-\Phi$ is exploited, i.e. both switches, S_1 and S_2, in Fig. 2.3 are permanently open. The low-frequency limit depends predominantly upon the desired input and output powers. It may be shown that

$$P'_{\text{in}} = P'_{\text{out}} \sim \frac{\omega}{S_{\max}}(V_B + \Phi)^2 = \frac{\omega}{\omega_c}\frac{(V_B + \Phi)^2}{R_S} = \frac{\omega}{\omega_c}P_{\text{norm}} \tag{2.29}$$

In other words the series impedance $1/\omega C_{\min}$ must be reasonably low. However, when this impedance becomes comparable with that of the diode bulk resistance R_s the efficiency drops off rapidly and the high-frequency limit is attained. The cut-off frequency is defined as

$$\omega_c = \frac{S_{\max}}{R_s} \tag{2.30}$$

At present, the useful frequency range of varactor frequency multipliers is below 100 GHz. The highest cut-off frequencies reported are in the range from 100 to 400 GHz though exceptionally up to 1000 GHz with the latest technology.

Overdriving of a varactor in the conduction region (switch S_1 temporarily closed) results in larger capacity changes and consequently in larger powers. At the same time the efficiency is not reduced appreciably if the input frequency is $\omega > 1/\tau_d$. Overdriving is the only possible working mode for step recovery diodes (see below) because the junction capacitance in the reverse bias region is almost voltage independent.

In Fig. 2.6(a) the equivalent circuit of the nonlinear capacitance diode is redrawn with all stray impedances and admittances added. In actual conditions the inductances L'_s and L''_s, of wires leading to the semiconductor, and the capacitance C_s introduced by the container may be tuned out by input and output filters. The resistance R_B, which accounts for the losses due to the reverse and leakage currents, is ordinarily so high that it may be neglected. Finally, combining the series stray resistance R'_s, due to connexions, with R_s we get the simple equivalent circuit in Fig. 2.6(b) on which all mathematical treatments of varactor multipliers are based.

Introducing time t as the independent variable into (2.10) leads to

$$\frac{dv(t)}{dt} = s(t).i(t) \tag{2.31}$$

where $s(t)$ is the instantaneous elastance. Noting the resemblance to (2.12), the solution of (2.31) may proceed according to the same ideas. Introducing the

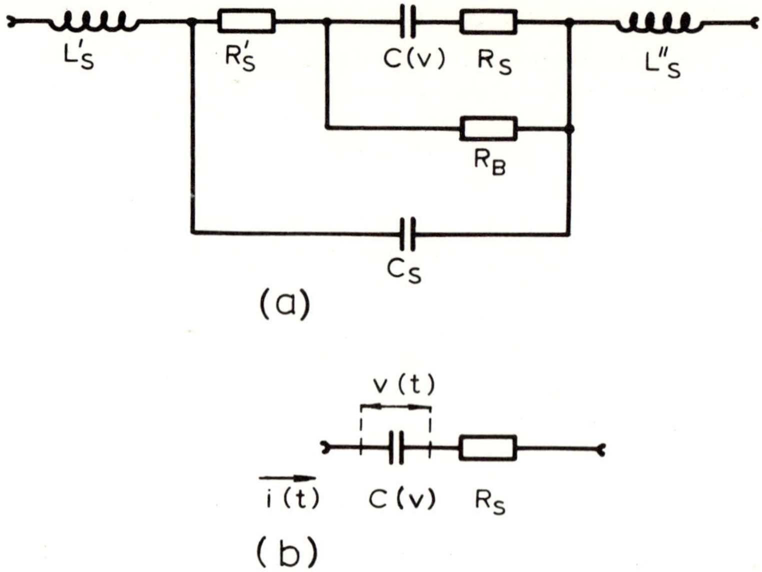

Fig. 2.6 Equivalent circuit of nonlinear capacitance diode: (a) with all stray impedances and admittances, and (b) simplification on which all mathematical treatments of varactor multipliers are based.

Fourier expansions (2.13), (2.14), and

$$s(t) = \sum_{r=-\infty}^{\infty} S_r \exp(jr\omega t) \tag{2.32}$$

and comparing the components with equal frequency we get

$$jk\omega V_k = \sum_{m=-\infty}^{\infty} S_{k-m} I_m \tag{2.33}$$

The use of charges instead of currents may sometimes be desirable.

$$i(t) = \frac{dq(t)}{dt} = \frac{d}{dt} \sum_{m=-\infty}^{\infty} Q_m \exp(jm\omega t) = \sum_{m=-\infty}^{\infty} jm\omega Q_m \exp(jm\omega t)$$

$$= \sum_{m=-\infty}^{\infty} I_m \exp(jm\omega t) \tag{2.34}$$

Thus

$$kV_k = \sum_{m=-\infty}^{\infty} mS_{k-m} Q_m \tag{2.35}$$

Both (2.33) and (2.35) may be expressed as nonlinear matrix equations

$$(jk\omega V_k) = (S_{k-m})(I_m) \tag{2.36}$$

or

$$(kV_k) = (S_{k-m})(m.Q_m) \tag{2.37}$$

The particular value of γ in (2.9) and the limited number of currents† in actual multiplier circuits are the constraints making possible the solution of the matrix equation (2.36) or (2.37). The basic circuit is shown in Fig. 2.7. The network N_1 provides a match at the fundamental frequency and blocks all other frequencies; similarly N_2 provides a match at the second harmonic and blocks all other frequencies. Nevertheless, even in the simplest cases the use of a

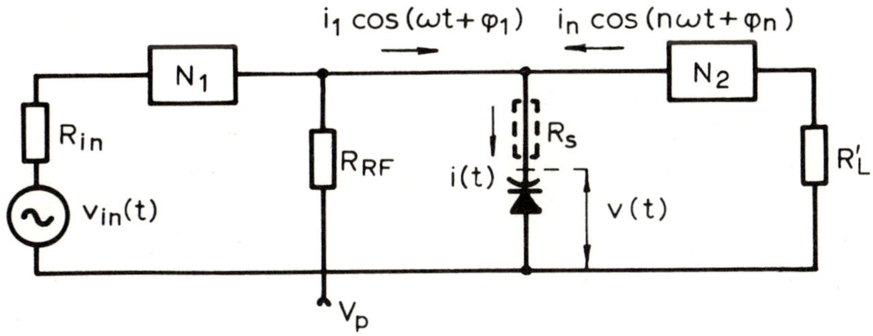

Fig. 2.7 The basic circuit of the abrupt junction varactor doubler; the d.c. biasing voltage V_p gets to the diode through R_{RF} which has, however, an infinite impedance for all a.c. involved.

computer is essential for establishing optimum working conditions. The results are published in the form of charts in an early textbook [2.5]. Extension of the analysis of varactor frequency multipliers for arbitrary capacitance variation and drive level has been worked out, e.g. in [2.6], whereas microwave frequency multipliers are discussed in [2.7].

Two-signal abrupt-junction varactor multipliers cannot be realized for $n > 2$. To enable the power exchange between input and output circuits to take place, the auxiliary second harmonic signal and, eventually, the others as well, must be allowed to circulate in the nonlinear capacitance with the help of an 'idling' circuit which passes only the respective auxiliary signal and blocks all other signals, see the basic circuit of the quadrupler in Fig. 2.8.

Another way to realize high-order varactor multipliers without idlers is the use of diodes having doping profiles other than the abrupt junction configuration. The step recovery diodes are particularly suitable for this purpose.

† Note the duality form of (2.12) and (2.31) in respect to the admittance–impedance and consequently the series and the shunt connexion of the diodes in Fig. 2.4 and 2.7.

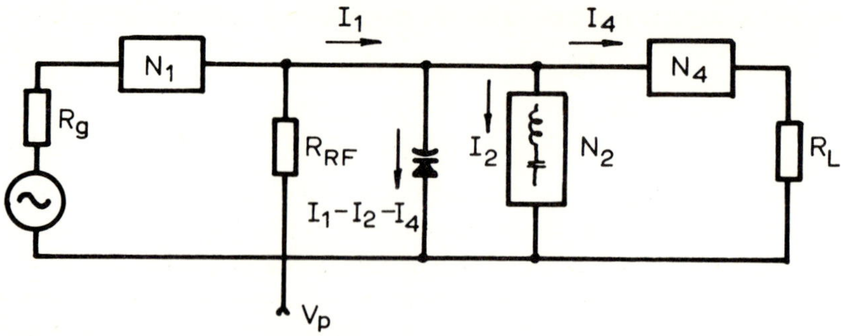

Fig. 2.8 The basic circuit of the abrupt-junction varactor quadrupler with one idling circuit N_2.

(D) Step-Recovery-Diode Operation

When a step recovery diode (SRD) is reverse-polarized the junction capacitance $C(v)$ is nearly voltage-independent, and to a reasonable approximation may be considered equal to C_{min}. Whereas during the conduction period (if much shorter than τ_d) – switch S_1 closed – the charge of minority carriers is responsible for the equivalent capacitance C_d which is effectively infinite with respect to C_{min}. Evidently in the piecewise linearization the SRD may be replaced by the elastance $s(t)$ alternating between $S_B = 1/C_{min}$ and zero. For simplicity in analysis, the SRD will be allowed to enter the forward bias region once only per fundamental cycle; further the time origin will be shifted so that $s(t)$ is an even function, i.e. $S_k = S_{-k}$.

The basic frequency-multiplying circuit is that of Fig. 2.7. There are only two harmonic currents

$$i(t) = i_1 \cos(\omega t + \varphi_1) + i_n \cos(n\omega t + \varphi_n) \tag{2.38}$$

and the diode series resistance R_s may be thought of as a resistor in the series with input and load resistances, i.e.

$$R'_{in} = R_{in} + R_s \quad \text{and} \quad R_L = R_s + R'_L \tag{2.39}$$

Consequently the SRD may be replaced with a pure nonlinear elastance, in which no power can be dissipated, and

$$P_{in} = P_{out} \tag{2.40}$$

From (2.39) and (2.40) one readily derives the power conversion efficiency of the frequency multiplier

$$\eta = \frac{\text{harmonic power output}}{\text{fundamental power input}} = \frac{R'_L/R_L}{R'_{in}/R_{in}} = \frac{1 - R_s/R_L}{1 + R_s/R_{in}} \tag{2.41}$$

In the following we shall proceed in the same way as in the case of the nonlinear resistor above. Inserting (2.38) into (2.36), the fundamental and the

nth harmonic voltages will be calculated. Then, employing (2.20) we get the input power to the idealized SRD.

$$P_{in} = \frac{1}{j\omega}[S_{n-1}(I_{-1}I_n - I_1 I_{-n}) + S_{n+1}(I_{-1}I_{-n} - I_1 I_n) + S_2(I_{-1}I_{-1} - I_1 I_1)] \tag{2.42}$$

Because all the input power must be delivered to the load (see eq. (2.40)) the last term in the above equation must be zero, which is achieved when

$$2\varphi_1 = k\pi \quad (k = 0, \pm 1, \ldots)$$

Further optimization gives:

$$\left.\begin{aligned} \varphi_1 &= \pi/2 \\ \varphi_n &= \pi, 2\pi \end{aligned}\right\} \tag{2.43}$$

and finally the maxima of $(S_{n-1} - S_{n+1})$ are found for the non-conducting diode angles

$$\frac{\theta}{2} = \frac{m}{n}\pi \quad (m = 1, 2, \ldots < n) \tag{2.44}$$

From now on, only results that are of importance for the designer of an SRD frequency multiplier will be considered. The reader interested in the respective derivations is referred to [2.8].

One readily gets from (2.40)

$$\frac{R_L}{R_{in}} = \left|\frac{I_1}{I_n}\right|^2 \tag{2.45}$$

and from the condition of one conduction period only per fundamental cycle

$$\left(\frac{R_L}{R_{in}}\right)_{min} = MRR = \frac{1}{\sin^2(m\pi/n)} \tag{2.46}$$

Further (2.42) to (2.45) give

$$R_{in}R_L = \left(\frac{S_{n-1} + S_{n+1}}{\omega}\right)^2 = \left(\frac{K_G}{\omega C}\right)^2 \quad (C = C_{min}) \tag{2.47}$$

The load resistance for the maximum power conversion efficiency is

$$R_L = \frac{K_G}{\omega C}\left[\left(\frac{\omega^2}{\omega_{cn}^2} + 1\right)^{1/2} - \frac{\omega}{\omega_{cn}}\right]^{-1} \quad (\omega_{cn} = \omega_c K_G) \tag{2.48}$$

The maximum efficiency at low frequencies

$$\eta_{LF} = 1 - \frac{\omega}{\omega_c}K_{LF} \tag{2.49}$$

and the approximate value for high frequencies

$$\eta_{HF} \approx \left(\frac{\omega_{cn}}{2\omega}\right)^2 \qquad (\omega/\omega_{cn} \geqslant 2) \tag{2.50}$$

Three different diode properties may limit the power-handling capability of the multiplier: reverse breakdown voltage, maximum power dissipation, and peak current flow. In a conventional varactor multiplier, the reverse breakdown voltage is usually the limiting factor.

Case of the Reverse Breakdown Limitation

From (2.29)

$$P_{0,\,\max} = \frac{(V_B + \Phi)^2}{R_s} \cdot \frac{\omega}{\omega_c} K_G K'_{PO} = P_{\text{norm}} \frac{\omega}{\omega_c} K_G K'_{PO} \tag{2.51}$$

At low frequencies the power-handling capability of the SRD multiplier can be increased if the cut-off frequency ω_c is artificially reduced through a parallel capacity, C_{ext}, to the diode. Naturally the efficiency will be reduced but (2.49) will give a rather pessimistic value (some of the current is shunted through C_{ext}).

$$\frac{P_{0,\,\max}}{1 - \eta_{LF}} = P_{\text{norm}} \cdot K_{FM} \tag{2.52}$$

Case of the Maximum Power Dissipation

At very high frequencies, where the conversion efficiency of the multiplier is low, the power-handling capability of the circuit is likely to be limited by maximum power dissipation of the SRD. In this situation maximum output power will always be obtained if the diode is operated in the most efficient mode; i.e.

$$P_{0,\max} = \frac{P_{\text{dis}} \cdot \eta}{1 - \eta} \tag{2.53}$$

where P_{dis} is the maximum permitted power dissipation of the diode.

All constants which have been introduced in equations (2.46) to (2.53) are tabulated, for different values of n and m, in Table 2.1 which relies on more extensive data (for n up to 20) published in [2.8].

The two-current (charge) SRD frequency multiplier is advantageous when the circuit simplicity is important and where the output frequency is very high. It has been found that the conversion efficiency is maximum if the conduction period is equal to one half of the fundamental period. On the contrary, at low frequencies the maximum output of the circuit is greatest if the conduction angle is large (i.e. as $m \to 0$).

Table 2.1

Constants for the design of the SRD frequency multipliers

n	m	K_G	MRR	K_{LF}	$K_G K'_{PO}$	K_{FM}
2	1	0·2122	1·0	9·42	$0{\cdot}629 \times 10^{-1}$	$0{\cdot}677 \times 10^{-2}$
3	1	$0{\cdot}689 \times 10^{-1}$	1·333	$2{\cdot}93 \times 10^{1}$	$0{\cdot}722 \times 10^{-1}$	$0{\cdot}246 \times 10^{-2}$
4	1	$0{\cdot}3001 \times 10^{-1}$	2·00	$7{\cdot}069 \times 10^{1}$	$0{\cdot}879 \times 10^{-1}$	$0{\cdot}124 \times 10^{-2}$
	2	$0{\cdot}4244 \times 10^{-1}$	1·00	$4{\cdot}712 \times 10^{1}$	$0{\cdot}150 \times 10^{-1}$	$0{\cdot}319 \times 10^{-3}$
5	1	$0{\cdot}1559 \times 10^{-1}$	2·894	$1{\cdot}468 \times 10^{2}$	0·105	$0{\cdot}718 \times 10^{-3}$
	2	$0{\cdot}2523 \times 10^{-1}$	1·106	$7{\cdot}938 \times 10^{1}$	$0{\cdot}179 \times 10^{-1}$	$0{\cdot}225 \times 10^{-3}$
6	1	$0{\cdot}9095 \times 10^{-2}$	4·00	$2{\cdot}749 \times 10^{2}$	0·124	$0{\cdot}450 \times 10^{-3}$
	3	$0{\cdot}1819 \times 10^{-1}$	1·00	$1{\cdot}10 \times 10^{2}$	$0{\cdot}703 \times 10^{-2}$	$0{\cdot}639 \times 10^{-4}$
7	1	$0{\cdot}5755 \times 10^{-2}$	5·31	$4{\cdot}759 \times 10^{2}$	0·142	$0{\cdot}299 \times 10^{-3}$
	3	$0{\cdot}1293 \times 10^{-1}$	1·052	$1{\cdot}547 \times 10^{2}$	$0{\cdot}807 \times 10^{-2}$	$0{\cdot}522 \times 10^{-4}$
8	1	$0{\cdot}3867 \times 10^{-2}$	6·828	$7{\cdot}747 \times 10^{2}$	0·161	$0{\cdot}208 \times 10^{-3}$
	4	$0{\cdot}1011 \times 10^{-1}$	1·00	$1{\cdot}979 \times 10^{2}$	$0{\cdot}412 \times 10^{-2}$	$0{\cdot}208 \times 10^{-4}$
9	1	$0{\cdot}2722 \times 10^{-2}$	8·549	$1{\cdot}200 \times 10^{3}$	0·180	$0{\cdot}150 \times 10^{-3}$
	4	$0{\cdot}7837 \times 10^{-2}$	1·031	$2{\cdot}552 \times 10^{2}$	$0{\cdot}461 \times 10^{-2}$	$0{\cdot}181 \times 10^{-4}$

(Extracted from Johnston & Boothroyd [2.8])

2.3 CLASS C FREQUENCY MULTIPLIERS

The advantage of class C frequency multipliers is the loose coupling between the input and output circuits and the power gain at the desired harmonic. The theory of operation is the same as that of class C amplifiers with the exception that the loading resonant circuit is tuned to the desired harmonic and not to the fundamental frequency.

Class C frequency multipliers using electron tubes are still frequently used in radio transmitters. The design procedure was worked out in the late thirties and has since formed part of many textbooks, e.g. [1.12, Section 13-5].

The treatment of transistor class C amplifiers and multipliers may be based upon approaches originally applied to analogous electron tube circuits, particularly for frequencies far below f_α. However, the large signal validity of the Ebers-Moll transistor model [2.9] provides both a much better insight, and also means for the analytical solution [2.10]. The basic Ebers–Moll transistor equations are:

$$i_e = I_{ES}[\exp(\lambda v_{be}) - 1] - \alpha_R I_{CS}[\exp(-\lambda v_{cb}) - 1] \tag{2.54a}$$

$$i_c = \alpha_F I_{ES}[\exp(\lambda v_{be}) - 1] - I_{CS}[\exp(-\lambda v_{cb}) - 1] \tag{2.54b}$$

where λ is that of eq. (1.18); α_F and α_R are the forward and reverse short-circuit current gains, and I_{ES} and I_{CS} are the emitter and collector saturation currents mutually connected through the reciprocity relation

$$\alpha_F I_{ES} = \alpha_R I_{CS} \tag{2.55}$$

The sign conventions are as in Fig. 2.9.

Assuming that saturation is avoided ($V_{cn} < V_{CC}$) then the collector current is given by a simple exponential law:

$$i_c = \alpha_F I_{ES}[\exp(\lambda v_{be}) - 1] + I_{CS} \tag{2.56}$$

As long as the respective amplifier-multiplier can be considered as voltage driven, and as long as the source resistance R_g and the lumped base resistance

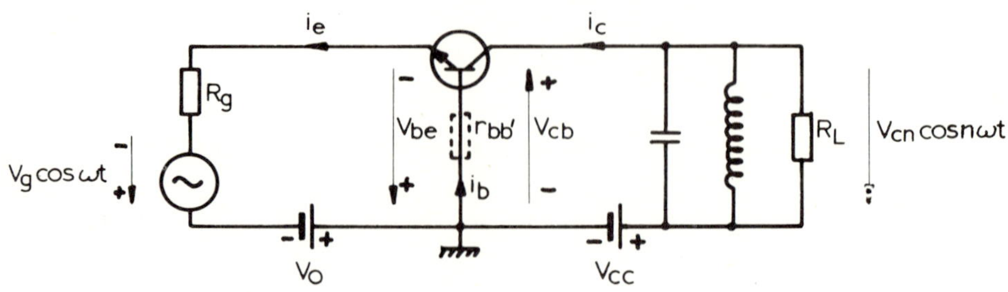

Fig. 2.9 Common base large-signal (Class C) transistor frequency multiplier. (From Harrison [2.10].)

$r_{bb'}$ are small compared with the intrinsic input impedance of the stage, then

$$v_{be} = V_0 + V_g \cos \omega t \tag{2.57}$$

and harmonic components I_{cn} of the collector current can easily be found by using the Jacobi-Anger formula (1.71)

$$I_{cn} = \alpha_F I_{ES} \exp(\lambda V_0) 2\bar{I}_n(\lambda V_g)^{\dagger} \tag{2.58}$$

The d.c. power supplied to the amplifier-multiplier is

$$P_s = V_{CC} I_{c0} = V_{CC}\{\alpha_F I_{ES}[\exp(\lambda V_0)\, \bar{I}_0(\lambda V_g) - 1] + I_{CS}\}$$

$$= V_{CC}\, \alpha_F I_{ES}\left[\exp(\lambda V_0)\, \bar{I}_0(\lambda V_g) + \frac{1}{\beta_R}\right] \tag{2.59}$$

where $\beta_R = \alpha_R/(1 - \alpha_R)$ and β_R^{-1} usually may be neglected. Evidently the d.c. component in the collector current is practically

$$I_{c0} = \alpha_F I_{ES} \exp(\lambda V_0)\, \bar{I}_0(\lambda V_g) \tag{2.60}$$

This is a very important relation because the quantity $\alpha_F I_{ES} \exp(\lambda V_0)$ may be expressed as a simple function of the d.c. collector current which is easily measured.

The power dissipated in the collector loading circuit tuned to the nth harmonic is

$$P_n = \frac{1}{2} V_{cn} \cdot I_{cn} = V_{cn} \cdot I_{c0} \cdot \frac{\bar{I}_n(\lambda V_g)}{\bar{I}_0(\lambda V_g)} \tag{2.61}$$

When exciting the amplifier-multiplier to the threshold of saturation, the a.c. peak amplitude V_{cn} is nearly equal to V_{CC}, and

$$P_{n,\max} \approx V_{CC} \cdot I_{c0} \cdot \frac{\bar{I}_n(\lambda V_g)}{\bar{I}_0(\lambda V_g)}, \tag{2.62}$$

the maximum collector efficiency is

$$\eta_{\max} = \frac{P_{n,\max}}{P_s} \approx \frac{\bar{I}_n(\lambda V_g)}{\bar{I}_0(\lambda V_g)}, \tag{2.63}$$

and the optimum resonance impedance of the collector loading circuit is

$$R_{L,\mathrm{opt}} \approx \frac{1}{2} \cdot \frac{V_{CC}}{I_{c0}} \cdot \frac{\bar{I}_0(\lambda V_g)}{\bar{I}_n(\lambda V_g)}. \tag{2.64}$$

† To avoid confusion with currents, the modified Bessel functions are denoted here by $\bar{I}_n(\lambda V_g)$.

The input a.c. power in the common base connexion is

$$P_{i,B}=\frac{1}{2}V_g.\frac{I_{c1}}{\alpha_F}=\frac{1}{\alpha_F}V_g.I_{c0}.\frac{\bar{I}_1(\lambda V_g)}{\bar{I}_0(\lambda V_g)} \tag{2.65}$$

From (2.61) and (2.65) the power gain may be calculated

$$G_{p,B}=\frac{P_n}{P_{i,B}}=\alpha_F\frac{\bar{I}_n(\lambda V_g)}{\bar{I}_1(\lambda V_g)}\cdot\frac{V_{cn}}{V_g} \tag{2.66}$$

and in the common emitter connexion respectively

$$G_{p,E}=\frac{G_{p,B}}{1-\alpha_F}=\beta_F\cdot\frac{\bar{I}_n(\lambda V_g)}{\bar{I}_1(\lambda V_g)}\cdot\frac{V_{cn}}{V_g} \tag{2.67}$$

Similarly the input impedances at the fundamental frequency are

$$R_{i,B}=\alpha_F\cdot\frac{V_g}{I_{c1}}=\frac{1}{2}\alpha_F\cdot\frac{V_g}{I_{c0}}\cdot\frac{\bar{I}_0(\lambda V_g)}{\bar{I}_1(\lambda V_g)} \tag{2.68}$$

$$R_{i,E}=\frac{1}{2}\beta_F\frac{V_g}{I_{c0}}\frac{\bar{I}_0(\lambda V_g)}{\bar{I}_1(\lambda V_g)} \tag{2.69}$$

In equations (2.61) to (2.69) the higher-order Bessel functions $\bar{I}_n(\lambda V_g)$ normalized to the *zero*th order $\bar{I}_0(\lambda V_g)$ are encountered. Because of their evident importance these normalized quantities are given in Fig. 2.10. With its help, when designing the transistor multiplier stage, the normalized input voltage λV_g may be so chosen as to ensure both a reasonable collector efficiency (for the respective harmonic) and a safe base-emitter voltage v_{be}.

The other important design parameter is the d.c. collector current I_{c0} which is finally adjusted through the emitter-base bias voltage V_0 in such a way that neither the collector dissipation

$$P_d=V_{CC}\cdot I_{c0}\left[1-\frac{V_{cn}}{V_{CC}}\cdot\frac{\bar{I}_n(\lambda V_g)}{\bar{I}_0(\lambda V_g)}\right] \tag{2.70}$$

nor the admitted maximum peak collector current are exceeded. According to eq. (2.54(b)) and (2.60)

$$\frac{I_{c,\,\mathrm{peak}}}{I_{c0}}=\frac{\exp(\lambda V_g)}{I_0(\lambda V_g)} \tag{2.71}$$

This last relationship is drawn in Fig. 2.11.

To understand the behaviour of transistor amplifier–multiplier stages at higher frequencies Harrison suggests in [2.10] the use of Koehler's [2.4] charge-control concept of the large-signal transistor model, see Fig. 2.12. Solution in closed form is possible but for limited cases only; the general solution is of the form

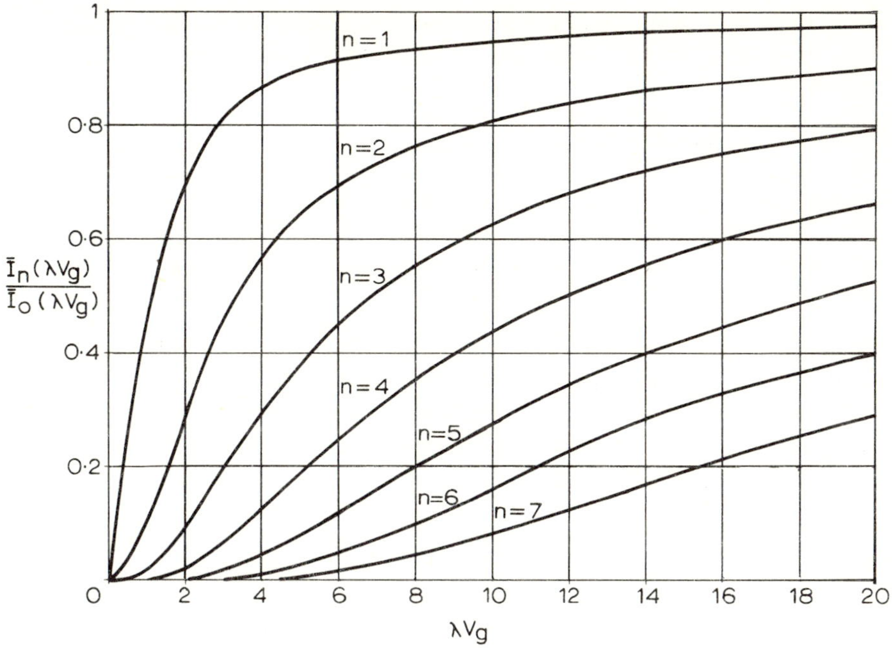

Fig. 2.10 Plot of modified Bessel functions, of higher orders, $\bar{I}_n(\lambda V_g)$ normalized to that of the *zero*th order $\bar{I}_0(\lambda V_g)$ vs. the argument λV_g.

$$\sum_{-\infty}^{+\infty} I_{cp} . \exp(jp\omega t) = \alpha_F . I_{ES} . \exp\left(\lambda \sum_{-\infty}^{+\infty} \frac{V_{gq} - (Z_1 + jq\omega Z_2) . I_{cq}}{1 + jq\omega T_1} \exp(jq\omega t)\right) \tag{2.72}$$

where the left-hand side is the Fourier expansion of the collector current $[i_c(t) - I_{CS}]$, and the constants on the right-hand side are

$$V_{g,q} = 0, \quad q \neq 0, \quad \text{and} \quad \pm 1 \tag{2.73a}$$

$$V_{g,0} = V_0 + \alpha_F I_{ES} \left[\frac{r_{bb'}}{\beta_R} + \frac{R_g + r_{bb'}}{\beta_F}\right] \tag{2.73b}$$

$$V_{g,1} = V_{g,-1} = \frac{1}{2} V_g ; \tag{2.73c}$$

$$T_1 = C_{TE}(R_g + r_{bb'}) ; \tag{2.74}$$

(C_{TE}, the emitter transition capacitance is a function of the junction voltage; when deriving (2.72) C_{TE} was approximated with a voltage-independent capacitance.)

$$Z_1 = \frac{R_g + (1 - \alpha_F) r_{bb'}}{\alpha_F} \tag{2.75}$$

$$Z_2 = \frac{1}{\omega_{\alpha F}} \cdot \frac{R_g + r_{bb'}}{\alpha_F}; \quad \frac{\omega_{\alpha F}}{2\pi} \text{ is the forward } \alpha\text{-cut-off frequency} \tag{2.76}$$

Examination of (2.72) proves useful also when looking for methods which should be adopted for the solution of the current-driven transistor frequency multiplier, i.e. when Z_1 resp. R_g is large. In this case the resemblance between (1.20) and (2.72) is evident and the actual working conditions may be simplified to those of an ideal switch in series with R_g. The current flow angles are chosen according

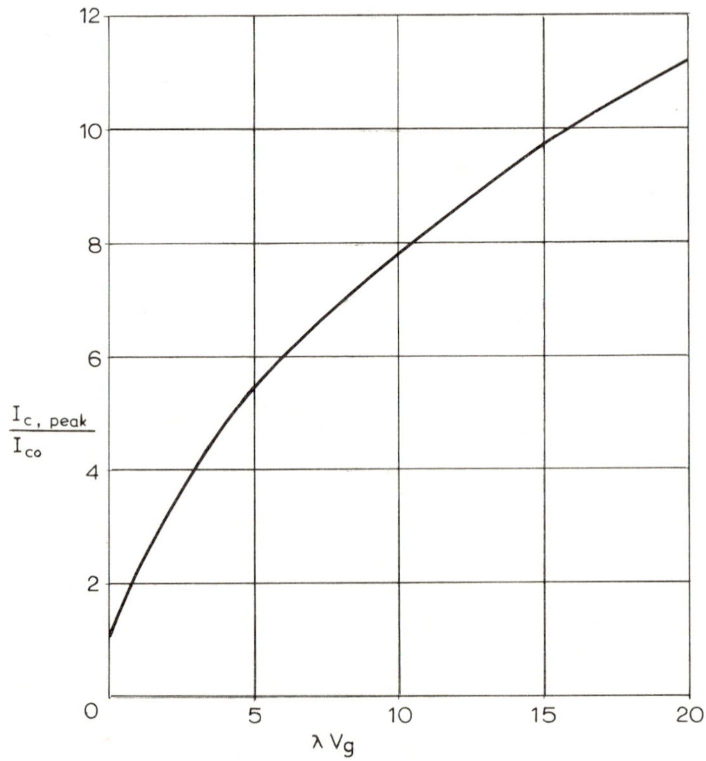

Fig. 2.11 Ratio of the peak collector current to the respective d.c. component for different values of λV_g in the voltage-driven transistor frequency multiplier.

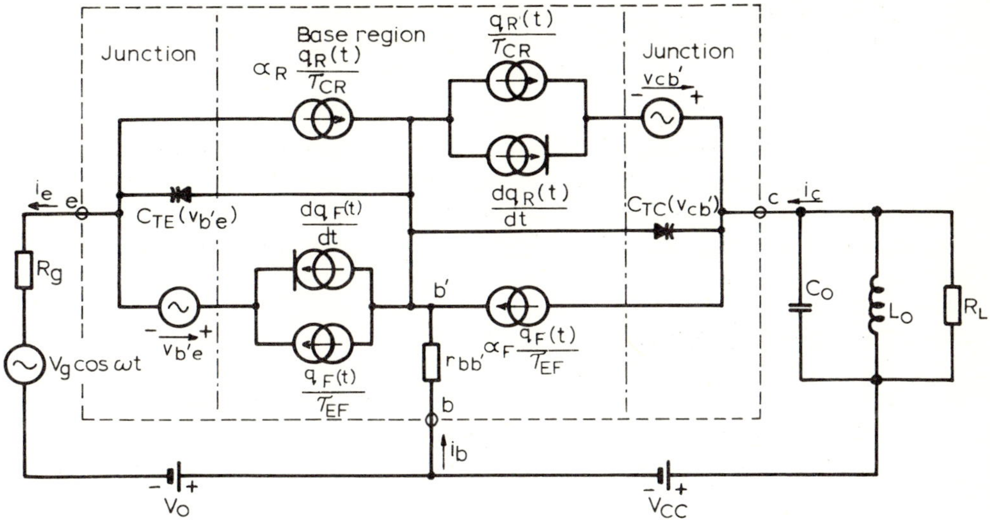

Fig. 2.12 Charge-control concept of the large-signal transistor model. Time constants τ_{EF} and τ_{CR} are inverses of the forward and reverse α-cut-off frequencies α_F and α_R respectively. Charges $q_F(t)$ and $q_R(t)$ are related to the transistor currents by $q_F(t)/\tau_{EF} = I_{ES}[\exp(\lambda v_{b'e})-1]$ and $q_R(t)/\tau_{CR} = I_{CS}[\exp(-\lambda v_{cb'})-1]$. Current generators ⊕| are charge stores which cannot sustain any voltage drop. (Harrison [2.10] and Koehler [2.4]. Copyright, 1967, American Telephone & Telegraph Co., reprinted by permission.)

to the maxima in Fig. 1.3(a) and the pertinent i_n/i_p are used for calculating the desired collector harmonic current I_{cn}. The modification of the above equations (2.59) to (2.70) is obvious. The limiting factors are either the collector dissipation, the collector peak current, or the base-emitter reverse voltage.

Now directing attention to the higher frequencies, (2.72) reveals for the voltage-driven case that the amplitude of the input voltage is reduced to an effective value $V_g/\sqrt{(1+\omega^2 T_1^2)}$ and shifted in phase by $\theta = \arctan \omega T_1$.†

Finally it should be mentioned that, with simultaneous operation of the transistor as a power amplifier and the base-collector junction capacitance, C_{TC} in Fig. 2.12, as a varactor frequency multiplier, useful harmonic power may be gained well above the transistor cut-off frequency, with satisfactory efficiency. The simplified circuit analysis together with the experimental results are given in [2.11].

---

† Thus it is possible to proceed with the calculation by methods analogous to those adopted in the low-frequency case. Precise results are given by Harrison [2.10].

2.4 SIGNAL-TO-NOISE AND PHASE-STABILITY DEGRADATION IN FREQUENCY MULTIPLIERS

The frequency multiplication process deteriorates the spectral purity and the long-term phase stability with respect to the parameters of the input signal. If very high multiplication factors are involved and inadequate precautions are taken, the output signal-to-noise ratio may become insufficient for many applications in frequency synthesis, frequency measurement, communications, etc.

(A) Degradation of the Signal-to-Noise Ratio

Considering that amplitude and phase disturbances (due to discrete signals with frequencies $\omega_s \approx \omega$, noise, etc.) of the sinusoidal input signal are small and band-limited, expressions (9.4) or (9.15) derived in Chapter 9 may be applied, i.e.

$$v_{\text{in}}(t) = v_1(t)\cos[\omega t + \varphi_1(t)] = \frac{v_1(t)}{2}[\![\exp\{j[\omega t + \varphi_1(t)]\}$$

$$+ \exp\{-j[\omega t + \varphi_1(t)]\}]\!] \qquad (2.77)$$

where both the amplitude $v_1(t)$ and the phase $\varphi_1(t)$ are slowly varying functions of time.

Assuming the power series description of the nonlinear device

$$i(t) = \sum_{k=0}^{\infty} a_k v^k \qquad (2.78)$$

and applying the binomial theorem on $[\![\exp\{j[\omega t + \varphi_1(t)]\} + \exp\{-j[\omega t + \varphi_1(t)]\}]\!]$ one readily sees that the nth harmonic component is

$$v_{\text{out}}(t) = v_n(t)\cos n[\omega t + \varphi_1(t)] \qquad (2.79)$$

In other words the slowly varying phase perturbation increases in the same proportion as the frequency.†

The behaviour of frequency multipliers in respect of spurious amplitude modulation is different. In the large-signal diode multipliers, discussed in Section 2.2, the small amplitude-modulation indices do not change, i.e.

$$m_{a1} = m_{an} \qquad (2.80)$$

For varactor multipliers the above statement follows immediately from (2.45), and for the nonlinear-resistor diode multipliers from (2.25) if (2.27) and (2.28)

---

† For the exact proof one would have to make use of the general intermodulation product theorem as described in Appendix 4.1.

are fulfilled. Also transistor class C multipliers, if current driven and optimized in respect to current flow angle θ, behave similarly.

If the noise power in the input signal is equally distributed between the amplitude and phase components, the signal-to-noise ratio is

$$\left(\frac{S}{N}\right)_{in} = \frac{P_S}{P_{NA} + P_{NP}} \qquad (P_S \gg P_{NA} = P_{NP}) \tag{2.81}$$

At the output of these types of frequency multiplier the situation is changed to

$$\left(\frac{S}{N}\right)_{out} = \frac{P_S}{P_{NA} + n^2 P_{NP}} \tag{2.82}$$

which may also be written as

$$\left(\frac{S}{N}\right)_{out} = \left(\frac{S}{N}\right)_{in} + 3 - 10 \log (n^2 + 1) \qquad \text{dB} \tag{2.83}$$

If the transistor class C multiplier is voltage-driven the small amplitude-modulation index increases more than n-times, as shown in Appendix 2.1, and the output signal-to-noise ratio is nearly

$$\left(\frac{S}{N}\right)_{out} = \left(\frac{S}{N}\right)_{in} - 20 \log n \qquad \text{dB} \tag{2.84}$$

When comparing (2.83) and (2.84) it is clear that no appreciable error is made if one uses simply the latter expression (2.84) for the signal-to-noise ratio deterioration.

Example 2.1 – Evaluation of the signal-to-noise degradation in the single-frequency synthesizer as used in the caesium atomic clock.

According to Example 5.2 and Fig. 5.8 the overall multiplication factor is nearly 2000; assuming an ideal multiplier

$$\left(\frac{S}{N}\right)_{out} = \left(\frac{S}{N}\right)_{in} - 20 \log 2000 = \left(\frac{S}{N}\right)_{in} - 66 \qquad \text{dB}$$

For a 100-kHz fundamental oscillator the overall multiplication factor would be 10^5 and the degradation

$$\left(\frac{S}{N}\right)_{out} = \left(\frac{S}{N}\right)_{in} - 20 \log 10^5 = \left(\frac{S}{N}\right)_{in} - 100 \qquad \text{dB}$$

The advantage of the higher frequency of the driving crystal oscillator is evident. Another fact is that the excellent $(S/N)_{in}$ at the input of long chains of frequency multipliers is of prime importance. This is achieved very often by an additional crystal filter at the multiplier input [2.14].

(B) The Additive Noise Generated in Frequency Multiplying Circuits

The noise figure of transistor amplifiers in the white-noise range is small (few decibels) and therefore no appreciable deterioration of the input S/N ratio need be feared. However, conditions are different with respect to the flicker ($\sim 1/f$) noise. Halford *et al.* in [2.12] report that the one-sided spectral density of the phase noise was about $10^{-11\cdot 2}$ radian$^2/f$ for the surveyed range 5 to 100 MHz, quite independent of the multiplication factor and the transistor type. Laboratory experiments proved that the direct-phase modulation of the RF carrier by the transistor was responsible for the flicker noise of the output phase. The improvement – typically more than 30 dB – has been achieved by local negative feedback (emitter degeneration).

In varactor multipliers it is the onset of either the forward or reverse current that is responsible for the increase of the additive noise. Watanabe *et al.* [2.13] have ascertained by measurement the $1/f$ noise-characteristic in the case of the considerable flow of the forward current, and the frequency independent noise-characteristic for the reverse current flow (avalanche noise).

(C) Phase Drifts Due to Temperature and Supply-Voltage Variations

Selective circuits, which are inevitable in frequency multipliers for suppressing the undesired harmonic signals, are generally temperature dependent. Consequently any change in ambient temperature causes detuning and shifts the phase of the signal passed. Small random changes enhance the phase noise already present. Whereas large and long temperature drifts, such as are encountered in warm-up periods, result in change of the average frequency $\overline{f(t)}_{\text{out}}$.

$$\overline{f(t)}_{\text{out}} = \frac{1}{t_2 - t_1} \int_{t_1}^{t_2} f(t)\,dt = \frac{\Phi(t_2) - \Phi(t_1)}{t_2 - t_1} = \frac{\Delta\Phi}{t_2 - t_1} + \overline{f(t)}_{\text{input}} \tag{2.85}$$

where $(t_2 - t_1)$ is the observing time.

When a simple resonant circuit is detuned by a small amount the consequent phase shift is [2.14]:

$$\Delta\Phi = K \,.\, Q \,.\, \alpha_T \,.\, \Delta T \tag{2.86}$$

where K is a function of the initial detuning and is plotted in Fig. 2.13(a); $K = 2/(1 + a^2)$, $a = Q[(\omega/\omega_0) - (\omega_0/\omega)] \approx Q[2(\omega - \omega_0)]/\omega_0$, $\omega_0^2 = 1/LC$; Q is circuit Q of the respective resonant circuit;

$$\alpha_T = -\frac{1}{2}\left(\frac{1}{L}\frac{\partial L}{\partial T} + \frac{1}{C}\frac{\partial C}{\partial T}\right) \tag{2.87}$$

and ΔT is the average change of the effective temperature.

The same equation (2.86) is valid for a pair of coupled resonant circuits, when assuming $Q_1 = Q_2 = Q$ and $\alpha_{T1} = \alpha_{T2} = \alpha_T$.

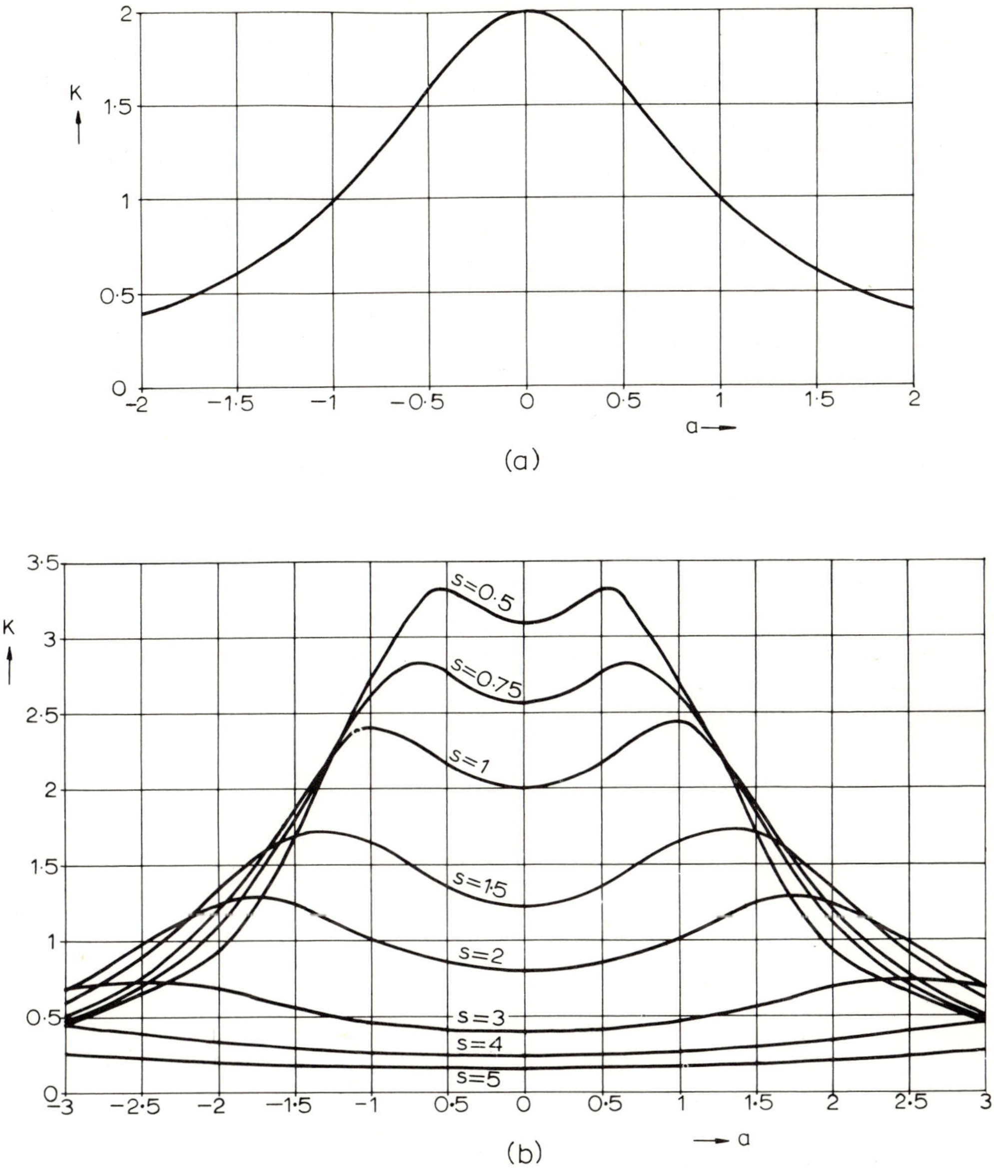

Fig. 2.13 Detuning function K vs. normalized detuning a:

$$a = Q\left(\frac{\omega}{\omega_0} - \frac{\omega_0}{\omega}\right) \approx Q\,\frac{2\Delta\omega}{\omega_0}$$

(a) for simple resonant circuit; (b) for a pair of coupled resonant circuits having $Q_1 = Q_2 = Q$. (From Sabury [2.14].)

The detuning function K, plotted in Fig. 2.13(b), is

$$K = \frac{4(1 + s^2 + a^2)}{(1 + s^2)^2 + (1 - s^2)a^2 + a^4} \quad \text{where } s = kQ,$$

k being the coupling coefficient between the primary and secondary circuits.

Evidently some sort of temperature stabilization and temperature compensation of tuned circuits is desirable. However, in the latter case the different thermal capacity of the coil and of the capacitor may spoil the effect when rapid changes in the ambient temperature are not effectively suppressed.

Example 2.2

Assume a simple resonant circuit in which $a = 0$, $Q = 100$, $\alpha_T = 0{\cdot}5 \times 10^{-4}/°\text{C}$, $\Delta T = 5 \times 10^{-3}$ °C, and $t_2 - t_1 = \tau = 10$ seconds, $(\Delta\Phi)_\tau = 5 \times 10^{-5}$ rad, and $(\Delta\omega)_\tau = 5 \times 10^{-6}\ 2\pi\text{Hz}$.

In long chains of frequency multipliers the most critical are resonant circuits (including crystal filters) in the first and second stages. A profound study of the problem illustrated with many experimental results, may be found in [2.14].

In the case where resonant-circuit parameters are voltage-dependent (the most common example is when semiconductor junction capacity forms part of the total tuning capacity) the phase variations are calculated in the same way as above, except that the temperature coefficient α_T must be replaced by the voltage coefficient α_v

$$\alpha_v = \frac{1}{\omega} \cdot \frac{\partial \omega}{\partial v} = \frac{K_0(v)}{\omega} \tag{2.88}$$

where $K_0(v)$ is as defined in chapter 6, see eq. (6.105) and (6.106).

There could be other environmental conditions, such as humidity etc., which might cause additional phase shifts in frequency multipliers.

2.5 CONCLUSIONS

Surveying the properties of the different types of frequency multiplier one may notice some important common features:

(i) There is generally a decrease in efficiency at the larger multiplying factors; this is particularly true for the low-frequency range where the non-linear resistor-diodes and transistor class C frequency multipliers are used.

(ii) Selective circuits (filters) are becoming more complicated with the increase in the multiplying factor because the relative distance with respect to the undesired harmonic signals is reduced; in varactor multipliers this tendency is increased by the presence of the idling circuits.

(iii) There is an inherent degradation of the signal-to-phase-noise (perturbations) ratio in proportion to the square of the multiplying factor, i.e. to n^2.

Thus in respect of the first and second points one may conclude that, in one stage only, multiplication with small prime numbers $n = 2$, 3, and 5 is a simple task.

The third point is restrictive as to the overall multiplication factor n, and this must be kept in mind when designing frequency synthesizers, particularly for higher frequencies, in which long chains of frequency multipliers are used.

APPENDIX 2.1
AMPLITUDE MODULATION ENHANCEMENT IN VOLTAGE-DRIVEN TRANSISTOR FREQUENCY MULTIPLIER

The behaviour in respect of small-index amplitude modulation may be evaluated by applying the Taylor expansion to eq. (2.58) and neglecting higher-order terms.

Assuming that

$$V_g = V_{g0}[1 + m_{a1} \cos \nu t]; \quad (m_{a1} \ll 1) \tag{2.89}$$

one gets

$$\begin{aligned} \bar{I}_n(\lambda V_{g0}[1 + m_{a1} \cos \nu t]) &= \bar{I}_n(\lambda V_{g0}) + \frac{\partial \bar{I}_n(\lambda V_{g0})}{\partial(\lambda V_{g0})} . \lambda V_{g0}\, m_{a1} \cos \nu t \\ &= \bar{I}_n(\lambda V_{g0})\left[1 + \frac{m_{a1}}{2} \lambda V_{g0} \frac{\bar{I}_{n-1}(\lambda V_{g0}) + \bar{I}_{n+1}(\lambda V_{g0})}{\bar{I}_n(\lambda V_{g0})} \cos \nu t\right] \end{aligned} \tag{2.90}$$

after applying the Bessel function recursion formula (1.68) on the derivative. Using the first part of the recursion formula of (1.68) on the denominator in (2.90) reveals that the amplitude modulation index on the nth carrier is

$$m_{an} = n . m_{a1} \frac{\bar{I}_{n-1}(\lambda V_{g0}) + \bar{I}_{n+1}(\lambda V_{g0})}{\bar{I}_{n-1}(\lambda V_{g0}) - \bar{I}_{n+1}(\lambda V_{g0})} > n . m_{a1} \tag{2.91}$$

Evidently the enhancement of the small amplitude-modulation index is more than n-times.

REFERENCES

[2.1] BARNES, F. S. and EIBER, G. F. An ideal harmonic generator. *Proc. I.E.E.E.*, **53**, 693-5.

[2.2] KEISER, B. E. The cycle splitter – A wide-band precision frequency multiplier. *I.R.E. Nat. Con. Rec. 1959*, 7, P.4, 275-81.

[2.3] VAN DUZER, V. E. 500 kc/s–500 Mc/s frequency doubler, *Hewlett-Packard Journal*, **17**, Oct. 1965.

[2.4] KOEHLER, D. The charge-control concept in the form of equivalent circuits, representing a link between the classic large signal diode and transistor models, *Bell System Technical Journal*, **46**, 523-76, March 1967.

[2.5] PENFIELD, P. and RAFUSE, R. P. *Varactor applications.* The M.I.T. Press, Cambridge, Massachusetts, 1962.

[2.6] SCANLAN, D. O. and LAYBOURN, M. A. Analysis of varactor harmonic generators, with arbitrary drive levels. *Proc. I.E.E.*, **114**, 1598-1604, 1967.

[2.7] WATSON, H. A. (ed.). *Microwave Semiconductor Devices and Their Circuit Applications.* New York: McGraw-Hill, 1969.

[2.8] JOHNSTON, R. H. and BOOTHROYD, A. R. Charge storage frequency multipliers. *Proc. I.E.E.E.*, **56**, 167-76, 1968.

[2.9] EBERS, J. J. and MOLL, T. L. Large signal behavior of junction transistors. *Proc. I.R.E.*, **42**, 1761-72, 1954.

[2.10] HARRISON, R. G. A nonlinear theory of class C transistor amplifiers and frequency multipliers. *I.E.E.E. Journal of Solid-State Circuits*, **SC-2**, 93-102, 1967.

[2.11] CAULTON, M. *et al.* Generation of microwave power by parametric frequency multiplication in a single transistor, *RCA Review*, **26**, 286-311, June 1965.

[2.12] HALFORD, D., WAINRIGHT, A. E. and BARNES, J. A. Flicker noise of phase in RF amplifiers and frequency multipliers: Characterization, cause, and cure. *Proc. of the 22-nd Annual Frequency Control Symposium*, Fort Monmouth, N.Y., 1968.

[2.13] WATANABE, T. and YOSHIHARU, F. Characteristics of semiconductor noise generated in varactor frequency multipliers, *Rev. of Electrical Communication Laboratory*, **15**, 752-68, Nov.-Dec. 1967 (Tokyo).

[2.14] SABURI, Y., YASUDA, Y. and HARADA, K. Phase variations in the frequency multiplier. *J. Radio Research Labs.*, **10**, 137-75.

[2.15] KROUPA, V. Nonlinear-resistance-diode frequency multipliers (*publication pending*).

3 Frequency Division

Frequency division cannot be achieved by a mere distortion of the original signal (with the exception of certain types of nonlinear reactance dividers). The mechanism involved is quite different and there are several ways of approach:

(a) Synchronization of a subharmonic oscillator rich in higher harmonic signals of which at least one is near the original frequency;

(b) Timing circuits having a natural period slightly longer than the desired multiple of periods of the input signal;

(c) Dividers with regeneration and modulation comprising a frequency multiplier.

The dividing factor in all these systems is limited, however, to small numbers only (ordinarily below 10);

(d) Memory or counting dividers, particularly sequence-operated circuits, may have any arbitrary number of stable states and hence they may be applied for dividing by any arbitrary factor. These dividers are the most important frequency dividers today. One of the reasons is their potentiality for mass production in integrated-circuit form.

3.1 SYNCHRONIZATION OF SINUSOIDAL OSCILLATORS

Direct synchronization of sinusoidal LC or RC oscillators is easy when the division ratio does not exceed two or three; for larger factors, oscillators which generate waveforms rich in harmonics are necessary. Division ratios up to 10 : 1 in one stage in commercial equipment have been reported [3.1, 7.11].

The advantages are the simplicity of the circuit, the sinusoidal synchronizing signal, and the power gain. The last feature may be of importance at very high frequencies (GHz frequency bands) where the divider may be made for example as a synchronized tunnel diode oscillator.

The shortcomings, on the other hand, are somewhat poor phase stability, in some instances, and continuation of the output signal on a frequency only slightly different from the correct frequency when the synchronization is lost.

3.2 SYNCHRONIZATION OF RELAXATION OSCILLATORS

The main feature of relaxation oscillators is the alternation between two states (e.g. large current-small current) whereas during the transition a large negative conductance is present. The repetition period is determined predominantly by RC timing circuits. The basic properties of vacuum-tube devices are discussed in depth in [3.2].

Relaxation oscillators are easily synchronized, provided the harmonic of the natural frequency is slightly below the frequency of the synchronizing signal.

Two types of relaxation oscillators often encountered in frequency dividing techniques are multivibrators and blocking oscillators.

(A) Astable Multivibrator

The basic circuit configuration is shown in Fig. 3.1(a). The two distinct states are: transistor T_1 conducting and transistor T_2 off and vice versa. Suppose for example T_1 is conducting, then capacitor C_1 charges through R_1 and the base-emitter voltage of T_2 increases. When T_2 starts to conduct, the negative voltage-drop from its collector reaches through C_2 the base of T_1 which is consequently switched-off very rapidly. Now, capacitor C_2 charges through R_2 and after some time T_1 is again conducting. Evidently, the repetition period depends on time constants R_1C_1 and R_2C_2. The idealized collector and base voltage waveforms of a symmetrical multivibrator are shown in Fig. 3.1(b). Synchronization is achieved when, for example, positive pulses are applied to either base of transistors T_1 or T_2. In the case where both leading and trailing edges of the rectangular output waveform are synchronized, better phase stability is reached. Therefore for odd dividing factors it is desirable to choose the ratio of determining time constants to be approximately

$$\frac{R_1 C_1}{R_2 C_2} \approx \frac{n+1}{n-1} \tag{3.1}$$

Reliable frequency dividers with ratios 5:1 up to 10:1 may be made in this way.

Note: there exist circuit configurations which prevent multivibrators from oscillating when synchronization is lost [3.3].

(B) Blocking Oscillator

Blocking oscillators are transformer-coupled feedback oscillators in which collector current is permitted to flow for one half-cycle only. Through the feedback action a biasing voltage in the base circuit builds up which prevents further oscillation, see Fig. 3.2. (The diode across the collector winding suppresses the eventual

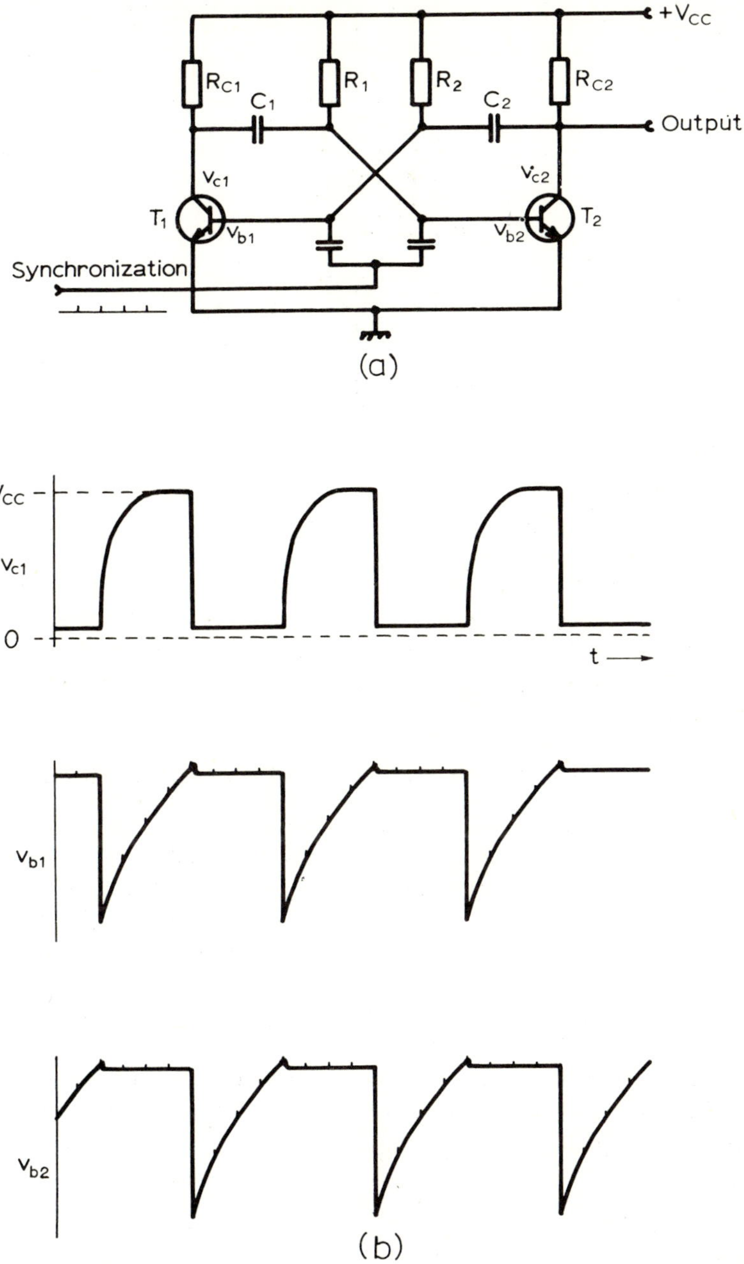

Fig. 3.1 Astable multivibrator: (a) basic configuration; (b) collector and base waveforms.

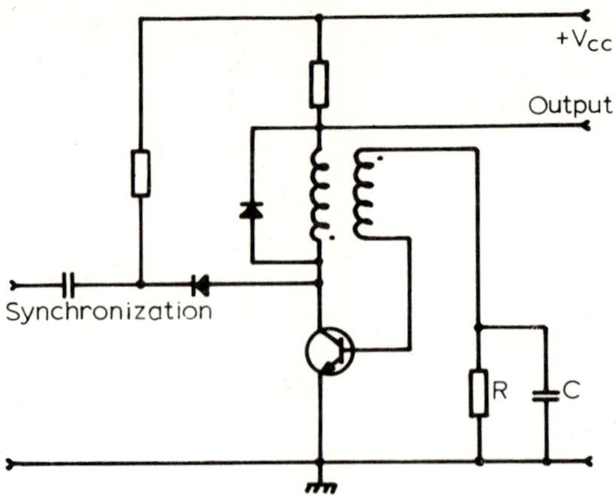

Fig. 3.2 Blocking oscillator frequency divider.

positive excursion following the abrupt interruption of the collector current.) Thereafter capacitor C starts slowly to discharge through R and finally the short-term regenerative conduction of the transistor is repeated. The onset of oscillations may be speeded up by suitably timed synchronizing pulses. This feature enables the blocking oscillator to act as the frequency divider.

The main drawback of this type of circuit is the transformer which is usually wound on a core of high permeability so that the feedback coupling is made as high as possible.

3.3 REGENERATIVE FREQUENCY DIVIDERS

A block diagram of this type of frequency divider is shown in Fig. 3.3.

Without the input signal $A_i \cos(\omega t + \varphi_i)$ the regenerative action cannot take place. There might even be difficulties with starting: when the circuit is switched on, a small signal $a_n \cos\{[(n-1)/n]\,\omega t + \psi\}$ is generated through the transient phenomenon at the second input of the mixer. After mixing with the input signal already present the component with the fractional frequency ω/n may be filtered out and amplified. From the output there is a feedback loop in which the multiplication $(n-1)$-times is performed. If the overall gain exceeds 'one' the regeneration takes place; finally saturation effects in either of the nonlinear circuits (inclusive even of the amplifier) limit the amplitude. The oscillation requires the zero phase shift around the loop, i.e.

$$\omega t + \varphi_i - \left(\frac{n-1}{n}\,\omega t + \varphi_{n-1}\right) + \varphi_a = \frac{\omega t}{n} + \varphi_0 \tag{3.2}$$

and

$$\left(\frac{\omega t}{n} + \varphi_0\right)(n-1) + \varphi_b = \varphi_{n-1} + \frac{n-1}{n}\,\omega t \tag{3.3}$$

The meaning of $\varphi_i, \varphi_{n-1}, \varphi_a, \varphi_b$, and φ_0 is evident from Fig. 3.3. After eliminating φ_{n-1} from (3.2) and (3.3) we get the relation between the output and the input phases φ_0 and φ_i respectively

$$\varphi_0 = \frac{\varphi_i + \varphi_a - \varphi_b}{n} \tag{3.4}$$

The maximum division ratio depends predominantly on the selectivity of the band-pass filter ($\omega_m \approx \omega/n$) following the mixer. The loop gain for the next possible working modes, i.e. $\omega/(n \pm 1)$ (and naturally for all others $\omega/(n \pm k)$ as well) must be safely below 'one'.

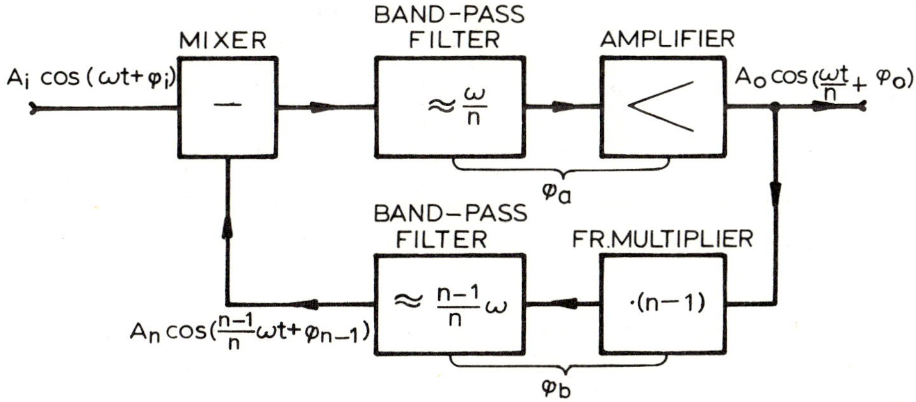

Fig. 3.3 Block diagram of a regenerative frequency divider.

Dividing circuits of this type may be very simple, as is shown in Fig. 3.4 which is taken from [3.4] where the mathematical theory is also given.

Frequency dividers of this type have good phase stability, but, the presence of resonant circuits prevents their use at very low frequencies.

3.4 PARAMETRIC FREQUENCY DIVIDERS

Any nonlinear device can generate harmonics of the driving frequency. Varactors have the peculiar property of being able also to supply subharmonics or rational fractions of the 'pumping' frequency. This working mode is much like that of an oscillator. The output current builds up with a phase that is at least partially arbitrary, accompanied perhaps by currents at other frequencies. As the output becomes larger, nonlinear effects reduce the rate of rise, and finally a steady state with constant output amplitude is reached. An extensive theoretical treatment may be found in [2.5, chapter 9].

For the abrupt junction varactors the small-signal theory predicts that the divide-by-n ($n > 2$) circuit will require an idler to start. However, the idler at frequency $(n - 1)\,\omega_0$ is not satisfactory because the resulting oscillations are not frequency stable; this problem is more profoundly dealt with in [3.5]. On the

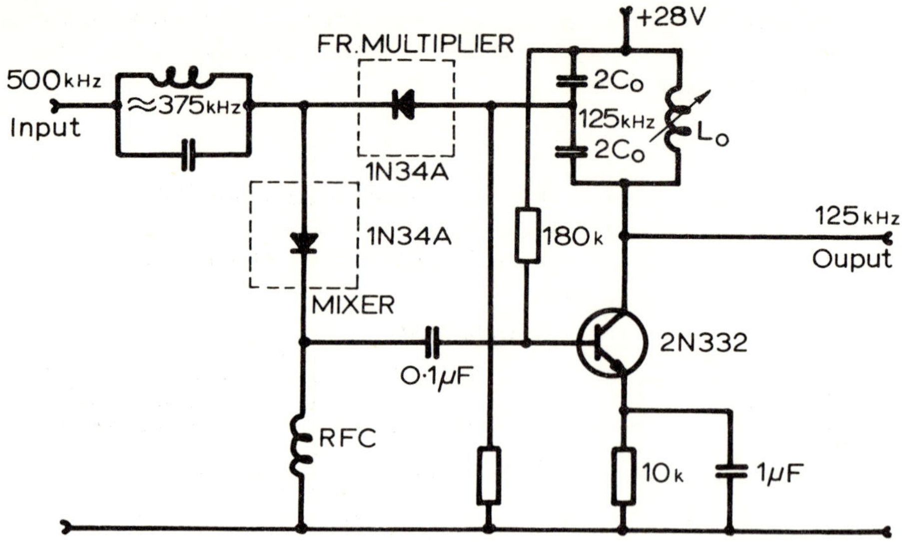

Fig. 3.4 A practical regenerative frequency divider. (From Plotkin & Lumpkin [3.4].)

other hand it is possible by assigning proper frequencies to idlers to generate phase-sensitive and therefore frequency-stable outputs at $1/2, 1/4, \ldots 1/2^k$ or at $m/2^k$ (m and k being integers) of the input frequency.

The advantage of varactor dividers is that they work at very high frequencies.

3.5 COUNTING DIVIDERS

Counting dividers which form a group of wide-band or even fully aperiodic frequency dividers may be constructed either as energy-storage devices or as sequence-operated multistate circuits. The latter are the most important frequency dividers today because they are fully aperiodic, have good phase stability, and can be produced on a large scale in integrated circuit (IC) form. Thus the complexity and great number of individual circuit elements encountered in conventional products are no longer a disadvantage.

(A) Energy-Storage Counting Dividers

The energy storage element in most electronic counters is usually a capacitor. Its advantages are stability with respect to temperature and ageing effects, and a relatively small leakage of the stored charge. The basic circuit is shown in Fig. 3.5. The negative-going pulse $-E_c$ charges the sampling capacitor C_1 and through the conducting diode D_1 the storage capacitor C_2 to the voltage

$$-E_1 = -E_c \frac{C_1}{C_1 + C_2} \tag{3.5}$$

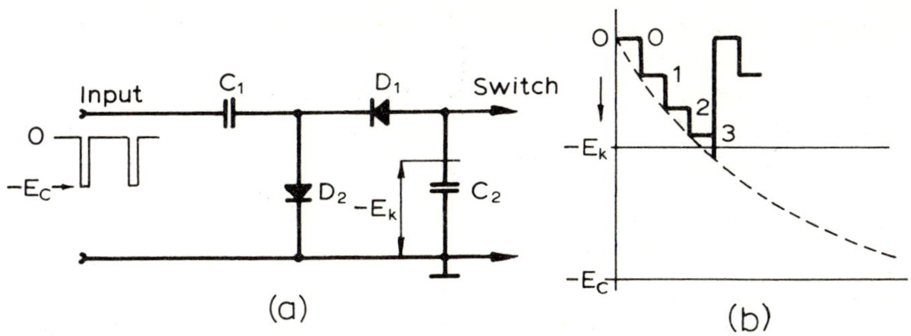

Fig. 3.5 Energy-storage counting divider: (a) basic configuration; (b) waveform on capacitor C_2.

When the input pulse ends, the input voltage returns to zero, D_1 closes, and C_1 is discharged through D_2. On succeeding pulses the input voltage must drop to $-E_1$, $-E_2$, etc. before the diode D_1 conducts. Consequently the voltage increments on C_2 get less and less following an exponential law [3.2, chapter 17]. However, this disadvantage can be overcome by substituting for D_2 an emitter follower (cf. transistor T_1 in Fig. 3.6). At a predetermined level E_k, i.e. k input pulses

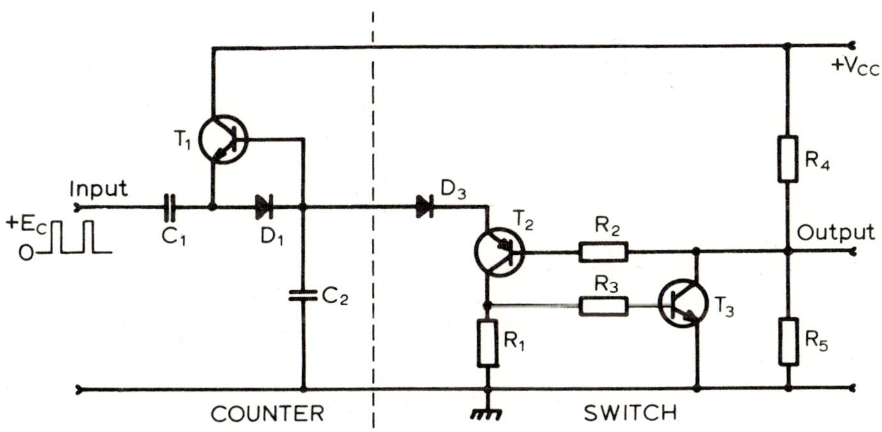

Fig. 3.6 Improved configuration of the energy-storage counting divider giving a linear 'staircase' waveform on C_2. (Preston [3.6]. Courtesy of Electronic Engineering.)

counted, C_2 is discharged through a suitable switch in a fraction of the input period. In this way the counter starts the new cycle without loss of any input pulse. In Fig. 3.6 the discharging is performed by the regenerative action of the complementary p-n-p/n-p-n transistor pair T_2 and T_3. As soon as the emitter potential of T_2 exceeds the trigger potential on its base, determined by the divider R_4-R_5, transistor T_2 starts conducting, in turn bringing T_3 into conduction, and thence positive feedback, causing a rapid discharge of the capacitor C_2 is formed.

In variable-frequency synthesizers described in chapter 7, a division ratio, changing over several decades, is desired. In [3.6] it is shown how this requirement can be fulfilled by using energy-storage counters.

(B) Sequence-Operated Counting Dividers

Memory circuits with two or several stable states form the basis of sequence-operated counting dividers.

With mechanical devices (gear boxes, etc.) the different space relationships of certain elements form the memory. However, when used as frequency dividers they have poor phase stability, and there is also the disadvantage of moving parts, and consequently large size, in comparison with the latest IC electronic circuits. This, and the very limited low-frequency range, are the reasons why mechanical devices have disappeared from frequency synthesis techniques.

The operation of counting tubes is dependent upon the space position of the current beam; but only the numerical indicator tubes are still of importance.

Amongst memory circuits with several electrical stable states, it is the bistable circuits that are most often encountered.

The basic circuit is the bistable multivibrator in Fig. 3.7, which is simply called a flip-flop. If transistor T_2 is in the conducting state, the collector voltage is low, and provided there is sufficient base drive, reaches the saturation value $V_{CE,sat}$

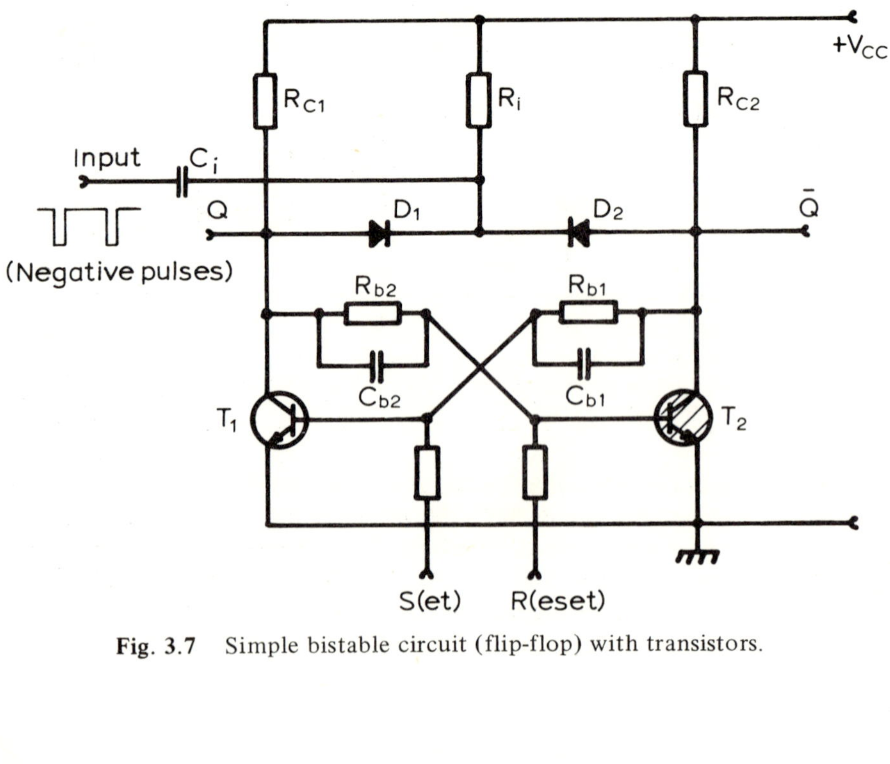

Fig. 3.7 Simple bistable circuit (flip-flop) with transistors.

(100 to 200 mV). Through the resistor R_{b1} the base of T_1 is also held at this low voltage and thus T_1 is cut-off. Its collector voltage is high, nearly V_{CC}, thus assuring a sufficiently large current through R_{b2} to maintain T_2 conducting. The circuit remains in this state until a regenerative triggering pulse reaches, via diode D_1 and (speed-up) capacitor C_{b2}, the base of the conducting transistor T_2. Consequently the collector current of T_2 is reduced, causing the collector voltage to rise. This voltage step is transferred via C_{b1} and R_{b1} to the base of the transistor T_1, which starts conducting, and through the regenerative action the bistable circuit gets into the other stable state, namely T_1 conducting and T_2 off. The following input pulse resets the flip-flop to the original state.

There are many design variations in flip-flop circuitry, depending upon working conditions. The most important parameter is the counting speed: the contemporary limit is in the range of 200 MHz for transistor bistable circuits and over 1 GHz for tunnel diode bistable circuits [3.7]; further, there are requirements of additional logical functions, reliability in the prescribed ambient temperature limits, etc. More information, together with design procedures, may be found in [3.8].

From the above description of the operation of the bistable element it is clear that, after a certain input pulse p_k, transistor T_2 in Fig. 3.7 is conducting and T_1 is off; the following input pulse p_{k+1} changes the state, namely, T_1 conducting and T_2 off; the next pulse p_{k+2} returns the bistable circuit to the state T_2 on, T_1 off, etc. Evidently the bistable memory circuit may be used as a frequency divider with the dividing factor $N = 2$. Both collector voltages are of the rectangular waveform from which, through the derivation action in a CR circuit (see C_iR_i in Fig. 3.7), the pulses necessary for driving another bistable circuit are easily generated.

(i) Binary Coded Dividers

The series connexion of two bistable circuits divides the input pulse rate frequency by four, the series connexion of three bistable circuits by eight, etc.

A schematic diagram of a four-stage counter-divider, which divides by 16, is shown in Fig. 3.8. The logical states of individual bistable circuits during one counting period are in Table 3.1.

In each bistable stage there is some delay between the arrival of the input pulse and the instant when the stage settles into its new state. Due to noise, temperature variations, voltage variations etc., there is a small time jitter of the leading and trailing edges of the output rectangular waveform as well (see Section 3.6). These disadvantages can be avoided by placing at the output of the dividing chain an additional bistable circuit, as illustrated in Fig. 3.9, activated, for example, by the trailing edge of the waveform supplied by the last bistable circuit and reset immediately by the next following input pulse (for the S and R inputs see Fig. 3.7). According to the function performed this circuit arrangement is sometimes called a pulse selector. There might be difficulties if the delay were

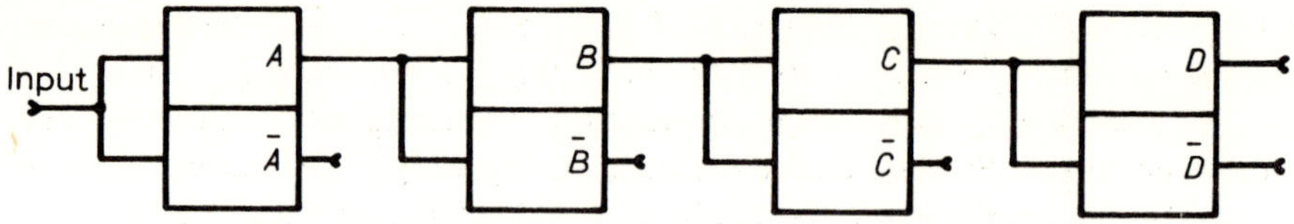

Fig. 3.8 Schematic diagram of the four-stage binary counter-divider

Table 3.1
Logical states in the four-stage binary counter-divider during one counting cycle

No. of the input pulse	Bistable circuit output			
P_k	D	C	B	A
0	0	0	0	0
1	0	0	0	1
2	0	0	1	0
3	0	0	1	1
4	0	1	0	0
5	0	1	0	1
6	0	1	1	0
7	0	1	1	1
8	1	0	0	0
9	1	0	0	1
10	1	0	1	0
11	1	0	1	1
12	1	1	0	0
13	1	1	0	1
14	1	1	1	0
15	1	1	1	1
16	0	0	0	0

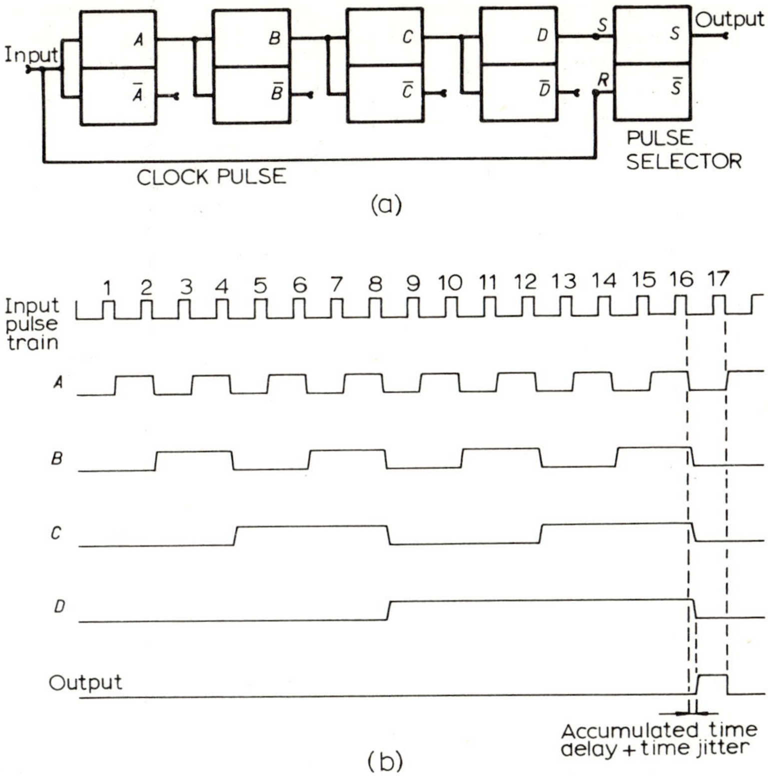

Fig. 3.9 Circuit arrangement for time delay and time jitter suppression in a binary counter-divider: (a) block diagram; (b) respective waveforms. (Kench [3.8]. Courtesy of Mullard.)

comparable with the input pulse train period. This problem is not present with synchronous counters, where each bistable circuit is triggered by the input pulse according to the logical state of all preceding stages, see Fig. 3.10. It may be seen that the loading on the trigger pulse source increases as the number of stages increases; the same is true for the loading on the output of individual bistable circuits, particularly those early in the chain. Another disadvantage is the need for a considerable number of extra components. For this reason the use of integrated circuits (IC) is desirable. A block diagram of a synchronous binary counter built from IC *J–K* flip-flops is in Fig. 3.11. (See Appendix 3.1 for the basic properties of *J–K* flip-flops.)

Any number of bistable circuits may be cascaded as shown above. The resulting dividing factor N is purely binary, i.e. $N = 2^n$ where n is the number of bistable elements in the chain.

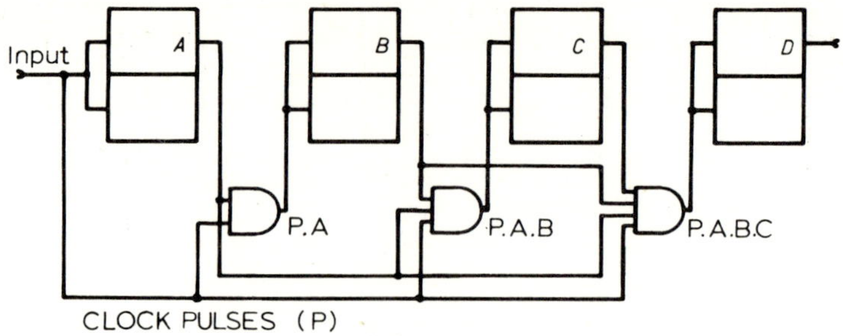

(a)

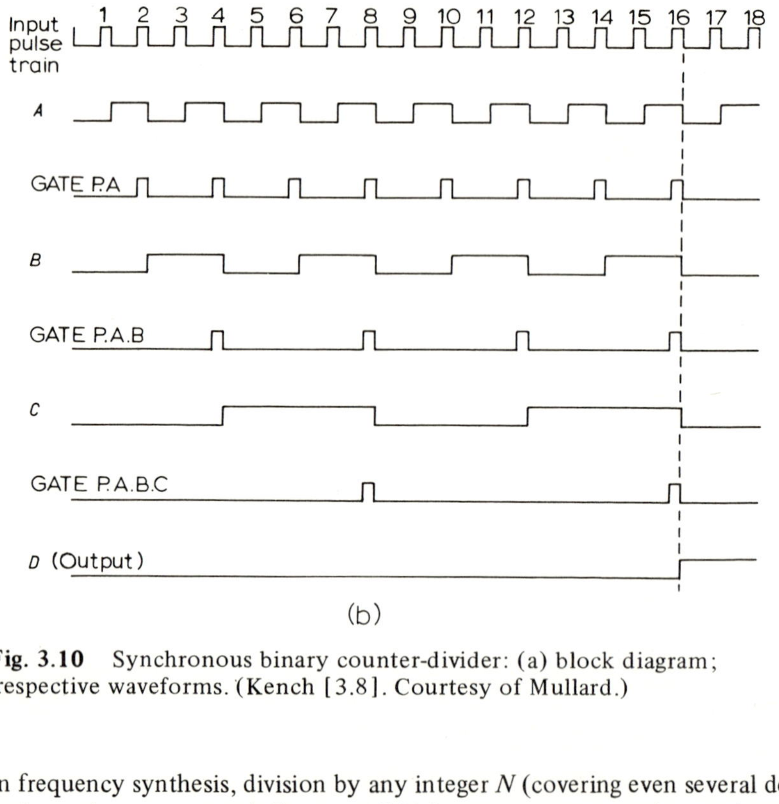

(b)

Fig. 3.10 Synchronous binary counter-divider: (a) block diagram; (b) respective waveforms. (Kench [3.8]. Courtesy of Mullard.)

In frequency synthesis, division by any integer *N* (covering even several decimal orders) may be encountered. For a small *N* this requirement can be met by modification of binary dividers by appropriate gating: during the counting cycle some configurations of stable states are omitted, or more precisely, prevented from forming. The same result may be achieved by introducing feedback to enable pulses derived from outputs of later stages to activate one or more former stages – the effect is that several dummy pulses are added. The disadvantage of this second configuration is reduction of maximum counting speed.

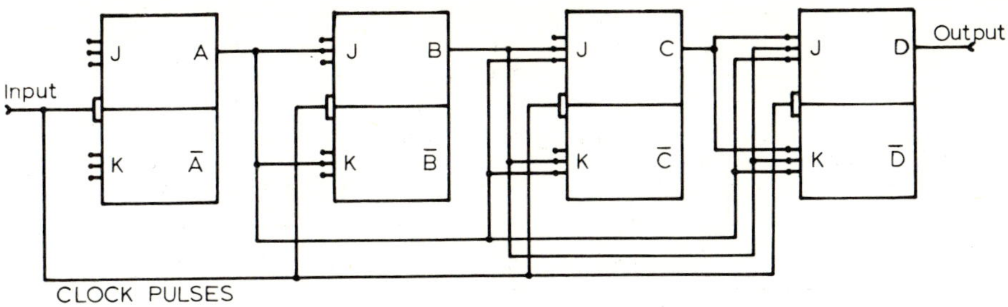

Fig. 3.11 Block diagram of the four-stage synchronous binary counter-divider built from *J–K* bistable elements.

Formal methods of deriving counters to produce given sequences are described in [3.9] and [3.10]. The following examples will assist the reader to understand the problem.

Generally the number of bistable circuits required for a divide-by-N counter is

$$2^{n-1} < N < 2^n \qquad (3.6)$$

Example 3.1 – Design of binary coded ternary counter

For $N = 3$ relation (3.6) gives $n = 2$; two bistable elements are required. Examination of Table 3.1 reveals that after the second pulse the two-stage binary chain (A,B) must return to the zero state. Thus the input pulse (P) may activate the first stage for logical combinations $P.\bar{A}.\bar{B}$ and $P.A.\bar{B}$ only. The respective logical equation is:†

$$P.\bar{A}.\bar{B} + P.A.\bar{B} = P.\bar{B} \qquad (3.7)$$

which says that at the input an *AND* gate must be placed. Similarly the activation of the second stage is governed by

$$P.A.\bar{B} + P.\bar{A}.B \qquad (3.8)$$

Because the combination of $P.A.B$ is prevented by (3.7), its addition is immaterial; it does however, assure that the counter is self-starting from every possible initial state. Thus the relation (3.8) may be simplified to

$$P.A.\bar{B} + P.A.B + P.\bar{A}.B + P.A.B = P.(A + B) \qquad (3.9)$$

which means that between the first and the second stage one *AND* gate and one *OR* gate must be placed. The block diagram of this ternary counter is shown in Fig. 3.12, and in Fig. 3.13 using the IC *J–K* flip-flops. In the latter case the necessary interconnexions are found from the truth table (Table 3.2).

† A summary of the laws of Boolean algebra is given in Appendix 3.1.

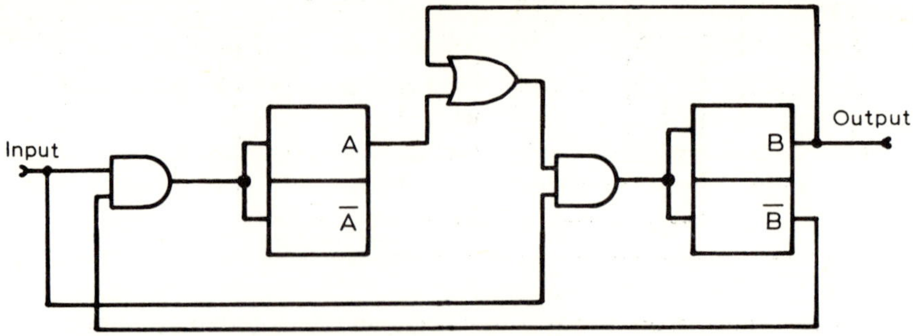

Fig. 3.12 Binary coded synchronous modulo-3 counter-divider.

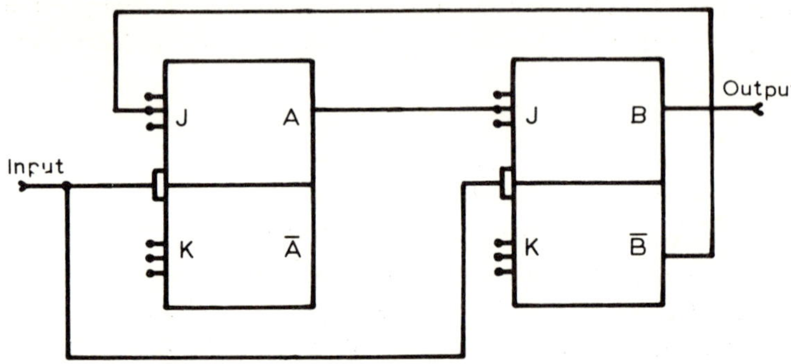

Fig. 3.13 Synchronous modulo-3 counter-divider built from *J–K* flip-flop elements.

Table 3.2
Truth table of the binary coded ternary counter

J_A	K_A	A_{k+1}	$\overline{A}_{k+1}$	J_B	K_B	B_{k+1}	$\overline{B}_{k+1}$
Assumed state before arrival of the starting clock pulse		0	1			0	1
1	1	1	0	0	1	0	1
1	1	0	1	1	1	1	0
0	1	0	1	0	1	0	1

Example 3.2 – Design of binary-coded decimal counter (BDC)-1248

According to (3.6) a four-stage counter is necessary. If the binary-decimal code – 1248 – is desired, the flip-flop state configurations corresponding to the numbers 10 to 15 must be inhibited. From Table 3.1 it may readily be deduced that after the 9th pulse both the first and the fourth stage must be 'reset' whereas the activation of the second binary circuit must be prevented. The logical condition of the latter requirement is $A.\bar{D}$ which is fulfilled when an *AND* gate is placed between the first and the second stage. A block diagram is shown in Fig. 3.14. It is supposed that the propagation delay through the gate, and through the *B* and *C* binary elements, is small enough to cause uncertainty in the transition from state '9' to state '0', see waveforms in Fig. 3.14. Otherwise an additional gate would be necessary.

Note: taking the output from the third binary stage results in a more regular space-mark ratio – see (b) in Fig. 3.14. However, the trailing edge of the fourth-stage waveform has a smaller time jitter because it is triggered directly from the first binary. Symmetrical rectangular output may be obtained by interchanging positions of the binary and modulo-5 counters.

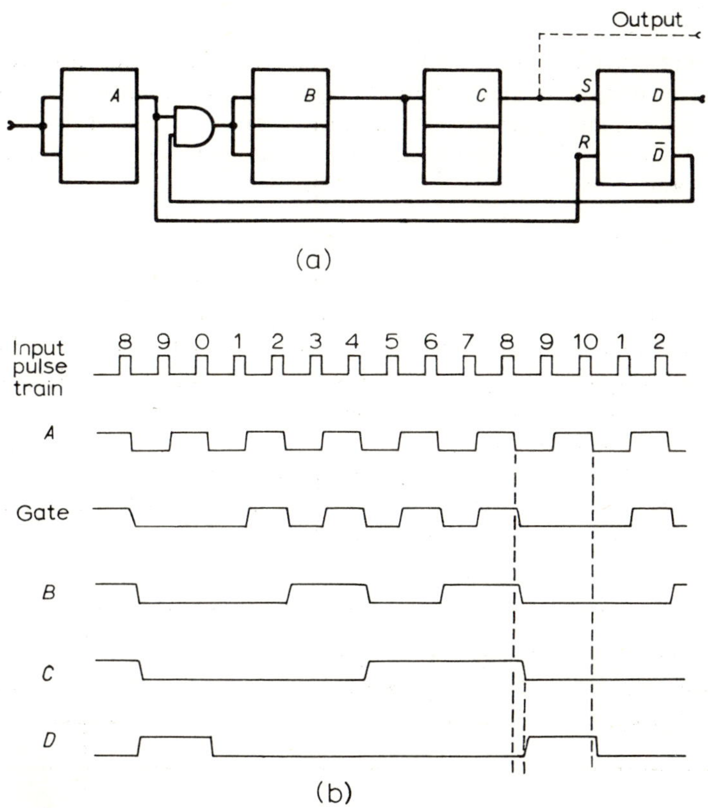

Fig. 3.14 Gated binary coded decimal counter-divider: (a) block diagram; (b) respective waveforms. (Kench [3.8]. Courtesy of Mullard.)

A logical diagram of the synchronous BDC-1248 decade counter using IC *J–K* bistable elements is shown in Fig. 3.15. The maximum operating frequency of the decade counter composed of contemporary commercial IC elements is above 10 MHz.

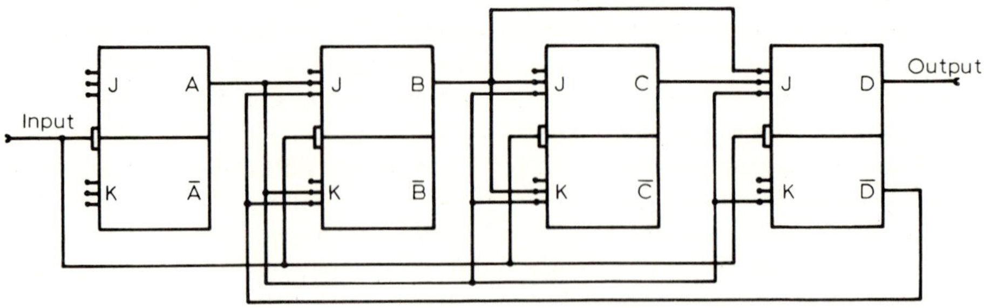

Fig. 3.15 Block diagram of the BCD (12̇48) decade counter-divider using *J–K* flip-flops.

The binary decimal code 1248 is not the only possible or the only used code when making decade counters. The 1242 code is often encountered, and sometimes the system economy requires an unconventional one, e.g. 1245, 148(-2), etc. These problems and the respective decade configurations are studied extensively in [3.8, chapters 7 and 8].

(ii) Ring Counters

A ring counter is composed of *N* identical binary circuits or switches from which ordinarily only one may be 'on'; all the others must be 'off'. The mutual interconnexions are arranged in such a way that each input pulse shifts the 'on' state to the next binary in the chosen 'direction'. Coming to the *N*th stage an output signal is delivered and the 'on' state is transferred back to the 'first' stage. A block diagram is shown in Fig. 3.16.

Ring counters having the working mode 'one-out-of-*N*' require more bistable elements than the binary-coded counters, see inequality (3.6). The building blocks must therefore be very simple to be competitive: for example, silicon controlled

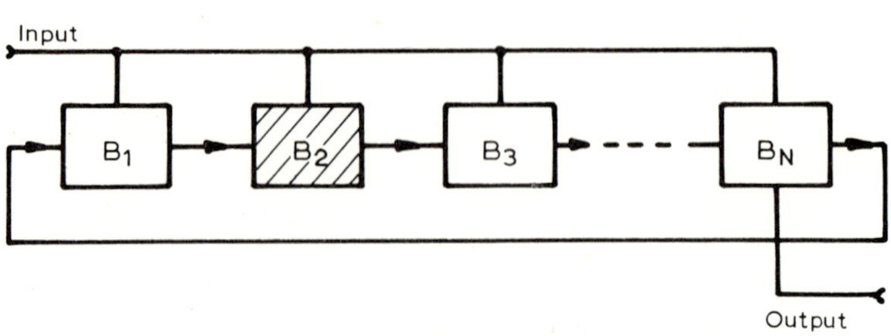

Fig. 3.16 Block diagram of a ring counter.

switches, p-n-p/n-p-n transistor pairs (see for example the switch shown in Fig. 3.6), etc.; a more detailed discussion of ring counters may be found in [3.8, chapter 6].

Vacuum and gas-filled counting tubes also belong to the class of ring counters.

(C) Variable-Ratio Counting Dividers

Digital frequency synthesizers described in Section 7.5 are based on dividers with large ratios generally variable over several decimal orders. Therefore when starting the design the point of departure will not be (3.6), which secures for a given divisor N the minimum of flip-flops, but a counter composed of BCD-subunits supplemented with coding and decoding logic circuits. The actual dividing process may be performed in different ways.

(i) Upcounter with Digit Recognition Logic

The principle is shown in Fig. 3.17. The input pulses accumulate in the counter until the programmed number N is reached; whereupon an output pulse is generated in the recognition logic circuits and the counter reset to the '0' state.

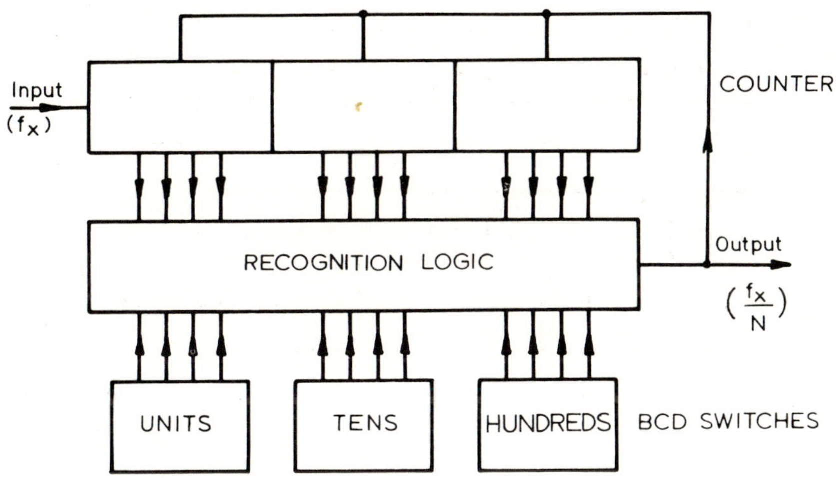

Fig. 3.17 Variable divider employing an up-counter and digit recognition logic.

It is evident that all these operations must be performed before the $(N + 1)$th (the 1st pulse of the next series) input pulse appears. Propagation delays of individual counting decades form a large recognition delay which if not fought against will cause the maximum allowable input frequency (f_x) to be reduced, eventually, far below the resolution capabilities of individual flip-flops. This disadvantage can be overcome by preventing a known number of input pulses from entering the counter, while at the same time retaining their information value (e.g. by resetting the counter to 0100 instead of 0000). The penalty one must pay for this compromise is that the minimum division factor is, in practice, much higher than 2. For an example of an actual divider of this type see [3.11].

(ii) Preset Downcounter

The idea is that the complement of N to the full counter capacity ($10^n - N$, when n is the number of counting decades) is inserted into the counter by the presetting logic circuits. After N input pulses, the zero state is reached and an output pulse generated. On its command the programmed number is transferred into the counter, and not until then can the new dividing cycle start (Fig. 3.18). The main disadvantage of this system is that the last, i.e. the Nth, pulse must 'propagate' through the counter, changing all the decades into the zero state before the output pulse is generated. This disadvantage can be overcome by combining both methods.

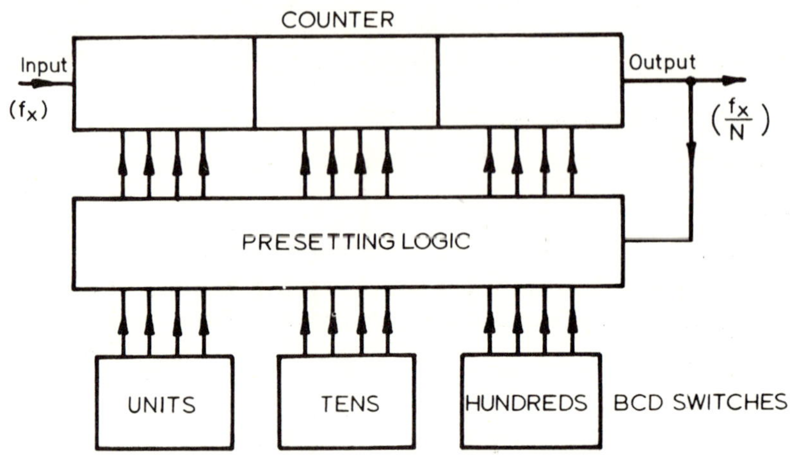

Fig. 3.18 Variable divider employing a down-counter and digit presetting logic.

(iii) Counter with Digit Presetting and Recognition Logic

In the simplest case the recognition state is moved from the '0s' to the '9s'. The advantage of this choice is evident. At the arrival of the last Nth pulse only the first counting decade is operated because all higher-digit decades are waiting in the '9' state long before. The counter must be preset to ($10^n - 1 - N$) or to the 9s complement in each decade.

When IC general purpose BCD counting decades (Fig. 3.19) are used as building blocks the variable divider may be realized with the minimum of additional logic circuits [3.12] – one two-input *NAND* gate is sufficient if three digits are programmed as illustrated in Fig. 3.20.

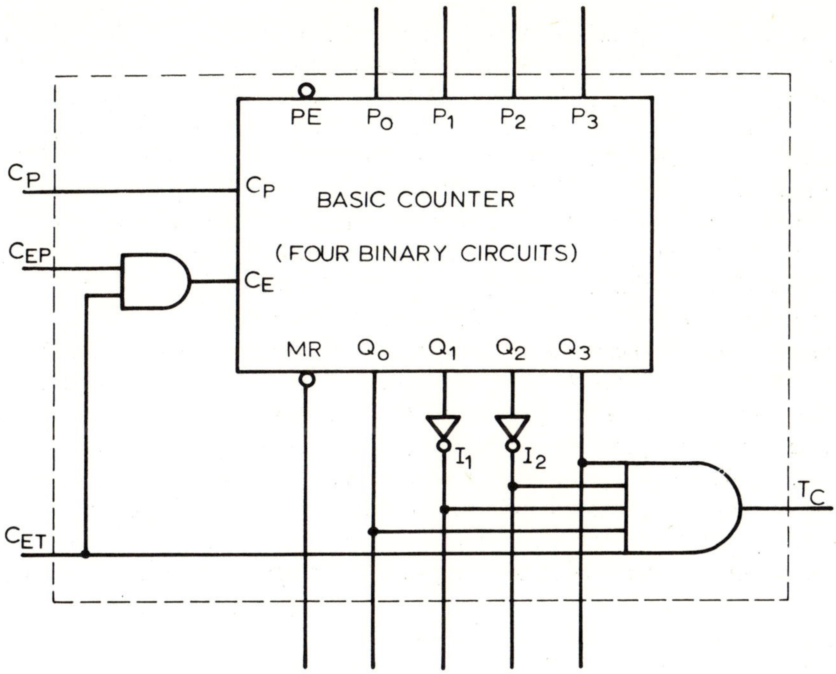

Fig. 3.19 General purpose integrated circuit (IC) *BCD* counting decade. C_P clock pulse; C_E count enable; C_{EP} count enable parallel; C_{ET} count enable trickle; *PE* parallel load (preset) enable (counter is loaded from $P_0 - P_3$ on the next clock pulse when *PE* is low). Inverters I_1 and I_2 secure the *BCD* operation (see Table 3.1). (From Stinehelfer & Nichols [3.12].)

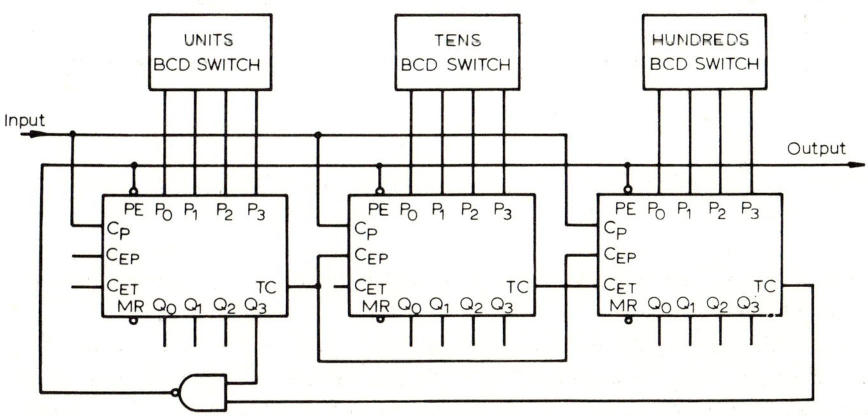

Fig. 3.20 Three digits programmable divider assembled from commercial IC elements. (From Stinehelfer & Nichols [3.12].)

To explain the operation we shall use a numerical example. Let the division ratio $N = 385$. The BCD switches must preset all the BCD counting units to the 9s complements, i.e. $\overline{N} = 614$. After 286 input (clock) pulses the 'hundreds' decade is in the '9' state. Adding another 90 pulses brings the 'tens' decade in the '9' state as well, and both TC outputs go high. From the remaining 9 pulses the penultimate imposes the '8' state on the 'units' decade and heightens Q_3 (see Table 3.1). Consequently the output of the *NAND* gate is low and the last pulse, i.e. the 385th, will again preset (or parallel load) the individual counting decades to 614 according to the information stored in the switches.

The programmable divider just described can operate up to 12 MHz.†

The input frequencies may be increased by an order if a special prescaler that divides by 11 or 10 is used [3.12]. The operation will be explained with reference to the above example and Fig. 3.21. At the beginning of the counting cycle the prescaler divides by 11. After 55 pulses the 'units' counting decade reaches the '9' state and its TC output disables its own C_{EP} and changes the prescaler ratio from 11 to 10. The 'tens-hundred' combination has already accepted 5 pulses and fills gradually up after each 10th input pulse until 98 (after the 375th pulse) when the presetting enable-state is prepared and ten pulses later performed. One

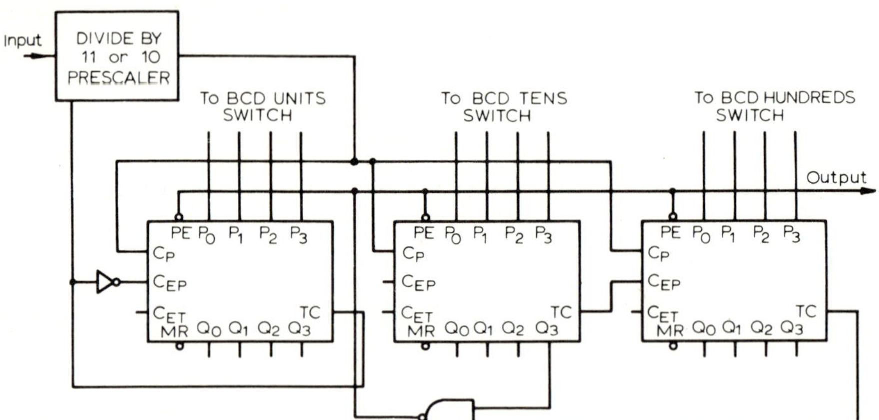

Fig. 3.21 Programmable divider with 11/10-ratio prescaler. (From Stinehelfer & Nichols [3.12].)

restriction imposed by this procedure is that the division ratio cannot be set less than 100.

Less restrictive systems with respect to the N_{min} have been devised. Jannazzo and Rustichelli [3.13] attempt to provide enough time for recognition and presetting by the accumulation of ten pulses at the beginning of each counting

† With Fairchild commercial counting units (9310 type).

cycle in an auxiliary counting decade, whereupon the information is transferred directly into the 'tens' stage of the main counter.

With more complicated recognition and presetting logic Frank [3.14] was able to reduce N_{min} to 3 without sacrificing the resolution capabilities of the counter on the one hand and still retaining slower flip-flops (power saving) at higher decimal places on the other hand.

3.6 PHASE INSTABILITY IN FREQUENCY DIVIDERS

The coherence condition requires that in any arbitrary time interval $(t - t_1)$ the number of input periods is N-times the number of output periods. Thus in the ideal case when neglecting the phase noise due to the dividing process

$$\frac{(1/2\pi)\,[\varphi_i(t) - \varphi_i(t_1)]}{(1/2\pi)\,[\varphi_0(t) - \varphi_0(t_1)]} = N \tag{3.10}$$

The above equation may be fulfilled for any arbitrary time interval $(t - t_1)$ only if

$$\varphi_0(t) = \frac{\varphi_i(t)}{N} \tag{3.11}$$

In the case of simple phase modulation of the input signal, (3.11) may be written as

$$\varphi_0(t) = \frac{\omega_i\, t + \Delta\varphi_i \sin \nu t}{N} \tag{3.12}$$

It is clear that the phase modulation index $\Delta\varphi_i$ is reduced in proportion to the division ratio N, but the modulation frequency remains unchanged. When differentiating (3.11) the same conclusion can also be drawn for the case of frequency modulation.

With a discrete spurious signal present at the input (with respective amplitudes $V_{sp} \ll V_i$) the resulting output phase modulation index is, according to (9.11) and (3.12),

$$\Delta\varphi_0 \approx \frac{1}{N} \cdot \frac{V_{sp}}{V_i} \tag{3.13}$$

In the case of narrow-band input noise eq. (9.12) gives for the input and output phase jitter

$$\varphi_{ni}(t) = \frac{n_s(t)}{V_i} \tag{3.14}$$

$$\varphi_{n0}(t) = \frac{\varphi_{ni}(t)}{N} = \frac{1}{N}\,\frac{n_s(t)}{V_i} \tag{3.15}$$

where $n_s(t)$ is the component of the noise voltage $n(t)$ which is in quadrature to the input signal.

The variance of (3.15) is

$$\overline{\varphi_{no}^2(t)} = \frac{\overline{\varphi_{ni}^2(t)}}{N^2} = \frac{1}{N^2} \cdot \frac{\overline{n^2(t)}}{V_i^2} = \frac{1}{N^2} \frac{1}{2(\text{SNR})_{\text{in}}} \tag{3.16}$$

If the frequency divider behaves as a perfect limiter it suppresses the amplitude noise completely and

$$(\text{SNR})_{\text{out}} = 1/\overline{\varphi_{no}^2(t)} = 2N^2(\text{SNR})_{\text{in}} \tag{3.17}$$

or

$$(\text{SNR})_{\text{out}} = (\text{SNR})_{\text{in}} + 20 \log N + 3 \quad (\text{dB}) \tag{3.18a}$$

$$\approx \text{SNR}_{\text{in}} + 20 \log N \quad (\text{dB}) \tag{3.18b}$$

In other words the ideal dividing process improves the signal-to-noise ratio by a factor $2N^2$ or approximately by $20 \log N$ (dB).

In actual dividing circuits the theoretical value given by (3.17) is not achieved because of the additive noise contributed by the dividing circuit itself.

(A) Regenerative Frequency Dividers

The output phase is given by (3.4)

$$\varphi_0 = \frac{\varphi_i + \varphi_a - \varphi_b}{N}$$

In a general case all the terms on the right-hand side are not constants and may contribute to additional output phase instability

$$\Delta\varphi_0(t) = \frac{\Delta\varphi_i(t) + \Delta\varphi_a(t) - \Delta\varphi_b(t)}{N} \tag{3.19}$$

where $\Delta\varphi_i(t)$ is in respect of the degradation of the input (SNR) due to the mixer, and $\Delta\varphi_a(t)$ and $\Delta\varphi_b(t)$ of the filter phase instabilities (see Section 2.4(C)) and also of the noise generated in the amplifier and multiplier.

The additional output phase instability may be further enhanced by intermodulation in the mixer (see Section 4.4), particularly when the division factor N is low.

(B) Time Jitter in Pulse-Synchronized and Counting Dividers

The triggering effect encountered in these types of frequency dividing circuit is released when the input waveform $u(t)$ reaches a predetermined level. The superimposed spurious signals and changes of the triggering level itself, due for example to the ambient temperature etc., shift the time of the triggering. In the first approximation the two terms of the Taylor expansion may be sufficient for its evaluation.

$$u(t_k) = u(kT) + u'(kT)(t_k - kT) \tag{3.20}$$

where kT is the ideal time of the kth triggering and t_k is the actual time. Evidently, the amount of the time jitter is

$$\Delta T_k = (t_k - kT) = \frac{u(t_k) - u(kT)}{u'(kT)} = -\frac{u_i(t_k)}{u'(kT)} \tag{3.21}$$

where $u_i(t)$ is the interfering signal (see Fig. 3.22(a)).

From eq. (3.21) it is readily seen that the smallest spurious time jitter ΔT_k requires placing of the triggering level at the steepest part of the input waveform.

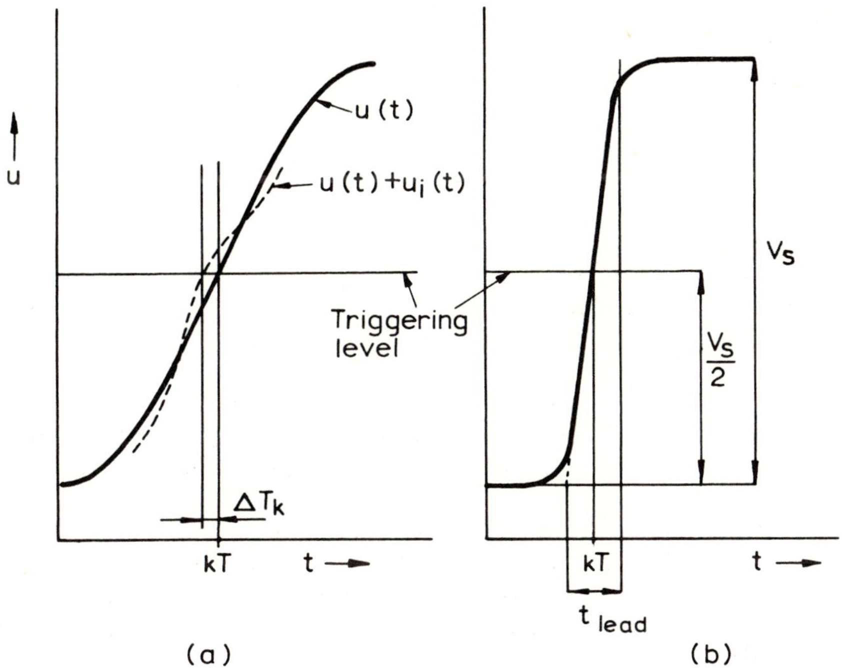

Fig. 3.22 Shift of the triggering time in the presence of an interfering signal: (a) considering a general input waveform $u(t)$; (b) considering a nearly rectangular or pulse input waveform.

A frequency dividing chain begins, generally, with a shaping circuit which is driven by a sinewave

$$u(t) = V_s . \sin \omega t; \quad u'(t)_{\max} = \omega V_s; \quad \text{and} \quad \Delta T_k = -\frac{u_i(t_k)}{\omega V_s} \tag{3.22}$$

The individual dividing circuits are driven by pulses or nearly rectangular waves with steep leading or trailing edges. The actual triggering slope may be expressed approximately by (see Fig. 3.22(b))

$$u'(t) = \frac{V_s}{t_{\text{lead}}} \quad \text{where} \quad t_{\text{lead}} \ll T \tag{3.23}$$

and

$$\Delta T_k = -\frac{u_i(t_k)}{V_s} t_{\text{lead}} \tag{3.24}$$

It is seen that the time jitter introduced in a pulse-synchronized or counting divider may be kept at a very low level, even in long chains of dividing circuits, and practically at the level given by eq. (3.22) when clocked binary systems are used or 'clocked pulse selectors' are placed at the output (Fig. 3.9, 3.10, and 3.11). This conclusion is supported by experimental results (cf. the accuracy of the time intervals in electronic counters used for frequency and time measurements).

3.7 CONCLUSIONS

In this chapter dealing with frequency division the emphasis has been on counting dividers or, more precisely, on sequence counting dividers which are the most important frequency dividers today. Their advantages are:

(1) Excellent output-phase stability (very small time jitter);
(2) Aperiodicity – they cover the whole frequency range from 0 to 200 MHz, and even above; extension to several GHz with the use of tunnel diodes has been reported;
(3) Possibility of achieving division by any positive integer;
(4) Use of solid-state techniques and RC circuits throughout;
(5) Mass production of individual bistable elements or even complete dividers in integrated circuit (IC) form for the whole frequency range from zero to several tens of MHz.
(6) Reliability and fail-safe operation.

All these properties are highly appreciated in frequency synthesis techniques. However, there are disadvantages also:

(1) Unsymmetrical rectangular or even pulse-shaped output waveform;
(2) Rather complex circuitry and a large number of individual elements when using the conventional form of construction.

Considering the above conclusions it is clear that frequency dividers of other types may be competitive on account of their simplicity. Therefore a short description of all important frequency dividing methods has been included in this chapter. However, the survey must not be regarded as complete; for example, the use of monostable multivibrators, unijunction relaxation oscillators, etc., for frequency division has not been discussed.

APPENDIX 3.1 PRINCIPLES OF LOGIC THEORY AND INTEGRATED LOGIC CIRCUITS

Integrated logic circuits provide an economic solution of many electronic switching and counting problems. Integrated circuits are more compact than the equivalent circuits made from discrete components (space and weight savings) and they are more reliable because they need so few external connexions. There is a further economic advantage in sparing the 'brain power' of designing engineers. No wonder the use of IC techniques, particularly in electronic counters, and in frequency synthesizers, is advancing rapidly (see chapters 7 and 8). It therefore seems appropriate to include a short summary of the basic properties of the integrated circuit elements. For deeper information the interested reader is referred to [3.9] or [3.15]; the latter contains a complete course in digital logic.

(A) Boolean Algebra

The basic IC elements are gates admitting two binary operations denoted as logical addition (+) and multiplication (.) Commutative, associative, and distributive laws are retained [1.15, Section 12.8-1] or [3.8, chapter 2]. Further properties are most easily derived and recalled with the use of the Venn diagrams shown in Fig. 3.23.

The complement to A, in respect to remaining space (the inverse statement), is $\bar{A}$ (see Fig. 3.23(a)), and

$$A + \bar{A} = I \qquad (3.25)$$

$$A + I = I \qquad (3.26)$$

$$A \,.\, I = A \qquad (3.27)$$

$$A \,.\, \bar{A} = 0 \qquad (3.28)$$

$$A + 0 = A \qquad (3.29)$$

$$A \,.\, 0 = 0 \qquad (3.30)$$

$$A \,.\, A = A \qquad (3.31)$$

$$A + A = A \qquad (3.32)$$

Further it may be easily verified that

$$A \,.\, (A + B) \equiv A + AB \equiv A \quad \text{(Laws of absorption)} \qquad (3.33)$$

DeMorgan's or dualization theorems

$$\overline{A + B} = \bar{A} \,.\, \bar{B} \qquad (3.34)$$

$$\overline{A \,.\, B} = \bar{A} + \bar{B} \qquad (3.35)$$

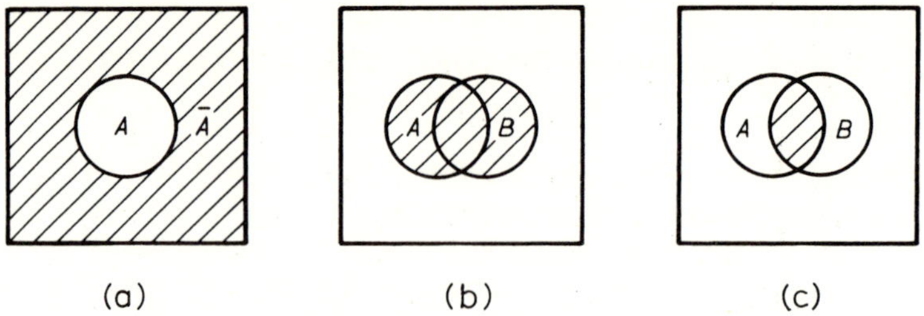

Fig. 3.23 Venn diagrams

(a) Inversion

The shaded area is the complement of A, i.e. the inverse statement is true.

(b) Logical addition

The shaded area is formed by A *or* B, i.e. the true statement is fulfilled by either of the partial statements.

(c) Logical multiplication

The shaded area is formed by parts common to A *and* B, i.e. the true statement requires the simultaneous fulfilment of both partial statements.

(B) Logic Levels and Symbols

In a binary system there are only two states, referred to as *H*(igh) or *L*(ow). The *H* is the relatively more positive level than *L*.

Positive-logic means that the voltage level assigned to the binary 'one' state is more positive than that assigned to the 'zero' state. In negative-logic the 'one' is less positive than 'zero'.

If the output of a gate is high only if all its inputs are high, then the gate performs the *AND* operation (logical multiplication) – the logical symbol is shown in Fig. 3.24(a). If the output of a gate is high if one of its inputs is high

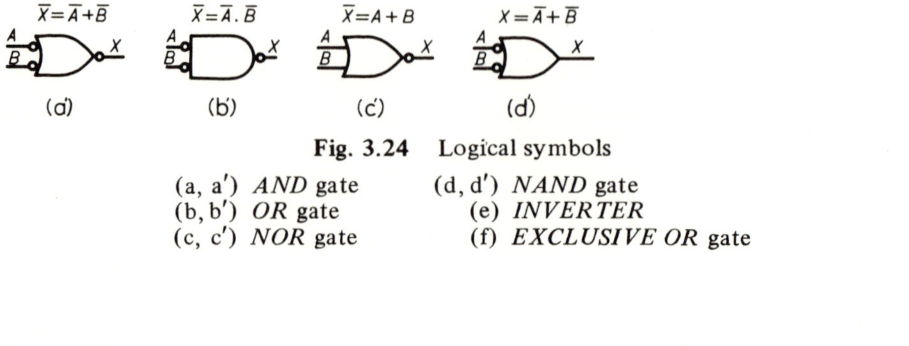

Fig. 3.24 Logical symbols

(a, a') *AND* gate
(b, b') *OR* gate
(c, c') *NOR* gate
(d, d') *NAND* gate
(e) *INVERTER*
(f) *EXCLUSIVE OR* gate

the gate performs the *OR* operation (logical addition) – the logical symbol is shown in Fig. 3.24(b). A small circle at the input line of a logic symbol indicates that a low (L) level activates the function (Fig. 3.24(c)); if it is at the output line then the low (L) level corresponds to the activated gate (Fig. 3.24(d)). Thus the small circle indicates inversion – the logical symbol of the *INVERTER* is in Fig. 3.24(e). The equivalence between the arrangements shown in Fig. 3.24(a) to (d) and 3.24(a′) to (d′), when using DeMorgan's theorems (3.34 and 3.35), is readily proved. Finally if the output of a gate is high only if one of the inputs is high and the other low (Fig. 3.24(f)) then the gate, or rather the system of gates, is called an *EXCLUSIVE OR* gate.

When positive-logic symbology is used to represent the negative-logic functions, the dual of the functions is produced. For example, the positive-logic *AND* *(NAND)* gate becomes a negative logic *OR (NOR)* gate.

(C) Integrated-Circuit Logic Families – *J-K* flip-flops

A logic family or series includes the basic gate elements, flip-flop elements, and more complex logic and storage elements (e.g. 4-bit binary counters, decade counters, decoders, shift registers etc.).

The basic building blocks of IC counters are generally *J-K* flip-flop elements and these deserve a more detailed description.

J-K flip-flops are versatile bistable elements performing simultaneously additional logical functions. Multiport *AND* gates are provided on the J and K inputs and the direct 'Set' and 'Reset' (also called 'Preset' and 'Clear') operations of the bistable element independent of the clock condition are possible. A functional diagram of a *J-K* master–slave flip-flop circuit is shown in Fig. 3.25. (This is not the only possible arrangement, nevertheless its properties as described below are representative). The complexity of the resultant discrete-element circuit is evident. For reasons of economy, therefore, only the integrated-circuit technique versions are practical.

From the gate interconnexion (Fig. 3.25) the exact logical operating characteristics can be deduced. Because of the *NAND* gate at the clock input the flip-flop responds to the positive transitions of the clock signal. The master latch and the slave latch operate together as follows: when the clock is high the output of the respective *NAND* gate is low and the synchronous inputs (J and K) are disabled; only the S(et) and R(eset) (asynchronous) inputs can affect the state of the flip-flop according to the truth table (Table 3.3).

When the clock goes low, the slave latch is locked in its present condition (due to the (5) and (5′) *AND* gates) independent of the master; further, either the J or K input gate is enabled according to the state of the slave latch. If all the data at the appropriate J and K inputs are high at any time when the clock is low, the master changes state to the opposite with respect to the slave. When the clock subsequently goes high the new information in the master is transferred to the slave independent of the synchronous inputs at that time.

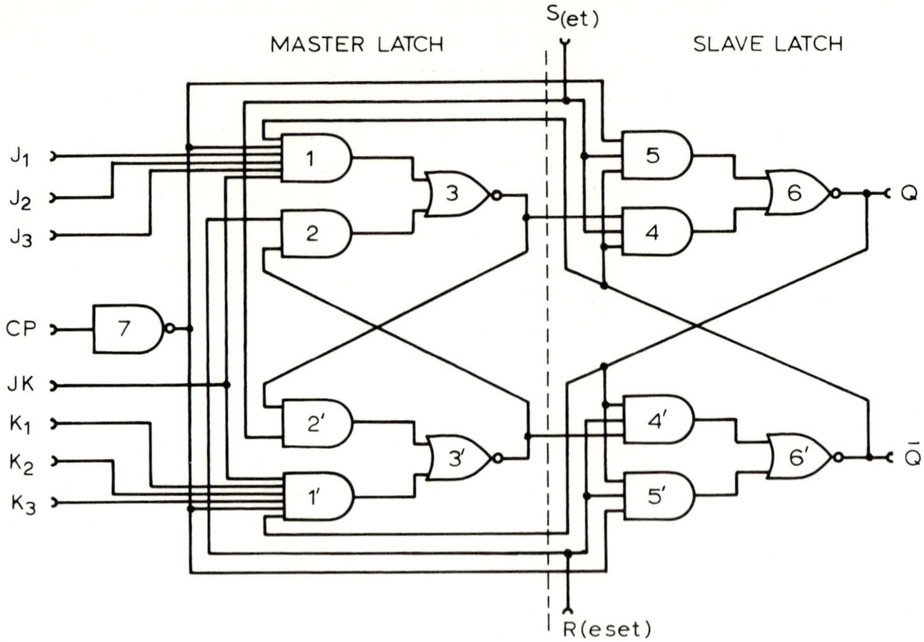

Fig. 3.25 Functional logic diagram of a *J–K* flip-flop bistable element in the 'master–slave' configuration.

Table 3.3
Set–Reset truth table for *J–K* flip-flop; asynchronous entry, independent of 'clock' and 'synchronous input'

S	R	Q_{n+1}	$\bar{Q}_{n+1}$	
0	0	1	1	
0	1	1	0	
1	0	0	1	
1	1	Q_n	$\bar{Q}_n$	(no change)

Evidently the *J–K* bistable element is a clocked device the state of which depends on the logical state of *J* and *K* inputs just before the arrival of the clock pulse according to the truth table (Table 3.4). The schematic representation used in this chapter is shown in Fig. 3.26. Contemporary counting speeds of individual *J–K* flip-flops range up to 50 MHz.

Table 3.4
Truth table of the *J–K* flip-flop; synchronous entry, *J-K* mode operation

J	K	Q_{n+1}	$\bar{Q}_{n+1}$	
0	0	Q_n	$\bar{Q}_n$	(no change)
0	1	0	1	
1	0	1	0	
1	1	$\bar{Q}_n$	Q_n	(Reversal of state)

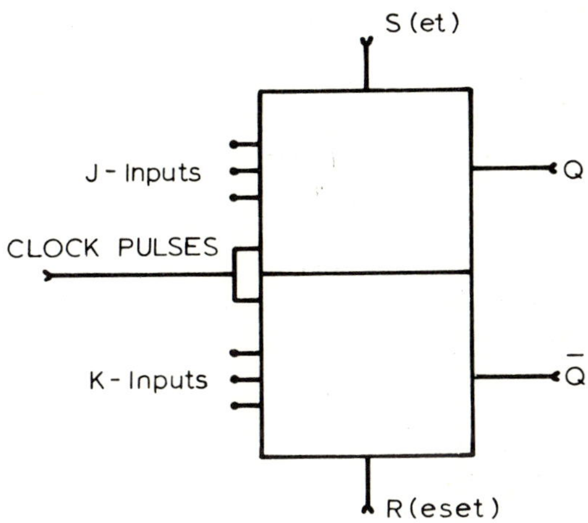

Fig. 3.26 Schematic representation of the *J–K* flip-flop adopted in this chapter.

REFERENCES

[3.1] NORRMAN, E. The inductance–capacitance oscillator as a frequency divider. *Proc. I.R.E.*, **34**, 799-803, 1946.

[3.2] CHANCE, B. (ed.) *et al. Waveforms.* New York, Toronto, London: McGraw-Hill, 1949.

[3.3] PLEVY, A. L. and MONACCHIO, E. N. Fail-safe frequency divider. *Electronics,* **39**, 127, Sept. 1966.

[3.4] PLOTKIN, S. and LUMPKIN, O. Regenerative fractional frequency generators. *Proc. I.R.E.*, **48**, 1988-97, 1960

[3.5] KAMP, Y. Amorçage des diviseurs de fréquence à capacité non linéaire. L'Onde Électrique, **48**, 787-93, Sept. 1968.

[3.6] PRESTON, B. A microelectronic frequency divider with a variable division ratio. *Electronic Eng.*, **37**, 240-4, April 1965.

[3.7] JUNGMEISTER, H. G. Eine bistabile Kippschaltung für den Gigahertz-Bereich. *Archiv der elektrischen Übertragung,* **21**, 447-58, Sept. 1967.
[3.8] KENCH, E. J. (ed.) *Electronic Counting: Circuits, Techniques, Devices,* London: Mullard, 1967.
[3.9] WICKES, W. E. *Logic Design with Integrated Circuits.* New York, London, Sydney: Wiley, 1968.
[3.10] KENCH, E. J. (ed.) *Integrated Logic Circuit Applications: Mullard FC Range.* London: Mullard, 1967.
[3.11] BLACHOVICZ, L. F. Dial any channel to 500 MHz. *Electronics,* **39**, 60-9, May 2, 1966.
[3.12] STINEHELFER, J. and NICHOLS, J. A digital frequency synthesizer for an *AM* and *FM* receiver. Trans. I.E.E.E., **BTR-15,** N.3, 235-43, 1969.
[3.13] JANNAZZO, S. and RUSTICHELLI, G. A variable-ratio frequency divider using micrologic elements. *Electronic Engineering,* **39**, 419, July, 1967.
[3.14] FRANK, R. W. The digital divider. *General Radio Experimenter,* **43**, 3-7, Jan.-Feb. 1969.
[3.15] HUGHES, J. L. *Computer Lab Workbook.* Maynard, Mass.: Digital Equipment Corp., 1969.

4 Frequency Mixing

The last algebraic operations used in the frequency synthesis process are adding and subtracting of two frequencies, performed in devices generally called mixers; however, the terms frequency converters and modulators are also used.

Almost every nonlinear element may be adopted for frequency mixing purposes. But the nonlinear resistors – semiconductor diodes – are still most often encountered, particularly at high and very high frequencies. Their advantages are simplicity and reliability of circuits. Their disadvantage is the conversion loss: typical values are from 6 to 10 dB. On the other hand transistor mixers exhibit conversion gain, but their frequency range is more limited, predominantly to lower frequencies.

Vacuum tube frequency converters, e.g. hexodes or pentagrid mixers in conventional radio sets, so extensively used in the past, should be mentioned, but for details earlier textbooks must be consulted, e.g. [1.12, Section 16.11 to 16.13].

4.1 DIODE MIXERS

In this section we shall limit our investigations to circuits comprising only a single diode as mixing element. Converters in VHF receivers are a typical example. However, there are other applications also, in frequency synthesis technique, in frequency measuring apparatus, etc.

In the following analysis we shall assume two input voltages: one much larger, generally designated as that of the local oscillator ($V_{LO} \cos \Omega t$), and the other called simply the signal voltage ($V_S \cos \omega t$). The desired output voltage $[V_L \cos (\Omega \pm \omega)t]$ is ordinarily the first upper or lower side-band. The diode mixing circuit is shown schematically in Fig. 4.1.

The general current voltage characteristic of a nonlinear resistor is

$$i = f(v) \tag{4.1}$$

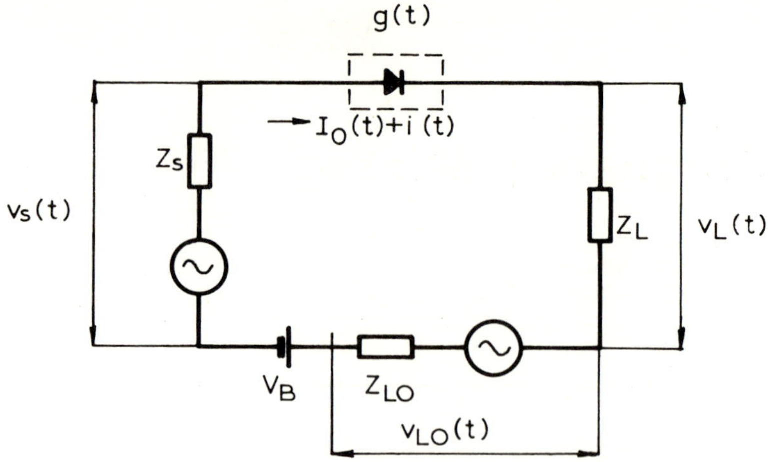

Fig. 4.1 Schematic arrangement of diode mixing circuit.

Assuming the signal and output voltages as small variations of the local oscillator voltage then the current variation is given by

$$\delta i = \left(\frac{di}{dv}\right)_{v_{LO}} . \delta v \tag{4.2}$$

where di/dv is the incremental conversion conductance $g[v_{LO}(t)]$. Introducing time as the independent variable and writing δi and δv simply as small a.c. quantities $i(t)$ and $v(t)$ respectively, eq. (4.2) changes to

$$i(t) = g(t) . v(t) \tag{4.3}$$

The conductance $g(t)$ is a function of the oscillator voltage and consequently a periodic function of time with the frequency Ω

$$g(t) = \sum_{n=-\infty}^{\infty} G_n . \exp(jn\Omega t) \tag{4.4}$$

Substituting (4.4) in (4.3), and using the signal voltage alone

$$v_S(t) = V_S \cos \omega t = V_{1,0}[\exp(j\omega t) + \exp(-j\omega t)]^{\dagger} \tag{4.5}$$

as a first approximation of the $v(t)$, we shall find that the current $i(t)$ is composed of the 'carrier' in ω and the side-bands in $\omega \pm n\Omega$. Equating the Fourier coefficients with the same frequency on both sides of (4.3) leads to (using complex Fourier series expansions)

$$I_{1,n} = \sum_{m=-\infty}^{\infty} G_{n-m} V_{1,m} \tag{4.6}$$

† If the frequencies Ω and ω are not coherent, which is generally the case, then $v_S(t) = v_S \cos(\omega t + \varphi_S)$, but by a suitable choice of the common time origin in (4.4) and (4.5) it is possible to reduce φ_S to a negligible value.

These equations may be written in the matrix form

$$\begin{vmatrix} \vdots \\ I_{1,2} \\ I_{1,1} \\ I_{1,0} \\ I_{1,-1} \\ I_{1,-2} \\ \vdots \end{vmatrix} = \begin{vmatrix} & \vdots & \vdots & \vdots & \vdots & \vdots & \\ \cdots & G_0 & G_1 & G_2 & G_3 & G_4 & \cdots \\ \cdots & G_{-1} & G_0 & G_1 & G_2 & G_3 & \cdots \\ \cdots & G_{-2} & G_{-1} & G_0 & G_1 & G_2 & \cdots \\ \cdots & G_{-3} & G_{-2} & G_{-1} & G_0 & G_1 & \cdots \\ \cdots & G_{-4} & G_{-3} & G_{-2} & G_{-1} & G_0 & \cdots \\ & \vdots & \vdots & \vdots & \vdots & \vdots & \end{vmatrix} \cdot \begin{vmatrix} \vdots \\ V_{1,2} \\ V_{1,1} \\ V_{1,0} \\ V_{1,-1} \\ V_{1,-2} \\ \vdots \end{vmatrix} \qquad (4.7)$$

The termination conditions which reduce substantially the number of voltages $V_{1,m}$ (or currents $I_{1,n}$) permit solution of (4.7). Generally only $V_{1,0}$, $V_{1,1}$, and $V_{-1,1} = V^*_{1,-1}$ i.e. the input signal, the desired side-band, and the image or undesired side-band, are of importance. Assuming further that $g(t)$ is an even function of time, (4.7) reduces to

$$\begin{vmatrix} I_{1,1} \\ I_{1,0} \\ I^*_{-1,1} \end{vmatrix} = \begin{vmatrix} G_0 & G_1 & G_2 \\ G_1 & G_0 & G_1 \\ G_2 & G_1 & G_0 \end{vmatrix} \cdot \begin{vmatrix} V_{1,1} \\ V_{1,0} \\ V^*_{-1,1} \end{vmatrix} \qquad (4.8)$$

In the simplest case Z_S and Z_L present impedances only to the signal and load currents respectively, whereas all other currents are by-passed. This situation is schematically illustrated in Fig. 4.2. Evidently only small-signal voltages $V_{1,0}$ and $V_{1,1}$ need be considered and the signal and the load currents are

$$I_{1,0} = G_1 \cdot V_{1,1} + G_0 \cdot V_{1,0} \qquad (4.9)$$

$$I_{1,1} = G_0 \cdot V_{1,1} + G_1 \cdot V_{1,0} \qquad (4.10)$$

Now the conversion loss, which is defined as the power available from the source at the signal frequency to the power delivered into the load at the desired sideband frequency is

$$L = \frac{P_{\text{in}}}{P_{\text{out}}} = \frac{I_{1,0} \cdot V^*_{1,0} + I^*_{1,0} \cdot V_{1,0}}{I_{1,1} \cdot V^*_{1,1} + I^*_{1,1} \cdot V_{1,1}} \qquad (4.11)$$

When optimizing (4.11) with respect to $V_{1,0}/V_{1,1}$ and $V^*_{1,0}/V^*_{1,1}$ we shall find that

$$L_{\min} = \left(\frac{G_0}{G_1}\right)^2 \left[1 + \sqrt{1 - \left(\frac{G_1}{G_0}\right)^2}\right]^2 \qquad (4.12)$$

$$\left(\frac{V_{1,1}}{V_{1,0}}\right)_{\text{opt}} = -\frac{G_0}{G_1}\left[1-\sqrt{1-\left(\frac{G_1}{G_0}\right)^2}\right] \tag{4.13}$$

$$R_{S,\text{opt}} = R_{L,\text{opt}} = \frac{1}{G_0\sqrt{1-(G_1/G_0)^2}} \tag{4.14}$$

Much profound theory of frequency conversion based on the linear approach, just explained, may be found in [4.1] where also other terminations with respect to the image sideband are considered. The effect of the nonlinear junction capacitance, which has been neglected in the above treatment, is clarified in [4.2].

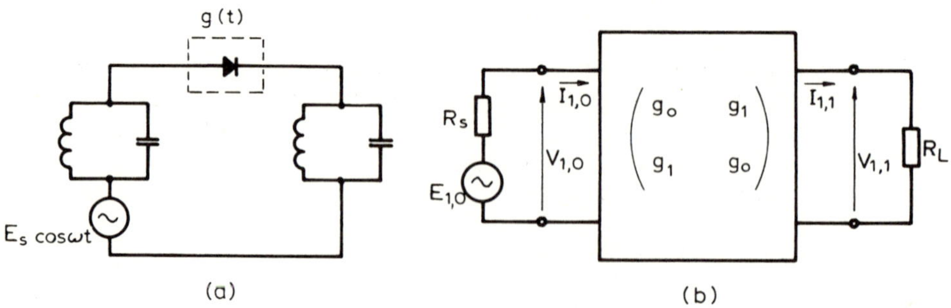

Fig. 4.2 A small-signal diode mixer presenting an infinite conductance to the image current i.e. $G_i = \infty$: (a) idealized circuit arrangement; (b) the equivalent 'mathematical' mixer.

When evaluating (4.12), (4.13), and (4.14) quite simple approximation of the nonlinear characteristic (4.1) will give sufficiently accurate results. Using piecewise linearization the diode conductance changes abruptly between G_d and zero (see Fig. 2.3) and consequently $g(t)$ has a rectangular waveform (cf. Section 2.2(B)). From its expansion, (see (1.30)),†

$$G_0 = \frac{G_d \theta}{2\pi} \tag{4.15}$$

and

$$\frac{G_1}{G_0} = \frac{\sin(\theta/2)}{\theta/2} \tag{4.16}$$

The Fourier expansion of the local oscillator current is given by (1.11), which reveals for the amplitude of the fundamental component

$$(I_{LO})_1 = \frac{1}{\pi} V_{LO} \cdot G_d \left(\frac{\theta}{2} - \sin\frac{\theta}{2}\cos\frac{\theta}{2}\right) \tag{4.17}$$

† Note the multiplication by 1/2 because G_1 is the complex Fourier coefficient.

From this equation the conductance presented by the mixing diode to the local oscillator is readily found:

$$\frac{G_{LO}}{G_d} = \frac{1}{\pi}\left(\frac{\theta}{2} - \sin\frac{\theta}{2}\cos\frac{\theta}{2}\right) \tag{4.18}$$

Note: the ratio G_{LO}/G_d is equal to the ratio i_1/i_m shown in Fig. 1.3(b).

Another simple way of solving diode mixers is the application of the exponential law, but the maximum positive peak voltage must be sufficiently small so that the series diode resistance R_s can be neglected (see Fig. 1.6). Then the diode current is given by (1.17) i.e.

$$i = i_0\,[\exp(\alpha v) - 1] \tag{4.19}$$

The incremental conductance is

$$\frac{di}{dv} = i_0\,\alpha \exp(\alpha v) \tag{4.20}$$

The effective diode voltage v, is, according to Fig. 4.1

$$v = V_B + V_{LO}\cos\Omega t + V_S\cos\omega t - V_L\cos(\Omega\,(\pm)\,\omega)\,t^{\dagger} \tag{4.21}$$

The instantaneous conductance is readily found, using the formula (1.71)

$$g(t) = i_0\,\alpha\exp(\alpha V_B)\sum_{n=-\infty}^{\infty}\bar{I}_n(\alpha V_{LO})\exp(jn\Omega t)\sum_{p=-\infty}^{\infty}\bar{I}_p(\alpha V_S)\exp(jp\omega t)$$

$$\times\sum_{q=-\infty}^{\infty}(-1)^q\,\bar{I}_q(\alpha V_L)\exp[jq(\Omega\,(\pm)\,\omega)\,t] \tag{4.22}$$

In the small-signal case the normalized signal and load voltage amplitudes are

$$\alpha V_S,\quad \alpha V_L \ll 1 \tag{4.23}$$

However, then all

$$\bar{I}_p(\alpha V_S),\ \bar{I}_q(\alpha V_L) \ll 1 \quad \text{except} \quad \bar{I}_0(\alpha V_S) \approx \bar{I}_0(\alpha V_L) \approx 1 \tag{4.24}$$

and the conductance expansion (4.22) simplifies to

$$g(t) = i_0\alpha\exp(\alpha V_B)\sum_{n=-\infty}^{\infty} I_n(\alpha V_{LO})\exp(jn\Omega t) \tag{4.25}$$

When introducing

$$I_{DC} = i_0\,.\exp(\alpha V_B)\,.\,\bar{I}_0(\alpha V_{LO})\,.\,\bar{I}_0(\alpha V_S)\,.\,\bar{I}_0(\alpha V_L)$$

$$\approx i_0\exp(\alpha V_B)\,.\,\bar{I}_0\,(\alpha V_{LO}) \tag{4.26}$$

---

† The cosinusoidal law of $v_L(t)$ is in accordance with results obtained above for the optimum conversion loss (see (4.12) to (4.14)).

the conductance presented to the local oscillator is

$$G_{LO} = \frac{2I_{DC}}{V_{LO}} \cdot \frac{\bar{I}_1(\alpha V_{LO})}{\bar{I}_0(\alpha V_{LO})} \tag{4.27}$$

and the normalized conversion conductance is

$$\frac{G_1}{G_0} = \frac{\bar{I}_1(\alpha V_{LO})}{\bar{I}_0(\alpha V_{LO})}; \tag{4.28}$$

the ratio (4.28) may be found in Fig. 2.10.

The optimum conversion loss L_{min}, the respective ratio of the load to the signal voltage $(V_L/V_S)_{opt}$ are given by (4.12) to (4.14) and drawn in Fig. 4.3 and 4.4 respectively as functions of the normalized L.O. voltage αV_{LO} (for the case of the exponential law) and simultaneously as functions of the effective

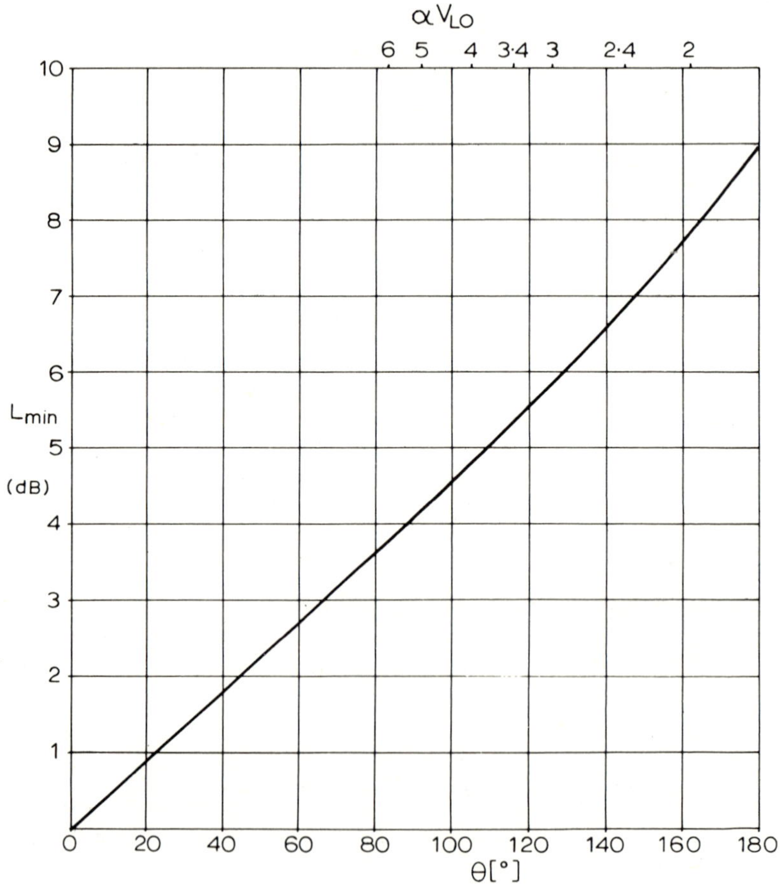

Fig. 4.3 Optimum conversion loss for different current flow angles θ and normalized L.O. voltages αV_{LO} respectively.

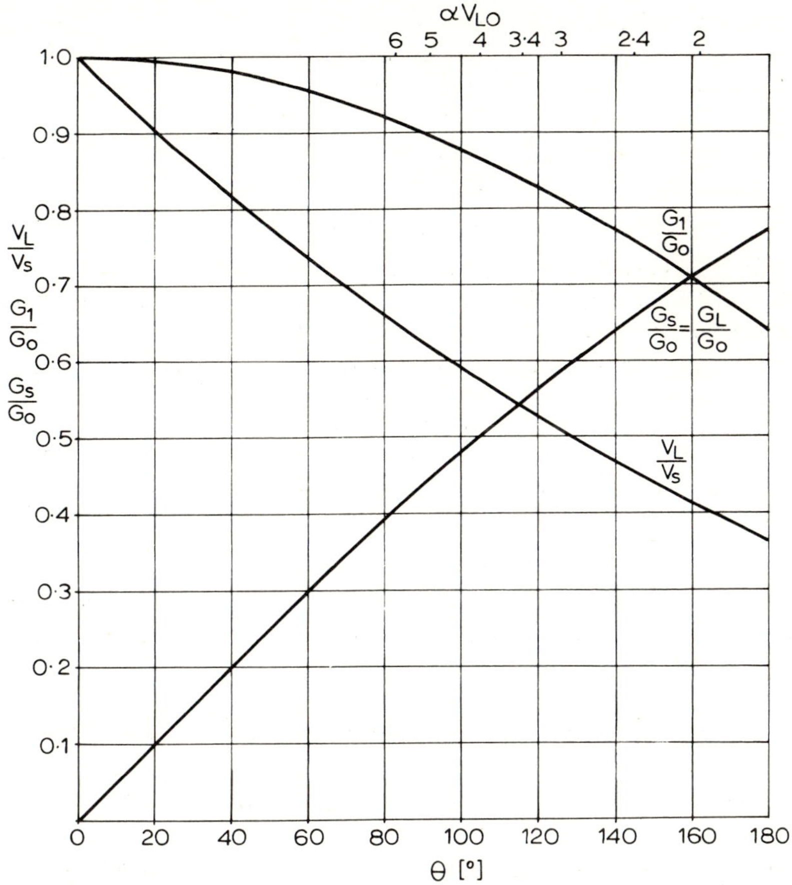

Fig. 4.4 Normalized source, load ($G_S/G_0 = G_L/G_0$), and conversion conductance (G_1/G_0) together with the load-to-signal voltage ratio ($V_{11}/V_{10} = V_L/V_S$) as function of θ and αV_{LO} respectively for the optimum conversion loss.

L.O. current flow angle θ (for the case of piecewise linearization). This procedure was originally suggested by Barber [4.3], the advantage being that the effect of both diode series resistance and junction capacitance can be simply allowed for by choosing a larger θ.

4.2 TRANSISTOR MIXERS

Mixers are generally large-signal devices, at least with respect to the *local oscillator* voltage. Therefore the large-signal transistor theory, recalled briefly in Section 2.3 when dealing with transistor multipliers, is also a suitable tool for designing transistor mixers. The basic common base configuration is shown in Fig. 4.5.

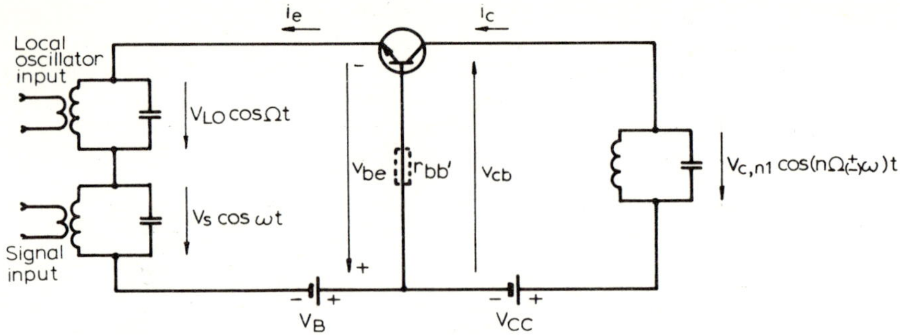

Fig. 4.5 Principle of the common-base transistor mixer.

Assuming that the *local oscillator* and *signal* tuned circuits are ideal, i.e. they have negligible impedance at frequencies other than their centre frequencies, and that the base resistance $r_{bb'}$ is small compared with the intrinsic input impedance of the stage, the instantaneous base-emitter voltage is:

$$v_{be} = V_B + V_{LO}\cos\Omega t + V_S\cos\omega t \tag{4.29}$$

(see footnote on p. 90)

When inserting (4.29) in (2.56) and applying (1.71) we get for the collector current

$$i_c(t) = \alpha_F I_{ES}\,.\exp(\lambda V_B)\sum_{n=-\infty}^{\infty}\bar{I}_n(\lambda V_{LO})\exp(jn\Omega t)\sum_{m=-\infty}^{\infty}\bar{I}_m(\lambda V_S)\exp(jm\omega t)$$

$$= \alpha_F I_{ES}\,.\exp(\lambda V_B)\sum_n\sum_m \bar{I}_n(\lambda V_{LO})\bar{I}_m(\lambda V_S)\exp[j(n\Omega + m\omega)t] \tag{4.30}$$

Amplitudes of individual sidebands in the collector current related to the d.c. component $I_{c,00}$, which is a quantity readily accessible to measurement (in contradistinction to I_{ES}), are

$$I_{c,nm} = 2I_{c,00}\frac{\bar{I}_n(\lambda V_{LO})\bar{I}_m(\lambda V_S)}{\bar{I}_0(\lambda V_{LO})\bar{I}_0(\lambda V_S)} \tag{4.31}$$

where

$$I_{c,00} = \alpha_F I_{ES}\exp(\lambda V_B)\bar{I}_0(\lambda V_{LO})\bar{I}_0(\lambda V_S) \tag{4.32}$$

Note: for the normalized Bessel functions see Fig. 2.10.

In the small-signal case when $\lambda V_S < 1$ and $m = 1$ eq. (4.31) simplifies to

$$I_{c,n1} \approx \lambda I_{c,00}\frac{\bar{I}_n(\lambda V_{LO})}{\bar{I}_0(\lambda V_{LO})}V_S\ . \tag{4.33}$$

The conversion transconductance $g_{con} = g_{n1}$, defined as the ratio of the current out at the sideband frequency $(n\Omega(\pm)\omega)$ to the input signal voltage V_S, follows immediately from (4.33)

$$g_{n1} = \lambda I_{c,00} \frac{\bar{I}_n(\lambda V_{LO})}{\bar{I}_0(\lambda V_{LO})} = g_m \frac{\bar{I}_n(\lambda V_{LO})}{\bar{I}_0(\lambda V_{LO})} \tag{4.34}$$

where g_m is the normal transconductance of the transistor at the quiescent current level $I_{c,00}$.

Other quantities of interest are: the L.O. input impedance

$$R_{LO,B} = \alpha_F \frac{V_{LO}}{I_{c,10}} = \frac{1}{2} \alpha_F \frac{V_{LO}}{I_{c,00}} \frac{\bar{I}_0(\lambda V_{LO})}{\bar{I}_1(\lambda V_{LO})} \tag{4.35}$$

and the signal input impedance

$$R_{S,B} = \alpha_F \frac{V_S}{I_{c,01}} = \frac{1}{2} \alpha_F \frac{V_S}{I_{c,00}} \cdot \frac{\bar{I}_0(\lambda V_S)}{\bar{I}_1(\lambda V_S)} \tag{4.36a}$$

$$\approx \alpha_F \frac{1}{\lambda I_{c,00}} = \frac{\alpha_F}{g_m} \quad \text{(in the small signal case)} \tag{4.36b}$$

In the common-emitter connexion, shown in Fig. 4.6, which is the most commonly used at low frequencies, the signal input impedance is

$$R_{S,E} = \beta_F \frac{1}{\lambda I_{c,00}} = \frac{\beta_F}{g_m} \tag{4.37}$$

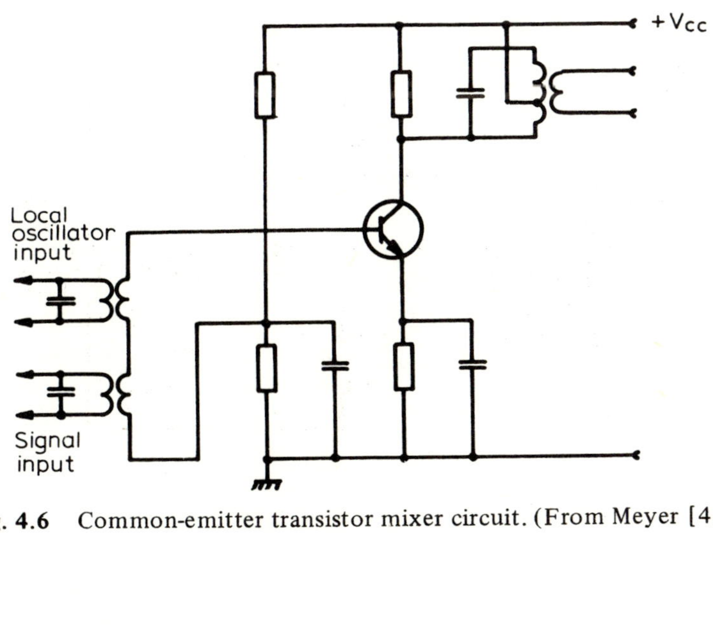

Fig. 4.6 Common-emitter transistor mixer circuit. (From Meyer [4.4].)

The situation when β_F is current-dependent is discussed in [4.4].

The optimum resonance impedance of the collector loading circuit in the large-signal case is

$$R_{L,\,\text{opt}} \approx \frac{1}{2} \cdot \frac{V_{CC}}{I_{c,00}} \cdot \frac{\bar{I}_0(\lambda V_{LO}) \cdot \bar{I}_0(\lambda V_S)}{\bar{I}_n(\lambda V_{LO}) \cdot \bar{I}_m(\lambda V_S)} \tag{4.38}$$

When looking for solutions of transistor mixers at high frequencies one must have recourse to eq. (2.72). However, its complexity requires the use of a computer, even if simplifications are adopted. Some practical results are worked out in references [4.5] and [4.6].

Returning to the fundamental analysis of transistor mixers, examination of (4.30) reveals that the collector current $i_c(t)$ comprises sidebands with all combinations $|n\Omega + m\omega|$, and not only the useful one, $(\Omega(\pm)\omega)$. The inconvenience of this fact will be explained later in Section 4.4; the inference is that the generation of undesired sidebands, often called intermodulation products, should be kept to a minimum. This may be, at least partially, achieved by a balanced or double-balanced configuration of mixing stages which, with respect to one, or other, or both input frequencies are effectively push-pull or push-push connexions.

A balanced transistor mixer is shown schematically in Fig. 4.7. Assuming an ideal symmetry of transistors, of input and output transformers, etc., the effective base-emitter voltages are:

$$v_{be,1} = V_B + V_S \cos \omega t + V_{LO} \cos \Omega t \tag{4.39a}$$

$$v_{be,2} = V_B + V_S \cos \omega t - V_{LO} \cos \Omega t \tag{4.39b}$$

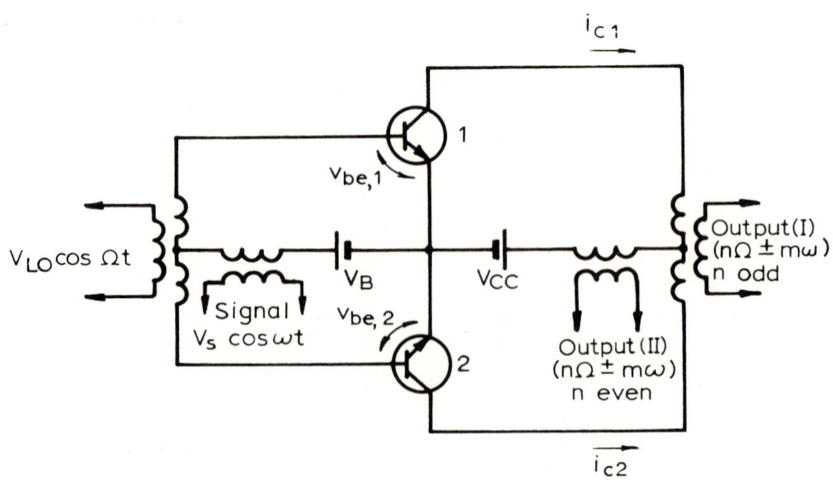

Fig. 4.7 The balanced transistor mixer circuit.

Substituting in (4.30) and taking into account the proper signs of $V_{LO}\cos\Omega t$ (see (1.71)) we get for individual collector currents

$$i_{c1} = I_{c,00} \sum_{n=-\infty}^{\infty} \sum_{m=-\infty}^{\infty} \frac{\bar{I}_n(\lambda V_{LO})\,\bar{I}_m(\lambda V_S)}{\bar{I}_0(\lambda V_{LO})\bar{I}_0(\lambda V_S)} \cos(n\Omega + m\omega)t \qquad (4.40a)$$

$$i_{c2} = I_{c,00} \sum_{n=-\infty}^{\infty} \sum_{m=-\infty}^{\infty} (-1)^n \frac{\bar{I}_n(\lambda V_{LO})\,\bar{I}_m(\lambda V_S)}{\bar{I}_0(\lambda V_{LO})\,\bar{I}_0(\lambda V_S)} \cos(n\Omega + m\omega)t \quad (4.40b)$$

Because the total output current at (I) is proportional to $(i_{c1} - i_{c2})$ and at (II) to $(i_{c1} + i_{c2})$ it may easily be proved that in the first case all components containing an even n, and in the second case all components containing an odd n, are balanced out.

Further groups of mixing products may be cancelled by using the double-balanced configuration shown in Fig. 4.8, the properties of which are summarized in Table 4.1. Transformerless versions which may be desirable in some applications have also been described (see for example [4.7] or [4.8]).

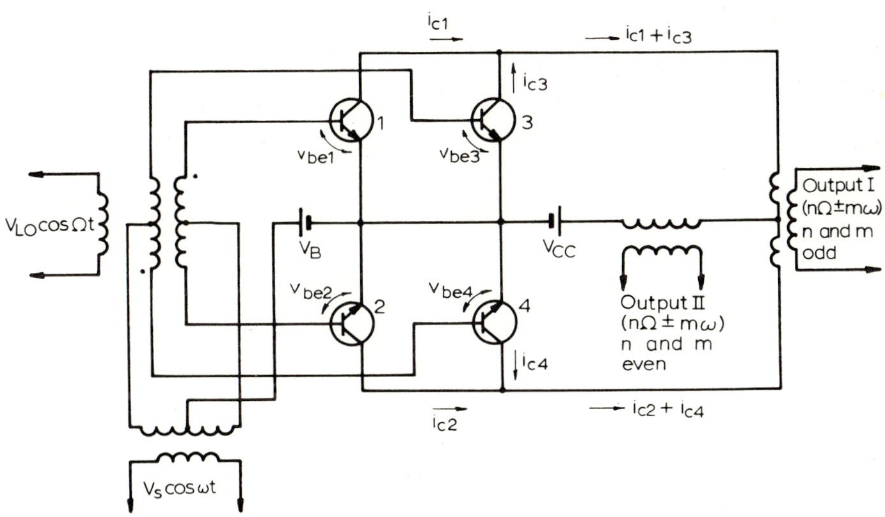

Fig. 4.8 Double-balanced transistor mixer circuit.

4.3 SWITCHING MIXERS

The benefits of balanced or double-balanced configurations of mixing circuits may also be realized with diodes in so-called switching modulators. There are two principal forms, each of which uses four diodes in a bridge arrangement (see Fig. 4.9 or 4.10, and 4.11).

Table 4.1

Properties of balanced and double-balanced transistor mixers

Transistor	$V_S \cos \omega t$	$V_{LO} \cos \Omega t$	$I_{c,nm}$	Balanced mixer		Double-balanced mixer	
No.	sign	sign	sign	Output (I) $i_{c1} - i_{c2}$	Output (II) $i_{c1} + i_{c2}$	Output (I) $i_{c1} - i_{c2} + i_{c3} - i_{c4}$	Output (II) $i_{c1} + i_{c2} + i_{c3} + i_{c4}$
1	+	+	$(+1)^{n+m}$	$\|n\Omega \pm m\omega\|$ n odd only	$\|n\Omega \pm m\omega\|$ n even only	$\|n\Omega \pm m\omega\|$ n and m odd only	$\|n\Omega \pm m\omega\|$ n and m even only
2	+	–	$(-1)^{n}$				
3	–	–	$(-1)^{n+m}$	n odd only	n even only		
4	–	+	$(-1)^{m}$				

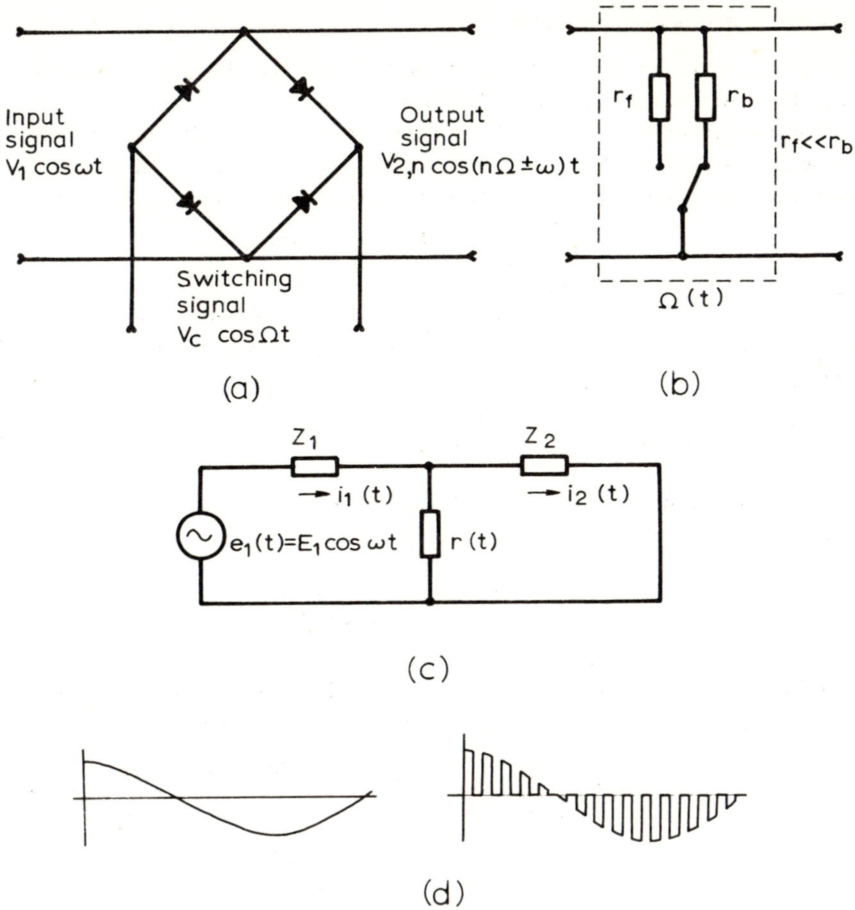

Fig. 4.9 Shunt or Cowan modulator: (a) arrangement of diodes; (b) simplification to ideal-switch operation; (c) the equivalent circuit with terminations; (d) idealized input and output waveforms. (From Belevitch [4.9].)

Assuming that the amplitude of the switching voltage is much larger than the signal amplitude, then the operation of rectifiers is controlled by the switching voltage alone – except for negligibly short intervals when this voltage is near zero. Thus the concept of the time-varying resistance may be applied, and the resulting linear theory is sufficient when solving problems of the impedance match between filters and mixers and when determining mixer losses, as in Section 4.1.

In the following paragraphs we shall make a short review of the theory of switching mixers as worked out by Belevitch [4.9], Tucker [4.10] and [4.11], Gensel [4.12], etc. and summarize the main results in surveying tables.

(A) Shunt and Series Modulators

The principle of the shunt modulator together with the idealized equivalent circuit and waveforms is shown in Fig. 4.9.

The analytic expression of the variable resistance $r(t)$, which takes the value r_f (nearly a short-circuit) during a half-period of the switching voltage and the value r_b (nearly an open-circuit) during the second half-period, can be written

$$r(t) = r_f \frac{1 - u(t)}{2} + r_b \frac{1 + u(t)}{2} \tag{4.41}$$

where $u(t)$ is a periodically discontinuous function of time alternating between (+1) and (−1) at the rate of the switching frequency $F = \Omega/2\pi$. Its Fourier expansion is, (see (1.30)),

$$u(t) = \frac{4}{\pi}(\sin \Omega t + \frac{1}{3} \sin 3\Omega t + \frac{1}{5} \sin 5\Omega t + \cdots) \tag{4.42}$$

Assuming the simplest case when Z_1 and Z_2 are pure resistances R_1 and R_2 respectively, the output current $i_2(t)$ is formally given by

$$i_2(t) = e_1(t) \frac{r(t)}{R_1 R_2 + (R_1 + R_2) r(t)} \tag{4.43}$$

The waveform is shown in Fig. 4.9(d).

Substituting (4.41) in (4.43) and eliminating the time function $u(t)$ from the denominator by multiplying and dividing by a suitable factor†

$$i_2(t) = \frac{e(t)}{2} \cdot \frac{2(R_1 + R_2) r_b r_f + R_1 R_2 (r_b + r_f) + R_1 R_2 (r_b - r_f) u(t)}{R_1^2 R_2^2 + R_1 R_2 (R_1 + R_2)(r_b + r_f) + (R_1 + R_2)^2 r_b r_f} \tag{4.44}$$

The first term represents the input leak $I_{2,0} \cos \omega t$, whereas the second gives the modulation products $I_{2,k} \sin (k\Omega (\pm) \omega t)$ where

$$\frac{I_{2,k}}{I_{2,k+2}} = \frac{k}{k+2} \quad (k \text{ odd}) \tag{4.45}$$

In the ideal case all components with the switching frequency and its harmonics are balanced out of the output.

The conversion loss L (defined as the ratio of the maximum available power $(E_1^2/8R_1)$ to the received power $(\frac{1}{2}R_2 I_{2,1}^2)$ in the useful sideband), is found to be minimum for

$$R_1 = R_2 = 2Z = 2\sqrt{r_f r_b} \tag{4.46}$$

$$L_{\min} = 20 \log \pi \frac{n+1}{n-1} \approx 9{\cdot}94 + \frac{17{\cdot}4}{n} \text{ dB} \tag{4.47}$$

† Because $u^2(t) = 1$, this factor is merely conjugate to the denominator with respect to the sign of the $u(t)$.

where

$$n = \sqrt{\frac{r_b}{r_f}} \tag{4.48}$$

The main properties of the shunt modulator with various termination conditions are summarized in Table 4.2.

Similarly the main properties of the series modulator in Fig. 4.10 are summarized in Table 4.3.

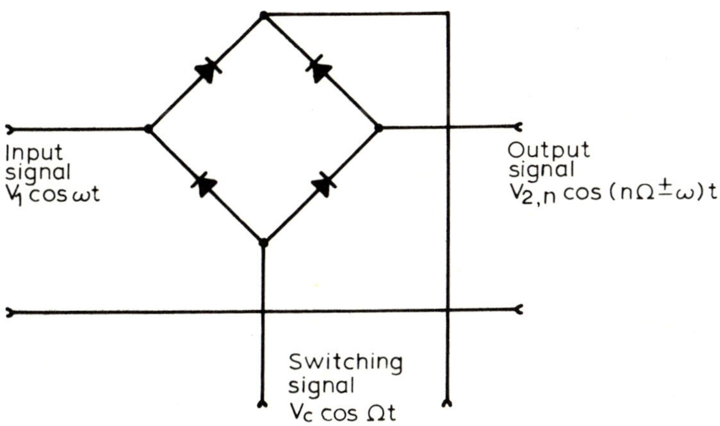

Fig. 4.10 Series modulator.

(B) Ring Modulator

According to assumptions made in the introductory part of this Section (4.3) the ring modulator represented in Fig. 4.11(a) is equivalent during a half-period of the switching signal to the lattice-resistive network in Fig. 4.11(b), and in the following half-period to a similar network with r_b and r_f interchanged. However, this is equivalent to a fixed lattice-resistive network followed by the periodic phase inverter (Fig. 4.11(c)). An ideal ring modulator ($r_f = 0, r_b = \infty$) reduces to the phase inverter alone. The transformers shown in Fig. 4.11(a) are only provided to supply the switching voltage and in the following discussions are assumed to have 1:1 ratio.

The input and output currents of the phase inverter in Fig. 4.11(c) are related by the discontinuous time function $u(t)$

$$i_2(t) = u(t)\, i_1(t) \tag{4.49}$$

The waveform is shown in Fig. 4.11(d).

Table 4.2
Main properties of the shunt modulator with various termination conditions

Case No.	Termination conditions		Matching conditions		Conversion loss (dB)
	Z_1	Z_2	R_1	R_2	
1	R_1 at all frequencies	R_2 at all frequencies	$2Z$	$2Z$	$9{\cdot}94 + \frac{17{\cdot}4}{n}$
2	R_1 at all frequencies	R_2 at all frequencies except ∞ at ω	$1{\cdot}73\,Z\left(1 + \frac{0{\cdot}38}{n}\right)$	$0{\cdot}86\,Z\left(1 - \frac{0{\cdot}77}{n}\right)$	$6{\cdot}94 + \frac{15{\cdot}02}{n}$
2′			$\sqrt{2}\,Z$	$\sqrt{2}\,Z$	$7{\cdot}46 + \frac{16{\cdot}4}{n}$
3	R_1 at all frequencies	R_2 at $\Omega + \omega$ ∞ at all other frequencies	$1{\cdot}73\,Z\left(1 + \frac{0{\cdot}38}{n}\right)$	$0{\cdot}86\,Z\left(1 - \frac{0{\cdot}77}{n}\right)$	$6{\cdot}94 + \frac{15{\cdot}02}{n}$
3′			$\sqrt{2}\,Z$	$\sqrt{2}\,Z$	$7{\cdot}46 + \frac{16{\cdot}4}{n}$
4	R_1 at ω ∞ at all other frequencies	R_2 at all frequencies	$0{\cdot}86\,Z\left(1 - \frac{0{\cdot}77}{n}\right)$	$1{\cdot}73\,Z\left(1 + \frac{0{\cdot}38}{n}\right)$	$6{\cdot}94 + \frac{15{\cdot}02}{n}$
4′			$\sqrt{2}\,Z$	$\sqrt{2}\,Z$	$7{\cdot}46 + \frac{16{\cdot}4}{n}$
5	R_1 at ω ∞ at all others	∞ at ω R_2 at all others	Z	Z	$3{\cdot}92 + \frac{17{\cdot}4}{n}$
6	R_1 at ω ∞ at all others	R_2 at $\Omega + \omega$ ∞ at all others	$0{\cdot}39\,nZ$	$0{\cdot}39\,nZ$	$8{\cdot}9 + \frac{22{\cdot}6}{n^2}$
7	∞ at $\Omega + \omega$ R_1 at all others	R_2 at $\Omega + \omega$ ∞ at all others	Z	Z	$3{\cdot}92 + \frac{17{\cdot}4}{n}$

(From Tucker [4.10] & Gensel [4.12])

Table 4.3
Main properties of the series modulator with various termination conditions

Case No.	Termination conditions		Matching conditions		Conversion loss (dB)
	Z_1	Z_2	R_1	R_2	
1	R_1 at all frequencies	R_2 at all frequencies	$\frac{1}{2}Z$	$\frac{1}{2}Z$	$9{\cdot}94 + \frac{17{\cdot}4}{n}$
2	R_1 at all frequencies	R_2 at all frequencies except 0 at ω	$0{\cdot}58\,Z\left(1 - \frac{0{\cdot}38}{n}\right)$	$1{\cdot}15\,Z\left(1 + \frac{0{\cdot}77}{n}\right)$	$6{\cdot}94 + \frac{15{\cdot}02}{n}$
2′			$\frac{Z}{\sqrt{2}}$	$\frac{Z}{\sqrt{2}}$	$7{\cdot}46 + \frac{16{\cdot}4}{n}$
3	R_1 at all frequencies	R_2 at $\Omega + \omega$ 0 at all other frequencies	$0{\cdot}58\,Z\left(1 - \frac{0{\cdot}38}{n}\right)$	$1{\cdot}15\,Z\left(1 + \frac{0{\cdot}77}{n}\right)$	$6{\cdot}94 + \frac{15{\cdot}02}{n}$
3′			$\frac{Z}{\sqrt{2}}$	$\frac{Z}{\sqrt{2}}$	$7{\cdot}46 + \frac{16{\cdot}4}{n}$
4	R_1 at ω	R_2 at all frequencies	$1{\cdot}15\,Z\left(1 + \frac{0{\cdot}77}{n}\right)$	$0{\cdot}58\,Z\left(1 - \frac{0{\cdot}38}{n}\right)$	$6{\cdot}94 + \frac{15{\cdot}02}{n}$
4′	0 at all other frequencies		$\frac{Z}{\sqrt{2}}$	$\frac{Z}{\sqrt{2}}$	$7{\cdot}46 + \frac{16{\cdot}4}{n}$
5	R_1 at ω 0 at all others	0 at ω R_2 at all others	Z	Z	$3{\cdot}92 + \frac{17{\cdot}4}{n}$
6	R_1 at ω 0 at all others	R_2 at $\Omega + \omega$ 0 at all others	$\frac{2{\cdot}60}{n}Z$	$\frac{2{\cdot}60}{n}Z$	$8{\cdot}9 + \frac{22{\cdot}6}{n^2}$
7	0 at $\Omega + \omega$ R_1 at all others	R_2 at $\Omega + \omega$ 0 at all others	Z	Z	$3{\cdot}92 + \frac{17{\cdot}4}{n}$

(From Tucker [4.10] & Gensel [4.12])

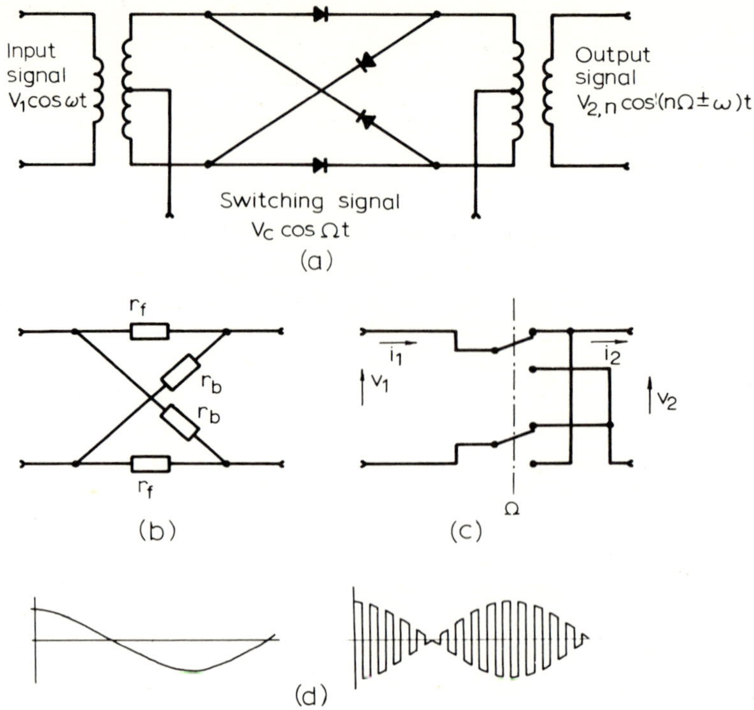

Fig. 4.11 Ring modulator: (a) arrangement of diodes and balanced transformers; (b) the equivalent lattice-resistive network during one half period of the switching signal; in the following half period r_b and r_f are interchanged; (c) the ideal phase inverter; (d) idealized input and output waveforms. (From Belevitch [4.9].)

If a phase inverter is connected between a generator of voltage $E_1 \cos \omega t$, with internal resistence R and a matched loading resistance R, the output voltage $v_2(t)$ will be

$$v_2(t) = i_2(t)\,R = u(t)\,.\,\frac{1}{2}E_1 \cos \omega t = \frac{1}{\pi}E_1 \sum_{k=1}^{\infty} \frac{1}{k} \sin(k\Omega \pm \omega)\,t^{\dagger} \quad (4.50)$$

Evidently, in the ideal case, described by (4.50), all components with the switching frequency and its harmonics as well as the input signal are balanced out of the output. We therefore often use the expression *double-balanced modulators.*

If only the first upper or lower sideband is considered as useful the conversion loss of the ideal modulator is

$$L = 20 \log \frac{\pi}{2} = 3{\cdot}92 \text{ dB} \quad (4.51)$$

† Because $v_2(t)$ is not a function of R, the source and load resistance may have any arbitrary value.

When considering the actual equivalent circuit including the lattice-resistive network (Fig. 4.11(b)) the image impedances of which are

$$Z = \sqrt{r_b \cdot r_f} \tag{4.52}$$

then the optimum terminating impedances for the ring modulator working between pure resistances are

$$R_1 = R_2 = Z \tag{4.53}$$

and the conversion loss for the sideband considered increases to

$$L = 3{\cdot}92 + 17{\cdot}4\sqrt{\frac{r_f}{r_b}} = 3{\cdot}92 + \frac{17{\cdot}4}{n}\ \text{dB} \tag{4.54}$$

In cases of selective terminating impedances the behaviour of the ring modulating circuit cannot be described in such a straightforward manner. Generally energy reflections both on source and load sides take place, and additional modulation products beyond those given by (4.50) may be generated. Matching conditions together with respective conversion losses are summarized in Table 4.4. It may be noted that a particularly unfavourable case is when input and output filters are of the same kind. The conversion loss increases to 8·9 dB; the undesired sideband is 1·3 dB above the useful one (Cases 4 and 5 – according to ref. [4.12]). On the other hand if the terminating filters have mutual inverse configuration the conversion loss tends to zero, with $n \to \infty$, and the useful sideband is stressed (6 dB) with respect to the conjugate (Cases 6 and 7).

4.4 INTERMODULATION PROBLEMS IN FREQUENCY MIXERS

In actual mixing circuits neither the signal nor load voltages are infinitesimal and they therefore affect the instantaneous conversion conductance. The result is that besides the desired sideband, ordinarily that of the sum or difference frequency $\Omega \pm \omega$, also many higher-order sidebands or intermodulation signals of the type $(n\Omega \pm m\omega)$ are generated. In frequency synthesis particular attention is paid to those intermodulation products which fall in the pass-band of the mixer output filters. The designer's task is to ensure that the spurious signal power in the load is below a prescribed useful level, say by –80, –90 dB, etc. Solution of the problem proceeds in several steps. First one assumes that all higher sidebands $n\Omega + m\omega$ $(m, n = 0, \pm 1, \pm 2, \ldots)$ may be generated, and from diagrams, as explained below, the lowest intermodulation products in the pass-band, which are the most troublesome, are found. Then using the actual nonlinear characteristic of the mixer the respective power levels are ascertained.†

† For mathematical treatment the actual nonlinear characteristic must be approximated as closely as possible, particularly when calculating higher-order terms. The simple piece-wise linearized models are unsuitable.

Table 4.4
Main properties of the ring modulator with various termination conditions

Case No	Termination conditions		Matching conditions		Conversion loss (dB)
	Z_1	Z_2	R_1	R_2	
1	R_1 at all frequencies	R_2 at all frequencies	Z	Z	$3{\cdot}92 + \frac{17{\cdot}4}{n}$
2	R_1 at ω; any value at all others	R_2 at all frequencies	Z	Z	$3{\cdot}92 + \frac{17{\cdot}4}{n}$
3	R_1 at all frequencies	R_2 at $\Omega + \omega$; any value at all others	Z	Z	$3{\cdot}92 + \frac{17{\cdot}4}{n}$
4	R_1 at ω ∞ at all others	R_2 at $\Omega + \omega$ ∞ at all others	$0{\cdot}39\, Zn$	$0{\cdot}39\, Zn$	$8{\cdot}9 + \frac{22{\cdot}6}{n^2}$
5	R_1 at ω 0 at all others	R_2 at $\Omega + \omega$ 0 at all others	$\frac{2{\cdot}60}{n} Z$	$\frac{2{\cdot}60}{n} Z$	$8{\cdot}9 + \frac{22{\cdot}6}{n^2}$
6	R_1 at ω	R_2 at $\Omega + \omega$	$0{\cdot}64\, Z$	$1{\cdot}57\, Z$	$\frac{27{\cdot}3}{n}$
6′	∞ at all other frequencies	0 at all other frequencies	$0{\cdot}92\, Z$	$0{\cdot}92\, Z$	$0{\cdot}87 + \frac{22{\cdot}7}{n}$
6″			Z	Z	$0{\cdot}87 + \frac{24{\cdot}8}{n}$
7	R_1 at ω	R_2 at $\Omega + \omega$	$1{\cdot}57\, Z$	$0{\cdot}64\, Z$	$\frac{27{\cdot}3}{n}$
7′	0 at all other frequencies	∞ at all other frequencies	$0{\cdot}92\, Z$	$0{\cdot}92\, Z$	$0{\cdot}87 + \frac{22{\cdot}7}{n}$
7″			Z	Z	$0{\cdot}87 + \frac{24{\cdot}8}{n}$

(From Tucker [4.10] & Gensel [4.12])

If the result is not satisfactory, and neither balanced nor double-balanced mixing device helps, the difficulty can only be overcome by changing the frequency synthesis plan.

(A) Intermodulation Diagrams

When only two signals with frequencies f_1 and f_2 are present at the input of the mixer, then intermodulation frequencies f_i encountered in the pass-band ($f_{p1} < f_i < f_{p2}$) of the output filter may be written symbolically

$$f_i = rf_1 + sf_2 = B(f_1 + f_2) \tag{4.55}$$

where r and s are integers either positive or negative ($r, s = 0, \pm 1, \pm 2, \ldots$) and

$$\frac{f_{p1}}{f_1 + f_2} = B_1 < B < B_2 = \frac{f_{p2}}{f_1 + f_2} \tag{4.56}$$

In the case where the lower sideband is the desired signal at the output, the smaller of both input frequencies, say f_1, must be considered as negative. Normalizing (4.55) with respect to f_1 and introducing the concept of the mixing ratio q

$$q = \frac{f_2}{f_1}^{\dagger} \tag{4.57}$$

we get

$$q = -\frac{r - B}{s - B} \tag{4.58}$$

This is an indetermined equation which may be solved for integer values of r and s only if both q and B are rational numbers. Here, the graphical approach which is easy and lucid, will be explained. We may consider (4.58) as a straight-line equation in the (r, s) plane having the slope $(-1/q)$ and going through the point $(+B; +B)$. The parallel straight lines through the points $(+B_1; +B_1)$ and $(+B_2; +B_2)$ bound the area where all integer pairs $(r; s)$ fulfil conditions for solving (4.58), and thus the general intermodulation relation (4.55) as well. See Fig. 4.12.

Note: In cases of narrow pass-bands there is in effect only one straight line.

Application of the graphical approach on actual intermodulation problems is given in examples below.

Example 4.1

Find the intermodulation signals in the mixer output of the regenerative frequency divider in Fig. 3.4.

<u>$f_1 = 375$ kHz; $f_2 = 500$ kHz</u>

---

† In agreement with the above statement q is positive in the case where the upper sideband is the desired signal at the output, and negative when the lower sideband is the desired signal.

Because the lower sideband is desired: $f_i = (-rf_1 + sf_2)$ and

$$q = -\frac{500}{375} = -\frac{4}{3} \quad \text{i.e.} \ \frac{1}{q} = -0{\cdot}75$$

Assuming a very narrow pass-band then $B_1 \approx B \approx B_2 \approx 1$. Drawing a straight line through points (−0·75) on the side axes of the diagram shown in Fig. 4.12 we find readily the pairs of coefficient $(r; s)$ for which the intermodulation signal f_i is in the output pass-band, or more precisely coincides with the desired sideband $(f_2 - f_1)$:

(1; 1) which represents the desired side-band
(5; 4), (9; 7), ...
(−3; −2), (−7; −5), ...

Example 4.2

Find the intermodulation signals in the output of the first mixer in the frequency synthesizer in Fig. 5.7.

$$f_1 = \frac{100}{36} \text{ kHz}; \ f_2 = \frac{100}{6} \text{ kHz}; \ B_1 \approx B_2 \approx 1$$

$$\frac{1}{q} = \frac{f_1}{f_2} = \frac{1}{6}$$

The straight line drawn through points +0·167 on both sides of the diagram intersects the following $(r; s)$ pairs:

. . . , (−5; 2), (1; 1), (7; 0), . . .

Example 4.3

Solution of the intermodulation problem at the output of the frequency synthesizer in Fig. 5.11.

Regardless of the fact that in the last mixer subtraction is performed we shall normalize the output pass-band 0 to 50 MHz with respect to $(f_1 + f_2)$ to obtain a smaller B_2 compared with the case $(f_2 - f_1)$ for which $B_2 \to \infty$ when $f_1 \to f_2$.

$$f_{1,\,\text{min}} = 260 \text{ MHz}; \ f_{2,\,\text{max}} = 310 \text{ MHz}; \ B_1 = 0; \ B_{2,\,\text{max}} = \frac{5}{56} \approx 0{\cdot}1$$

$$\left(\frac{1}{q}\right)_{\text{min}} = \frac{260}{310} = 0{\cdot}84 \ \text{and} \ \left(\frac{1}{q}\right)_{\text{max}} = 1$$

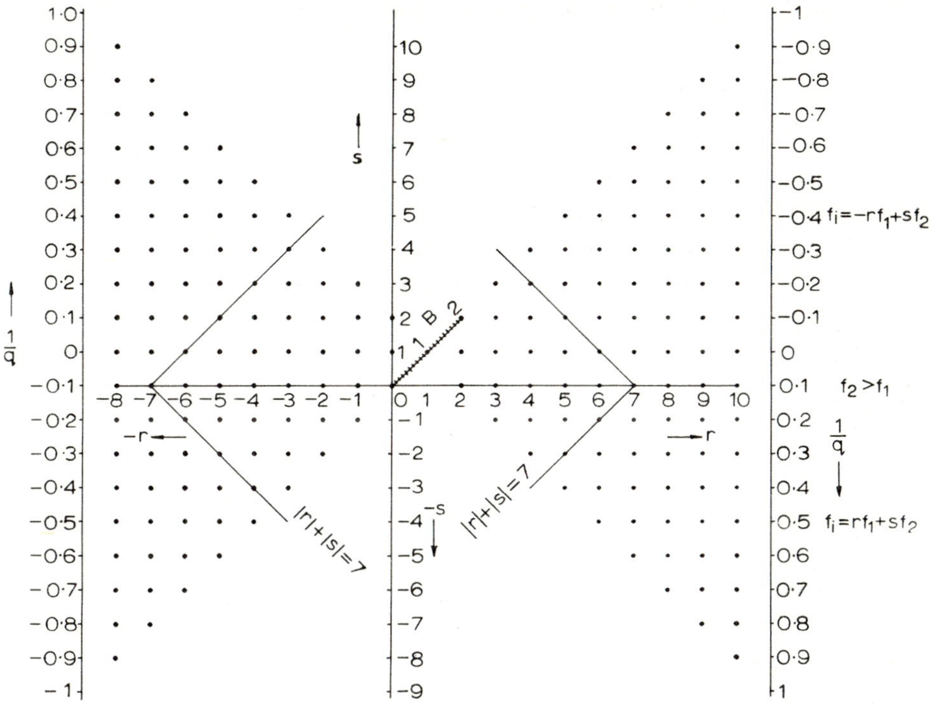

Fig. 4.12 The intermodulation 'point' diagram.

Integer pairs $(r; s)$ in the area bounded by the straight lines with slope parameters 0·84 and 1 and drawn through points $B = 0$ and $B = 0{\cdot}1$ respectively indicate the intermodulation signals $f_i = (rf_1 + sf_2)$ which may be encountered in the output pass-band: They are:

(−6; 5), (−7; 6), (−8; 7), . . .
(−1; 1) the desired sideband and its harmonics respectively (−2; 2), (−3; 3), . . . when $f_1 \to f_2$.

Solution of the intermodulation problem when using the 'point diagram' in Fig. 4.12 is very simple; all interfering pairs may be seen at once, particularly in the case of the narrow pass-band of the output filter. The only shortcoming is rather poor resolution which sometimes requires numerical verification of the 'suspected' pairs. Often therefore another kind of graphical solution is used [4.13].

Normalizing (4.55) with respect to f_2 ($f_2 > f_1$) we get

$$\frac{f_i}{f_2} = \frac{f_p}{f_2} = r\frac{f_1}{f_2} + s \tag{4.59}$$

This equation represents families of straight lines

$$y = rx + s \qquad (4.60)$$

with origins (0; s) and slopes equal to r, as shown in Fig. 4.13. The application for solving intermodulation problems will be explained with the following example.

Example 4.4

We shall reconsider the same intermodulation problem as in Ex. 4.3.

$f_{p1} = 0$, $f_{p2} = 50$ MHz; $f_1 = 260$ to 300 MHz, $f_2 = 300$ to 310 MHz

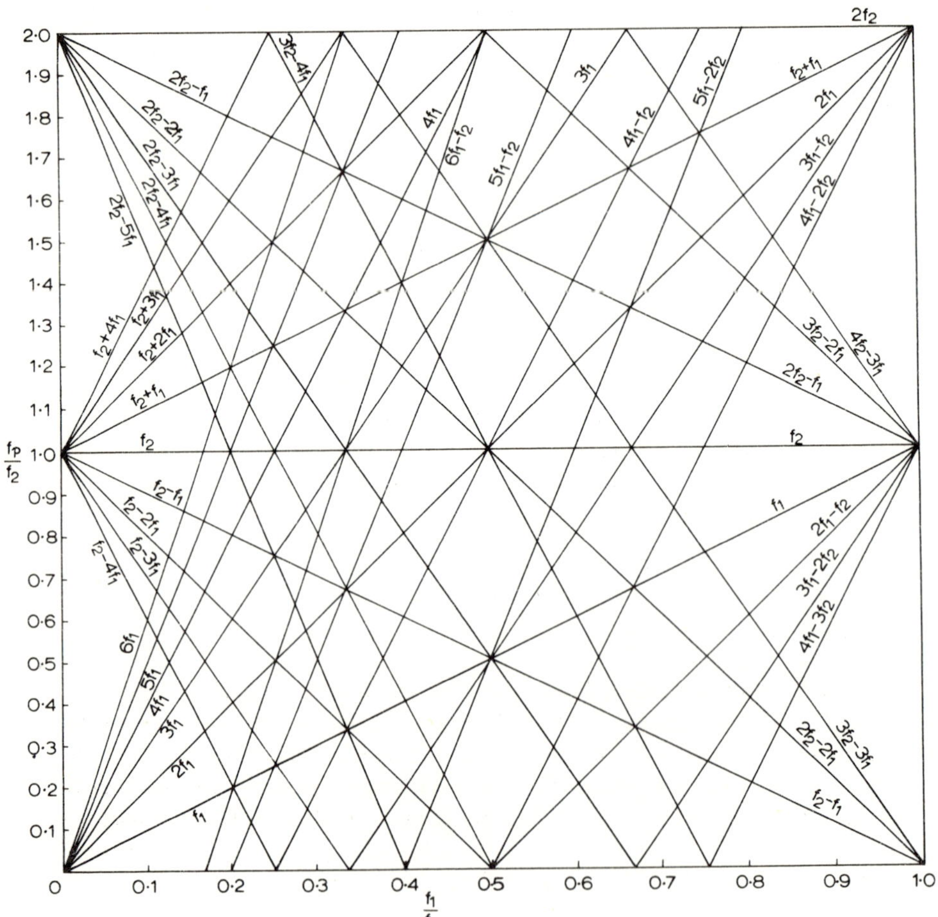

Fig. 4.13 The 'straight line' intermodulation diagram. (Brown [4.13].)

We shall find extremes of f_1/f_2 (0·84 and 1) and of f_p/f_2 (0 and 0·167), mark the respective points on axes, and finally draw through these points horizontal and vertical lines which bound a rectangle. Parameters $(r; s)$ of all straight lines intersecting its area indicate the order of the intermodulation signals which are present in the synthesizer output. Using the diagram in Fig. 4.13 we shall find $(f_2 - f_1), (2f_2 - 2f_1), (3f_2 - 3f_1), \ldots$ as in Example 4.3. But the higher-order terms are not indicated because no straight lines other than those of the type $|r| + |s| \leqslant 7$ are drawn in the diagram.

(B) The Smallest Order of Intermodulation Signals in the Output Pass-Band

When the frequency mixing ratio q is equal to an integer the lower bound of the intermodulation order

$$N = |r| + |s| \tag{4.61}$$

follows immediately from (4.58)

$$|q + 1| = \left| -\frac{r - B}{s - B} + 1 \right| \leqslant |-r + s| \leqslant |r| + |s| \tag{4.62}$$

for integer values of B.

In the case where

$$q = \pm \frac{u}{v} \tag{4.63}$$

(u and v are relatively prime positive integers) eq. (4.62) still applies, i.e.

$$u + v \leqslant |-r + s| \leqslant |r| + |s| \tag{4.64}$$

However,

$$u - v \leqslant |r - s| < |r| + |s| \tag{4.65}$$

because r and s have the same signs when the lower or difference sideband is the useful signal – see Fig. 4.12. Adding $2v = 2\,|s - B|$ to (4.65) gives for the lower bound of N a better approximation, namely

$$u + v - 2B \leqslant |r| + |s| \tag{4.66}$$

(C) Power Level of Intermodulation Signals

Solution of this problem is simple for transistor mixers at low frequencies when eq. (4.30) can be used. Then the intermodulation current is

$$I_{c,rs} = 2I_{c,00} \frac{\bar{I}_r(\lambda V_{LO})\,\bar{I}_s(\lambda V_S)}{\bar{I}_0(\lambda V_{LO})\,\bar{I}_0(\lambda V_S)} \tag{4.67}$$

If its frequency $(r\Omega + s\omega)$ is in the pass-band of the output filter the loading impedance is nearly the same as for the desired sideband $(\Omega(\pm)\omega)$ and the power level is simply

$$IM = 20 \log \frac{\bar{I}_r(\lambda V_{LO})\,\bar{I}_s(\lambda V_S)}{\bar{I}_1(\lambda V_{LO})\,\bar{I}_1(\lambda V_S)} \text{ (dB)} \tag{4.68}$$

Diagrams in Fig. 4.14 show that IM is smaller for larger $N = |r| + |s|$ and for higher orders of $|s|$ if $V_S < V_{LO}$.

Equation (4.68) is also often used when discussing properties of diode mixers [4.14]. However, one must bear in mind the limitations: namely the small signal cases, where the exponential law is actually valid, and in addition the output voltage V_L is small compared to the signal voltage V_S. When this last

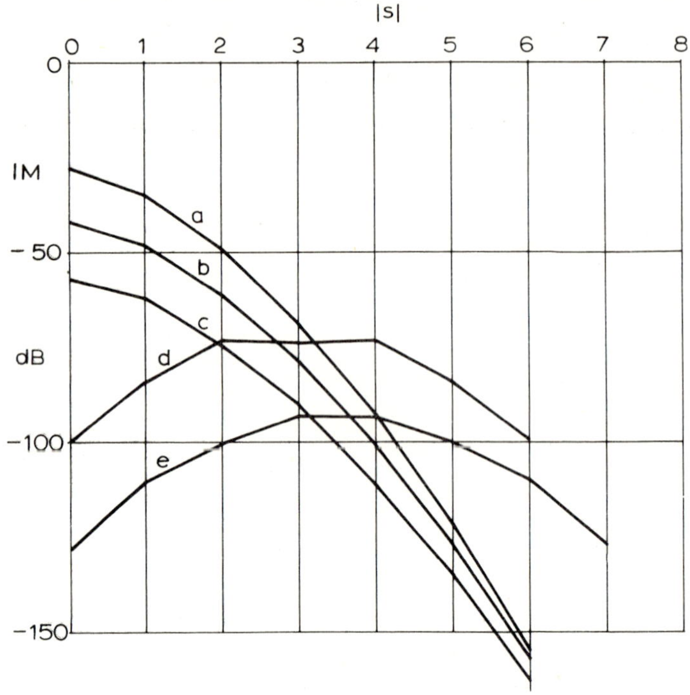

Fig. 4.14 Power levels of intermodulation signals in a transistor mixer:

(a) $|r| + |s| = 6$
(b) $|r| + |s| = 7$ $\quad \lambda V_{LO} = 2{\cdot}8$ and $\lambda V_S = 0{\cdot}2$
(c) $|r| + |s| = 8$
(d) $|r| + |s| = 6$ $\quad \lambda V_{LO} = \lambda V_S = 0{\cdot}6$
(e) $|r| + |s| = 7$

condition is not true we have to use the three-voltage current expansion of (4.19) which is the same as the conductance expansion (4.22), but with the factor α omitted. The main contributions to the $(r; s)$ intermodulation component defined by (4.55) are

$$I_{rs} \approx 2I_{DC} \sum_{q=0}^{s} (-1)^q \bar{I}_{r\mp q}(\alpha V_{LO}) \bar{I}_{s-q}(\alpha V_S) \bar{I}_q(\alpha V_L) \tag{4.69}$$

where the (−) or (+) sign in the factor $\bar{I}_{r \mp q}(\alpha V_{LO})$ is used when the first upper sideband or lower sideband is filtered at the output of the mixer.

As soon as the voltage drop on the resistance in series with the diode junction cannot be neglected, the simple exponential law must be modified in accordance with (1.20) i.e.

$$i = i_0 \{\exp[\alpha(v - iR_s)] - 1\} \tag{4.70}$$

The application of the Taylor expansion leads to the power series

$$i = a_0 + a_1 v + a_2 v^2 + \cdots \tag{4.71}$$

The coefficients a_k have been evaluated by Mills [1.2].

To find the amplitudes of intermodulation signals, the usual approach is to assume that the contribution of the lowest order term, i.e. $a_N v^N$, is the most significant for a given frequency, whereas only the local oscillator and the input signal are considered [4.15]. Using the binomial theorem we get

$$I_{rs} \approx \frac{a_N N!}{2^{N-1}(N - |r|)!|s|!} V_{LO}^{|r|} V_S^{|s|}; \qquad N = |r| + |s| \tag{4.72}$$

At high levels of operation the above assumptions no longer hold and intermodulation amplitudes should be evaluated using the general intermodulation formula (4.86) derived in Appendix 4.1.

If the intermodulation power level is too high it may be lowered by using a smaller signal-voltage V_S'. According to (4.68) or (4.72) the improvement is

$$IM' - IM \approx 20(|s| - 1) \log \frac{V_S'}{V_S} \tag{4.73}$$

Both equations (4.68) and (4.72) indicate that the larger is the intermodulation order N, the smaller the intermodulation interference might be expected to be. Therefore in accordance with (4.62) large mixing ratios $|q|$ should be preferred. But in this case there is the problem of the effective suppression of the 'carrier' and 'image sideband'. Consequently demands on the selectivity of the output filters in the mixing stages may become excessive. For this reason, predominantly economic considerations are decisive for the upper bound put on the mixing ratio. It may be shown that, if $|q + 1| \leqslant 19$, the necessary output filters may be very simple – one or two pairs of resonant circuits.

Another solution of this problem might be a mixing element, the power series expansion of which effectively ends with the second-order term, i.e. $a_2 v^2$. Inspection of the truncated series of (4.71) reveals that no intermodulation products of higher orders can be generated. Therefore the square law nonlinear characteristic is the most desirable for mixers.

Very often balanced or double-balanced mixing devices are an effective weapon when fighting the intermodulation problem. Properties of transistor circuits have been discussed in Section 4.2. Examination of diode switching

Table 4.5
Balancing properties of switching modulators

Modulator	Diode	Voltage	Sign	Port I	Port II	Port III
type	k	v_k	i_k	$i_I = i_1 - i_2$	$i_{II} = i_3 - i_4$	$i_{III} = i_2 + i_3$
Series modulator (Fig. 4.15(a))	1	$v_{LO} + v_S - v_L$	$(-1)^q$	$i_I = i_{II} = K(-1)^q[1-(-1)^s]$		$i_{III} = K'(-1)^q$
	2	$v_{LO} - v_S + v_L$	$(-1)^{s-q}$	components with s odd only		$\times [1 + (-1)^s]$
	3	$v_{LO} + v_S - v_L$	$(-1)^q$			components with s even only
	4	$v_{LO} - v_S + v_L$	$(-1)^{s-q}$			
				$i_I = i_1 + i_2 - i_3 - i_4$	$i_{II} = i_1 - i_2 - i_3 + i_4$	$i_{III} = i_1 - i_2 + i_3 - i_4$
Ring modulator (Fig. 4.15(b))	1	$- v_{LO} + v_S + v_L$	$(-1)^{r-q}$	$i_I = K_I(-1)^q$	$i_{II} = K_{II}(-1)^q$	$i_{III} = K_{III}(-1)^q$
	2	$+ v_{LO} + v_S - v_L$	$(-1)^q$	$\times [1 + (-1)^r]$	$\times [(-1)^r - 1]$	$\times [(-1)^r - 1]$
	3	$+ v_{LO} - v_S + v_L$	$(-1)^{s-q}$	$\times [1 - (-1)^s]$	$\times [1 + (-1)^s]$	$\times [1 - (-1)^s]$
	4	$- v_{LO} - v_S - v_L$	$(-1)^{r+s-q}$	r even s odd only	r odd s even only	r odd s odd only
Ring modulator (Fig. 4.15(c))	1	$+ v_{LO} + v_S - v_L$	$(-1)^q$	$i_I = K'_I(-1)^q$	$i_{II} = K'_{II}(-1)^q$	$i_{II} = K'_{III}(-1)^q$
	2	$- v_{LO} + v_S + v_L$	$(-1)^{r-q}$	$\times [1 + (-1)^r]$	$\times [1 - (-1)^r]$	$\times [1 - (-1)^r]$
	3	$+ v_{LO} - v_S + v_L$	$(-1)^{s-q}$	$\times [1 - (-1)^s]$	$\times [1 - (-1)^s]$	$\times [1 + (-1)^s]$
	4	$- v_{LO} - v_S - v_L$	$(-1)^{r+s-q}$	r even s odd only	r odd s odd only	r odd s even only

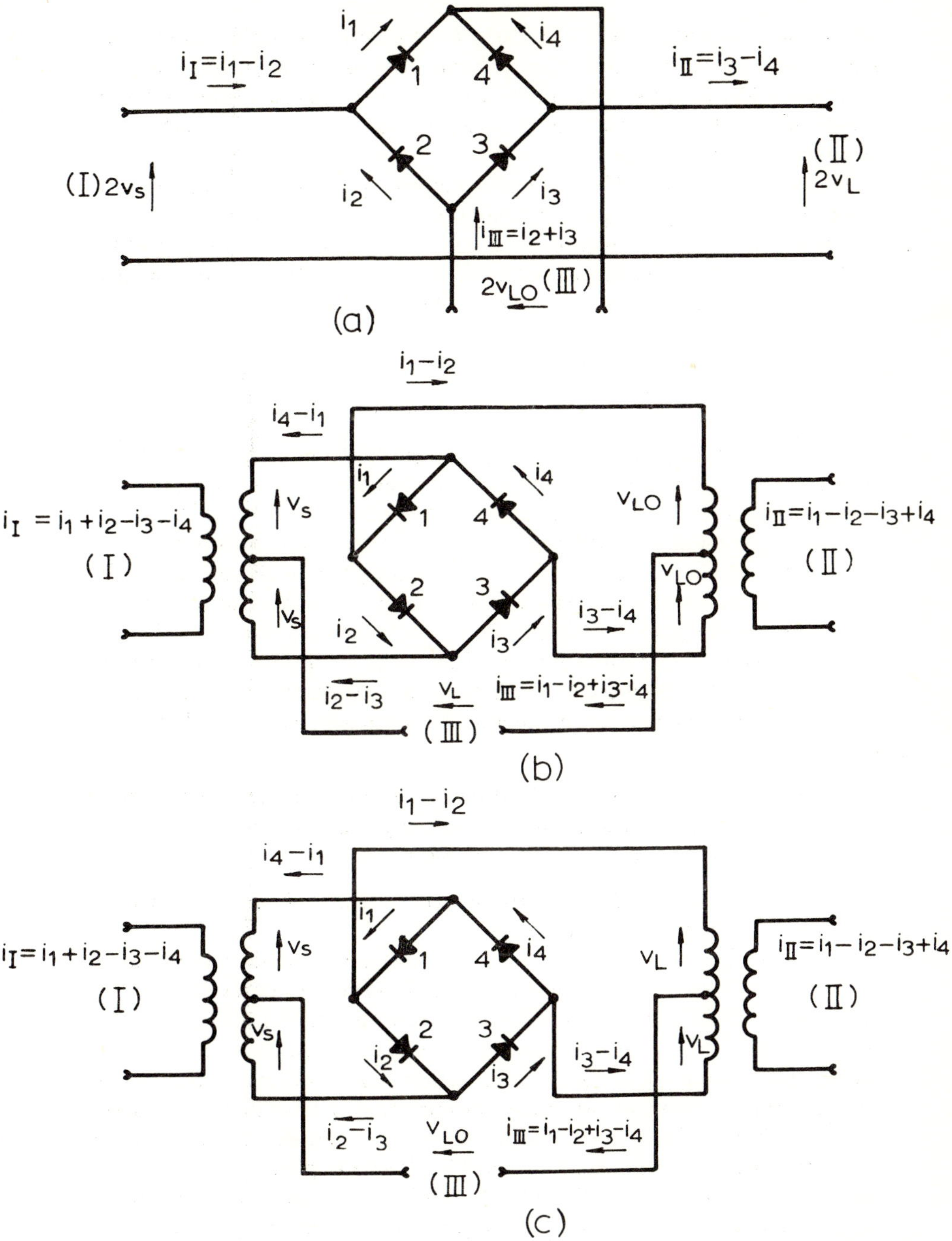

Fig. 4.15 Instantaneous voltages and currents in switching mixers: (a) series modulator; (b) ring modulator; (c) ring modulator, with v_{LO} and v_L interchanged.

mixers must be based either on (4.69) or (4.86). Both approaches lead to the same results which, for series and ring modulators, assuming the presence of three sinusoidal voltages, are summarized in Table 4.5. Balancing properties of the shunt modulator follow from the duality theorem with respect to the series modulator.

It must not be assumed that in real switching mixers all diodes will be identical and transformers perfect. Therefore neither of the intermodulation signals will be completely balanced out. Only a reduction of the order of 20 to 40 dB, and with special care 60 dB, may be expected. In this sphere modern diodes with very high reverse resistance are helpful [4.16].

4.5 CONCLUSIONS

In frequency synthesis one of the most appreciated features is the output frequency purity. From this viewpoint mixers must be considered as particularly dangerous because the indispensable nonlinearity affected by at least two signals with different frequencies (Ω, ω) generates all possible combinations $|r\Omega + s\omega|$ ($r, s = \pm 1, \pm 2, \ldots$). Recalling briefly the theory of simple diode and transistor mixing circuits it has been shown that the intermodulation power level -80 dB may be attained when $|r| + |s| \geqslant 7$. This conclusion agrees with former experimental results [5.2]. Further improvement may be provided by balanced configuration. Particularly important in this respect is the well-known ring modulator. Its advantages which should be emphasized are:

(1) suppression of all intermodulation products with either $|r|$ or $|s|$ even,
(2) a solid-state device available commercially as a whole in
(3) wide-band, low-noise, low-insertion-loss versions.

The disadvantage lies in balancing transformers which limit the useful range, particularly at the lower-frequency end.

Frequency mixers were the first electronic devices where the all-solid-state technique was applied, as early as in the 30s. Nevertheless improvements are still in progress. At least the introduction of Schottky barrier diodes with extremely fast turn-on turn-off times, and field-effect transistors (FET or MOSFET – metal-oxide-surface) with excellent noise properties, excellent carrier suppression and almost square-law operation, deserve to be mentioned.

Finally it should be stressed that the design of mixing circuits in frequency synthesis is a critical problem. Low signal levels lead to poor signal-to-noise ratios and phase noise. High signal levels may introduce spurious modulation at discrete frequencies as explained – an equally undesirable result. Mixers applied at the end of a synthesis sequence introduce less noise since no multiplication follows.

APPENDIX 4.1 GENERAL INTERMODULATION PRODUCT

The problem is completely covered in [4.17]. Here, only the main ideas will be recapitulated.

Assuming the power series description of the nonlinear device

$$i(t) = \sum_{k=0}^{\infty} a_k v^k \tag{4.74}$$

and the multisignal effective voltage

$$v(t) = \sum_{m=0}^{M} \hat{V}_m \cos(\omega_m t + \varphi_m) = \tfrac{1}{2} \sum_{m=-M}^{M} V_m \exp(j\varphi_m) \exp(j\omega_m t) \tag{4.75}$$

where

$$\hat{V}_m = V_m = V_{-m}, 2\hat{V}_0 = V_0, \omega_{-m} = -\omega_m, \varphi_{-m} = -\varphi_m, \text{and } \omega_0 = 0, \varphi_0 = 0 \tag{4.76}$$

the application of the multinomial theorem gives readily

$$v^k(t) = \frac{k!}{2^k} \sum_{(k)} \left[\prod_{m=-M}^{M} \frac{V_m^{n_m}}{(n_m)!} \exp(jn_m \varphi_m) \exp(jn_m \omega_m t) \right] \tag{4.77}$$

The summation is taken over all positive integers n_m such that

$$\sum_{m=-M}^{M} n_m = k \tag{4.78}$$

Introducing

$$n_m \omega_m + n_{-m} \omega_{-m} = (n_m - n_{-m}) \omega_m = p_m \omega_m \tag{4.79a}$$

$$n_m \varphi_m + n_{-m} \varphi_{-m} = p_m \varphi_m \tag{4.79b}$$

$$n_m + n_{-m} = q_m = |p_m| + 2h_m \tag{4.79c}$$

h_m being positive integers and

$$n_0 = p_0 = q_0 \tag{4.79d}$$

rearrangement of (4.77) results in a multiple Fourier series

$$v^k(t) = \frac{k!}{2^k} \left\{ \sum_{(k)} \left[\prod_{m=0}^{M} \frac{V_m^{q_m} \exp(jp_m \varphi_m)}{[(q_m + p_m)/2]!\,[(q_m - p_m)/2]!} \exp(jp_m \omega_m t) \right] \right\} \tag{4.80}$$

The summation $\sum_{(k)}$ is taken over all positive integers q_m such that

$$\sum_{m=0}^{M} q_m = k \tag{4.81}$$

Assuming a particular frequency

$$\Omega = r_1 \omega_1 + r_2 \omega_2 + \cdots + r_M \omega_M \tag{4.82}$$

and excluding, without loss of generality, $m = 0$, i.e. performing the expansion of $i(t)$ around the biasing voltage V_0, the modulus of the respective Fourier coefficient is

$$(v^k)_\Omega = \frac{k!}{2^k} \sum_{(\Sigma h_m)} \left(\prod_{m=1}^{M} \frac{V_m^{|r_m|+2h_m}}{(|r_m| + h_m)!\,(h_m)!} \right) \tag{4.83}$$

The summation is taken over all positive integers h_m such that

$$\sum_{m=1}^{M} h_m = \tfrac{1}{2}(k - N) \tag{4.84}$$

This simplification follows from (4.81) by introducing (4.79c) and the 'order' of the intermodulation frequency (4.82) defined by

$$N = \sum_{m=1}^{M} |r_m| \tag{4.85}$$

Considering (4.74), one readily finds the modulus of the Fourier coefficient of the total intermodulation current at frequency Ω

$$(i)_\Omega = \sum_{L=0} a_{N+2L} \frac{(N + 2L)!}{2^{N+2L}} \sum_{(L)} \left(\prod_{m=1}^{M} \frac{V_m^{|r_m|+2h_m}}{(|r_m| + h_m)!\,(h_m)!} \right) \tag{4.86}$$

In agreement with (4.84) summations $\sum_{(L)}$ are over all positive integers h_m such that

$$\sum_{m=1}^{M} h_m = L \tag{4.87}$$

The general case includes $m = 0$ and summations must be taken over (S) instead of (L)

$$S = k - N = r_0 + 2 \sum_{m=1}^{M} h_m \tag{4.88}$$

where r_0 are positive integers

$$r_0 = \begin{cases} 0, 2, \ldots \leqslant S & (S \text{ even}) \quad (4.89a) \\ 1, 3, \ldots \leqslant S & (S \text{ odd}) \quad (4.89b) \end{cases}$$

REFERENCES

[4.1] STRUM, P. D. Some aspects of mixer crystal performance. *Proc. I.R.E.*, **41**, 875-89, 1953.

[4.2] EDWARDS, C. F. Frequency conversion by means of a nonlinear admittance. *Bell Systems Technical Journal*, **35**, 1403-16, Nov. 1956.

[4.3] BARBER, M. R. Noise figure and conversion loss of the Schottky barrier mixer diode. *I.E.E.E. Transactions of the Microwave Theory and Techniques*, **MTT-15**, 629-35, 1967.

[4.4] MEYER, R. G. Noise in transistor mixers at low frequencies. *Proc. I.E.E.*, **114**, 611-18, 1967.

[4.5] MEYER, R. G. Signal processes in transistor mixer circuits at high frequencies. *Proc. I.E.E.*, **114**, 1605-12, 1967.

[4.6] MEYER, R. G. Noise in transistor mixers at high frequencies. *Proc. I.E.E.*, **115**, 487-95, 1968.

[4.7] HOIGAARD, J. C. Spurious frequency generation in frequency converters. *Microwave Journal*, **10**, 61-4, 78-82, July, Aug. 1967.

[4.8] JONES, H. E. On the transformerless full-wave modulator-demodulator, *I.E.E.E. Journal of Solid State circuits*, **3**, 198, June 1968.

[4.9] BELEVITCH, V. Linear theory of bridge and ring modulator circuits. *Electrical Communication*, **25**, 62-73, 1948.

[4.10] TUCKER, D. G. *Modulators and Frequency-changers.* London: MacDonald, 1953.

[4.11] TUCKER, D. G. *Circuit with periodically-varying Parameters.* London: MacDonald, 1964.

[4.12] GENSEL, J. Das Verhalten von Modulatorschaltungen bei komplexen, insbesondere selektiven Abschlüssen. *Frequenz*, **11**, 153-9, 175-85, May, June 1957.

[4.13] BROWN, T. T. Mixer harmonic chart. *Electronics*, **24**, 132, 134, April 1951.

[4.14] ORLOFF, I. M. Intermodulation analysis of crystal mixer. *Proc. I.E.E.E.*, **52**, 173-9, 1964.

[4.15] TRAMMEL, R.D., Jr. A method for determining mixer spurious response rejection. *I.E.E.E. Transactions on Electromagnetic Compatibility*, **EMC-8**, 81-9, 1966.

[4.16] DALLEMAGNE, R. Emploi des diodes au silicium dans les modulateurs en anneau. *Cables & Transmissions*, **14**, 12-29, 1960.

[4.17] KROUPA, V. Amplitude of the General Intermodulation Product. *Proc. I.E.E.E.*, **58**, 851-2, 1970.

5 Frequency Synthesis

In the preceding chapters the single frequency-changing procedures have been explained. The main feature, common to all, is that only one of the four fundamental algebraic operations is involved. When for the derivation of a new frequency from a given fundamental frequency, combinations of several additions, subtractions, multiplications, and divisions are necessary, the frequency-changing process is called frequency synthesis. The term was coined in the early 40s when the first measuring generators were devised, the variable output frequency of which was always harmonically related to a submultiple of the driving standard frequency [5.1].

It may seem surprising that this idea was not developed sooner, considering that the algebraic operations with frequencies are performed so easily and, at the same time, have been known for so long. The explanation is, without doubt, the complexity of the frequency synthesis system which requires a large number of passive and active elements. It therefore took almost twenty years for frequency synthesizers to become standard laboratory equipment and gain widespread use in communications.

5.1 THE MATHEMATICAL THEORY OF FREQUENCY SYNTHESIS

In this paragraph we shall follow the leading ideas of the 'Theory of Frequency Synthesis', published elsewhere [5.2], but the rigorous mathematical proof will be omitted.

The first generalization of the frequency synthesis problem is achieved when the desired output frequency f_x is normalized with respect to the input frequency f_s.

$$\frac{f_x}{f_s} = \frac{X_1}{Y_1} \tag{5.1}$$

If X_1 and Y_1 are products of small prime numbers, the task of deriving the output frequency f_x is an easy one and the synthesizer is a simple chain of frequency multipliers and dividers as shown in Fig. 5.1. In a more general case X_1 and Y_1 may also be products of small and large prime numbers. Then the simple scheme shown in Fig. 5.1 is not applicable in direct frequency synthesis systems because generation of the higher harmonic frequencies with sufficient power may be difficult, and the suppression of the undesired products to the prescribed level, especially of the neighbouring harmonic frequencies, becomes expensive.

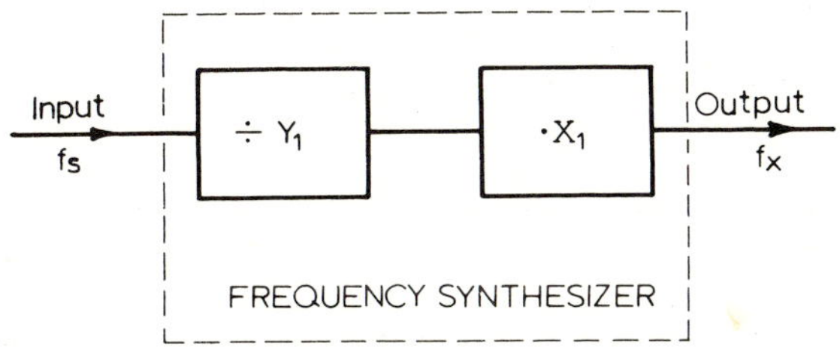

Fig. 5.1 Frequency synthesizer reduces to a simple chain of multipliers and dividers when both integers X_1 and Y_1 ($f_x/f_s = X_1/Y_1$) are products of small prime numbers. (Kroupa [5.2].)

Simple reasoning suggests that the solution of this problem may be achieved if the output frequency f_x is gained as the sum or difference of frequencies derived in separate circuits by the multiplication and division of the input frequency f_s. The corresponding mathematical analysis is as follows:

Let us suppose that

$$X_1 = z_1 \cdot X_1' \tag{5.2}$$

where z_1 is the product only of all small prime numbers in X_1, and X_1' is the product of all large prime numbers in the numerator X_1. Further it is possible to write the denominator Y_1 as the product of two suitable factors y_1 and Y_2

$$Y_1 = y_1 \cdot Y_2 \tag{5.3}$$

which may be chosen in such a way that after substituting relations (5.2) and (5.3) in equation (5.1) we get

$$\frac{f_x}{f_s} = \frac{z_1}{y_1} \cdot \frac{X_1'}{Y_2} = \frac{z_1}{y_1}\left(x_1 + \frac{X_2}{Y_2}\right) \tag{5.4}$$

where x_1 is the product of small prime numbers only.

The synthesizer corresponding to this mathematical model is composed of two divider–multiplier chains, the outputs of which are added or subtracted in a mixer according to the block diagram in Fig. 5.2. Multipliers x_1 and X_2 together with the mixer M_{12} and the divider Y_2 form a closed loop that may be called 'the synthesizing loop'. It has been shown in Section 4.4 that the output signal of a mixer contains, besides the desired sum or difference frequency, a

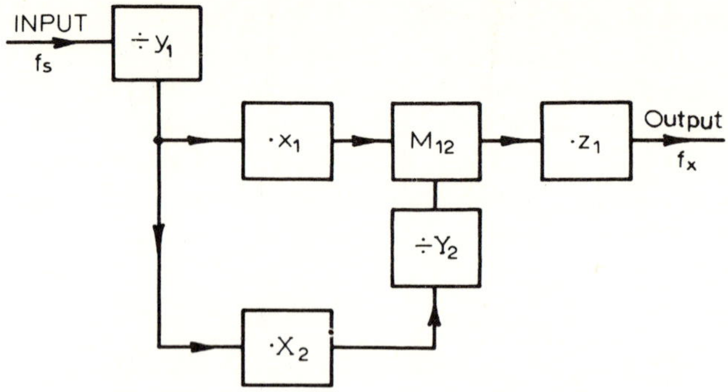

Fig. 5.2 Frequency synthesizer including one 'synthesizing loop' formed from two divider-multiplier chains and closed by a modulator. (Kroupa [5.2].)

large number of unwanted products; according to the analysis performed in Section 4.4 the suppression of the carrier and image frequencies is difficult when the ratio

$$|q_1| = \left|\frac{x_1}{X_2/Y_2}\right| \tag{5.5}$$

of the normalized input frequencies x_1 and X_2/Y_2 is too large; on the other hand if this ratio is small the combination of higher harmonics in the mixer may give rise to an intolerably high level of the interference signals in the pass-band of the following filters.

Evidently in algebraic operations with frequencies the following limitations must be imposed:

A. It is desirable to restrict the multiplication of frequencies only to products of small prime numbers, i.e. 2, 3, and 5.

B. The ratio $|q| > 0$ of frequencies added or subtracted in a mixer must not be larger than some upper bound and smaller than some lower bound.

Returning to equation (5.4), it is possible that the numerator X_2 is still the product of small prime numbers z_2 and of large prime numbers X_2'.

$$X_2 = z_2 \,.\, X_2' \tag{5.6}$$

In this case we must repeat with the residue X_2/Y_2 the same procedure which has been imposed upon the original normalized frequency X_1/Y_1. Thereafter the equation (5.4) changes to

$$\frac{f_x}{f_s} = \frac{z_1}{y_1}\left[x_1 + \frac{z_2}{y_2}\left(x_2 + \frac{X_3}{Y_3}\right)\right] \tag{5.7}$$

where x_2 is again the product of small prime numbers only according to the condition A, and the ratio between x_2 and the residue X_3/Y_3 is bounded in accordance with **B**. Finally for Y_2 the following substitution has been made

$$Y_2 = y_2 \cdot Y_3 \tag{5.8}$$

The same operation must be reiterated until the numerator in the last term becomes a product of small prime numbers only. Consequently the general equation which describes the synthesis of any arbitrary frequency f_x from a fixed standard frequency f_s may be written as

$$\frac{f_x}{f_s} = \frac{x_1 \cdot z_1}{y_1} + \frac{x_2 \cdot z_1 \cdot z_2}{y_1 \cdot y_2} + \cdots + \frac{x_n \cdot z_1 \cdot z_2 \ldots z_n}{y_1 \cdot y_2 \ldots y_n} \tag{5.9}$$

or

$$\frac{f_x}{f_s} = \frac{\sum\limits_{i=1}^{n} x_i \prod\limits_{k=1}^{i} z_k \prod\limits_{r=i+1}^{n} y_r}{\prod\limits_{r=1}^{n} y_r} = \frac{\sum\limits_{i=1}^{n} x_i \prod\limits_{k=1}^{i} z_k \prod\limits_{r=i+1}^{n} y_r}{Y_1}$$

and called the synthesizer series. All x_i and z_k $(i, k = 1, 2 \ldots n)$ are products of small prime numbers only or otherwise they are elements of the set $\{u\}$ of integers u_m

$$u_m = 2^{\rho} \times 3^{\sigma} \times 5^{\tau} \quad (m = 1, 2, \ldots; \rho, \sigma, \tau = 0, 1, 2, \ldots) \tag{5.10}$$

Integers u_m from the interval (0; 1000) are indicated in Table 5.1, and prime numbers from the same interval in Table 5.2.

Further all ratios

$$|q_i| = \left| x_i \cdot \frac{Y_{i+1}}{X_{i+1}} \right| \tag{5.11}$$

made between any term of the synthesizer series (5.9) and the appertaining residue are bounded as required by condition **B**. It is important to note that q_i is positive/negative if the respective residue X_{i+1}/Y_{i+1} has to be added/subtracted. This means, when considering the frequency synthesizer, that in the first case at the output of the mixer $M_{i,i+1}$ the upper sideband, and in the second case the lower sideband, has to be filtered out. However, the 'negative frequency' has no physical meaning, and therefore condition **B** refers only to the absolute values of q_i.

$$q_{\text{lower}} \leqslant |q_i| = \left| x_i \cdot \frac{Y_{i+1}}{X_{i+1}} \right| \leqslant q_{\text{upper}} \tag{5.12}$$

For the general discussion of the frequency synthesis problem it is desirable to have the lower bound as well as the upper bound in the condition **B** constant. Analysis of the mixing process with the use of the diagrams in Section 4.4 shows that a higher level of interfering signals in the pass-band must be

Table 5.1

Integers u_m in the interval (0; 1000)

m	u_m	$2^\rho \times 3^\sigma \times 5^\tau$	m	u_m	$2^\rho \times 3^\sigma \times 5^\tau$
1	1	$2^0 \times 3^0 \times 5^0$	44	180	$2^2 \times 3^2 \times 5$
2	2	2	45	192	$2^6 \times 3$
3	3	3	46	200	$2^3 \times 5^2$
4	4	2^2	47	216	$2^3 \times 3^3$
5	5	5	48	225	$3^2 \times 5^2$
6	6	2×3	49	240	$2^4 \times 3 \times 5$
7	8	2^3	50	243	3^5
8	9	3^2	51	250	2×5^3
9	10	2×5	52	256	2^8
10	12	$2^2 \times 3$	53	270	$2 \times 3^3 \times 5$
11	15	3×5	54	288	$2^5 \times 3^2$
12	16	2^4	55	300	$2^2 \times 3 \times 5^2$
13	18	2×3^2	56	320	$2^6 \times 5$
14	20	$2^2 \times 5$	57	324	$2^2 \times 3^4$
15	24	$2^3 \times 3$	58	360	$2^3 \times 3^2 \times 5$
16	25	5^2	59	375	3×5^3
17	27	3^3	60	384	$2^7 \times 3$
18	30	$2 \times 3 \times 5$	61	400	$2^4 \times 5^2$
19	32	2^5	62	405	$3^4 \times 5$
20	36	$2^2 \times 3^2$	63	432	$2^4 \times 3^3$
21	40	$2^3 \times 5$	64	450	$2 \times 3^2 \times 5^2$
22	45	$3^2 \times 5$	65	480	$2^5 \times 3 \times 5$
23	48	$2^4 \times 3$	66	486	2×3^5
24	50	2×5^2	67	500	$2^2 \times 5^3$
25	54	2×3^3	68	512	2^9
26	60	$2^2 \times 3 \times 5$	69	540	$2^2 \times 3^3 \times 5$
27	64	2^6	70	576	$2^6 \times 3^2$
28	72	$2^3 \times 3^2$	71	600	$2^3 \times 3 \times 5^2$
29	75	3×5^2	72	625	5^4
30	80	$2^4 \times 5$	73	640	$2^7 \times 5$
31	81	3^4	74	648	$2^3 \times 3^4$
32	90	$2 \times 3^2 \times 5$	75	675	$3^3 \times 5^2$
33	96	$2^5 \times 3$	76	720	$2^4 \times 3^2 \times 5$
34	100	$2^2 \times 5^2$	77	729	3^6
35	108	$2^2 \times 3^3$	78	750	$2 \times 3 \times 5^3$
36	120	$2^3 \times 3 \times 5$	79	768	$2^8 \times 3$
37	125	5^3	80	800	$2^5 \times 5^2$
38	128	2^7	81	810	$2 \times 3^4 \times 5$
39	135	$3^3 \times 5$	82	864	$2^5 \times 3^3$
40	144	$2^4 \times 3^2$	83	900	$2^2 \times 3^2 \times 5^2$
41	150	$2 \times 3 \times 5^2$	84	960	$2^6 \times 3 \times 5$
42	160	$2^5 \times 5$	85	972	$2^2 \times 3^5$
43	162	2×3^4	86	1000	$2^3 \times 5^3$

Table 5.2

Prime numbers in the interval (1; 1000)

2	109	269	439	617	811
3	113	271	443	619	821
5	127	277	449	631	823
7	131	281	457	641	827
11	137	283	461	643	829
13	139	293	463	647	839
17	149	307	467	653	853
19	151	311	479	659	857
23	157	313	487	661	859
29	163	317	491	673	863
31	167	331	499	677	877
37	173	337	503	683	881
41	179	347	509	691	883
43	181	349	521	701	887
47	191	353	523	709	907
53	193	359	541	719	911
59	197	367	547	727	919
61	199	373	557	733	929
67	211	379	563	739	937
71	223	383	569	743	941
73	227	389	571	751	947
79	229	397	577	757	953
83	233	401	587	761	967
89	239	409	593	769	971
97	241	419	599	773	977
101	251	421	601	787	983
103	257	431	607	797	991
107	263	433	613	809	997

expected when the sum frequency is filtered out, compared with the case when the difference frequency is desired. On the other hand, referring to the ratio $|q_i + 1|$ between the normalized output frequency X_i'/Y_{i+1} and the lower of the input frequencies X_{i+1}/Y_{i+1} the above-mentioned difficulty will be not encountered. But condition **B** must be slightly changed:

$$q_d \leqslant |q_i + 1| = \left| \frac{X_i'}{X_{i+1}} \right| \leqslant q_h \tag{5.13}$$

For larger values of $|q_i|$ the interfering frequencies are mainly the undesired sideband and the 'carrier' frequency, i.e. the higher of the input frequencies.

It may be easily shown that both definitions (5.12) or (5.13) of the upper bound are practically equivalent.†

Finally the rules for factorizing the denominator Y_1 into the sequence y_r ($r = 1, 2, \ldots n$) must be formed.

$$Y_1 = \prod_{r=1}^{n} y_r \tag{5.14}$$

Returning to eq. (5.4), which is the mathematical model of the first step in the frequency synthesis process, it is evident that the smallest multiplying factors x_1 and X_2 (fulfilling conditions **A** and **B**) are desirable from the viewpoint of the simplicity of the frequency synthesizer. Calculating x_1 from the equation

$$x_1 Y_2 + X_2 = X_1' \tag{5.15}$$

gives (with the help of (5.3))

$$x_1 = \frac{y_1}{Y_1}(X_1' - X_2) \tag{5.16}$$

and after inserting for X_2 from (5.13) we get

$$x_1 = \frac{y_1}{z_1} \cdot \frac{X_1}{Y_1} \cdot \frac{q_1}{q_1 + 1} \tag{5.17}$$

We shall see later that for q_1 it is always true that $|q_1| \gg 0$, therefore $q_1/(q_1 + 1) \approx 1$. Because X_1, Y_1, and z_1 are constants it is clear from eq. (5.17) that x_1 is proportional to y_1. This implies that for y_1 the smallest factor in Y_1 has to be chosen for which equation (5.4) may be solved (both conditions **A** and **B** being fulfilled). Similarly, following the expansion of (5.4) into (5.7), it may easily be proved that for y_2 the smallest factor in Y_2 enabling eq. (5.7) to be solved has to be chosen; in the same way all other $y_3, \ldots y_{n-1}$ are found.

Clearly, when expanding the normalized frequency X_1/Y_1 in the synthesizer series, the following condition must be taken into account:

---

† The ratio between the desired sideband and the 'carrier' frequency is

$$\frac{x_i + X_{i+1}/Y_{i+1}}{x_i} = 1 \pm \frac{1}{q_{\text{upper}}}$$

Substituting $q_{\text{upper}} = q_h - 1$ for the case of the lower sideband and $q_{\text{upper}} = q_h + 1$ for the case of the higher sideband according to relations (5.12) and (5.13) we get

$$1 \pm \frac{1}{q_{\text{upper}}} = 1 \pm \frac{1}{q_h \mp 1} = 1 \pm \frac{1}{q_h}\left(1 \pm \frac{1}{q_h} + \cdots\right)$$

For large values of q_h (this condition is always fulfilled) the higher terms in the expansion may be neglected, which leads to the conclusion that

$$1 \pm \frac{1}{q_{\text{upper}}} \approx 1 \pm \frac{1}{q_h}$$

C. For y_r, the smallest factor in Y_r ($r = 1, 2 \ldots n-1$) must be chosen for which the respective x_r and X_{r+1}/Y_{r+1} fulfil conditions **A** and **B**.

Application of condition **C** ensures that the frequency synthesis equation (5.9) is unique and simultaneously that the actual synthesizer is as simple as possible, because all unnecessary multiplying, dividing, and mixing operations are eliminated. The first statement about multiplying is evident from the fact that x_r is always proportional to y_r; the second statement is true as long as the relation (5.14) is true; the third statement will be proved later.

Generally Y_1 may be the product of small and large prime numbers; this means that sometimes dividing with large prime numbers is necessary. This is easily performed at low frequencies where counting dividers are common (see Section 3.5).

Condition **C** is in good agreement with this tendency because it favours the use of smaller dividing factors at the beginning of the frequency synthesis process where higher frequency levels are expected. Reverting to equation (5.9) it is evident that this is the general equation with which all problems of frequency synthesis may be solved. In accordance with this mathematical model the synthesizer has $(n - 1)$ loops or mixing levels. Its block diagram is shown in Fig. 5.3. As in equation (5.9) all those terms are multiplied by z_k and divided by y_r, for which the condition $i \geqslant k, r$ is fulfilled; the dividers y_i and the multipliers z_i may be placed either in the input or in the output branch of the respective synthesizing loop but not in those branches which are in common with other loops. Nevertheless the position of these dividers and multipliers determines the frequency levels at preceding modulators. This is important when division with large prime numbers is involved, as already discussed above.

Placing of dividers in the output branches tends to diminish the phase jitter of the preceding synthesizing loops. The other advantage is the possibility of using one and the same multiplier several times which enhances the simplicity of the synthesizer. In this case equivalence or simple proportionality between each individual x_i is desirable, and therefore smaller departures from the rules of condition **C** might be necessary.

Following the evolution of the frequency synthesis process from equation (5.1), to (5.4), and further to (5.7), etc., it is evident that from the purely mathematical point of view the procedure is equivalent to successive approximations of real numbers. In the number theory several ways are used to accomplish this task [5.3]. It has been shown in paper [5.2] that only Cantor series, including the systematic or g-adic fractions, are suitable for frequency synthesis purposes. Further, closer examination of the general frequency synthesis equation (5.9) reveals that this so-called synthesizer series is only a special type of the known Cantor series.

The only question which remains is proving that any positive rational number X_1/Y_1 may be expanded into the synthesizer series (5.9), all the terms of which

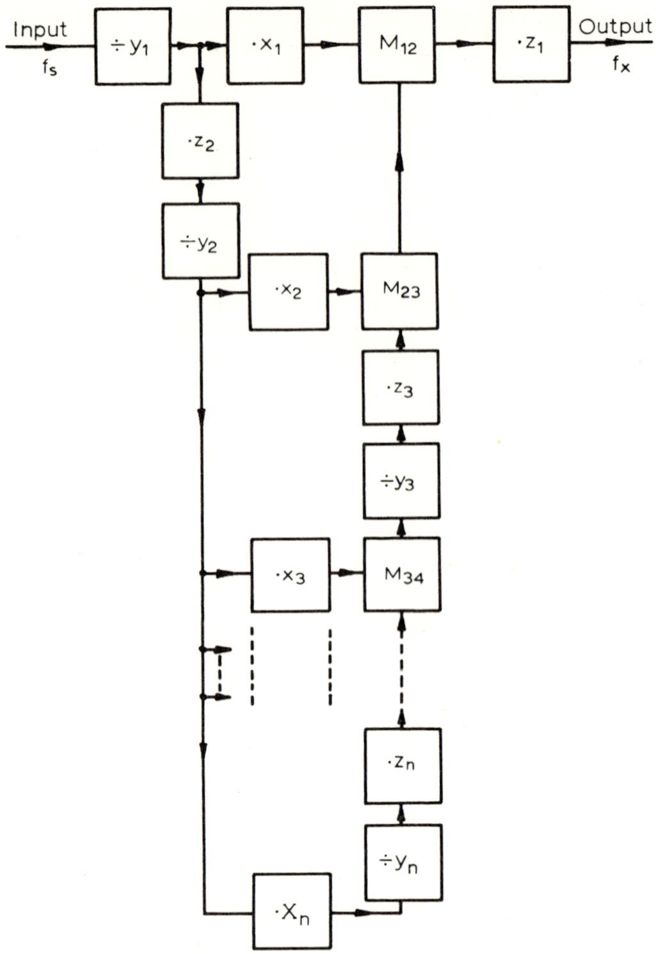

Fig. 5.3 Block diagram for the general case of frequency synthesis. (Kroupa [5.2].)

fulfil simultaneously both conditions **A** and **B**. The complete proof has been published in [5.2] but the process is rather lengthy; here, mentioning the leading ideas, and the quoting of the results, will be sufficient.

Determination of the lower bound q_d may be made, using the left-hand part of the expression (5.13)

$$q_d \leqslant \left| \frac{X_i'}{X_{i+1}} \right|_{\min} \leqslant \left| \frac{X_{n-1}'}{X_n} \right| \tag{5.18}$$

From the definition (5.2), (5.6) etc., X_i' is the product of large prime numbers which do not belong to the set of integers $\{u_m\}$, see (5.10). From Table 5.1 it is readily seen that $|X'_{n-1}|_{\min} = 7$; taking into account that $|X_n| \geqslant 1$ we get

$$q_d \leqslant 7 \tag{5.19}$$

similarly the expression (5.13) gives for the largest upper bound q_h

$$q_h \geqslant |X'_i|_{\max} = X'_1 \tag{5.20}$$

Conclusion: relations (5.19) and (5.20) prove the existence of the finite interval for all $|q_i + 1|$ $(i = 1, 2, \ldots n-1)$.

Preliminary estimation of n, i.e. of the number of terms in eq. (5.9), or of $n-1$, i.e. of the number of synthesizing loops in the block diagram (see Fig. 5.3), is desirable. For this purpose the product of expressions (5.13) for all i $(i = 1, 2, \ldots n-1)$ is suitable

$$q_d^{n-1} \leqslant \left|\frac{X'_1}{X_2}\right| \cdot \left|\frac{X_2}{z_2 X_3}\right| \cdots \left|\frac{X_{n-1}}{z_{n-1} X_n}\right| \leqslant q_h^{n-1}$$

or simply

$$q_d^{n-1} \leqslant \left|\frac{X'_1}{z_2 z_3 \ldots z_{n-1} X_n}\right| \leqslant q_h^{n-1} \tag{5.21}$$

because $|X_n| \geqslant 1$ and all $z_k \geqslant 1$ $(k = 2, 3, \ldots n-1)$ it is certainly true that

$$n - 1 \leqslant \frac{\log X'_1}{\log q_d} \tag{5.22}$$

and finally after substituting for q_d from (5.19) the following inequality may be written:

$$n - 1 \leqslant 1,2 \log X'_1 \tag{5.23}$$

The most important procedure in the frequency synthesis process is the finding of multiplying factors x_i $(i = 1, 2, \ldots n-1)$ fulfilling condition **A**, i.e. requirement (5.10). It is easy to verify that each x_i satisfies a similar expression to (5.17)

$$x_i = \frac{y_i}{z_i} \left|\frac{X_i}{Y_i}\right| \frac{q_i}{q_i + 1} \tag{5.24}$$

This equation simplifies to

$$u = v\,\frac{q_i}{q_i + 1} \tag{5.25}$$

after introduction of new variables u and v

$$u = x_i$$

$$v = \frac{y_i}{z_i}\left|\frac{X_i}{Y_i}\right| = \left|\frac{X_i'}{Y_{i+1}}\right|^{\dagger} \tag{5.26}$$

Because $|q_i|$ is bounded to an interval $(q_d; q_h)$ according to (5.13), equation (5.25) represents two families of lines. Inserting the lower and the upper bounds q_d and q_h respectively gives the equations of the envelope.

Boundary lines for negative values of q_i are

$$u = \frac{q_d + 1}{q_d} v \tag{5.27a}$$

$$u = \frac{q_h + 1}{q_h} v \tag{5.27b}$$

and boundary lines for positive values of q_i are

$$u = \frac{q_h - 1}{q_h} v \tag{5.28a}$$

$$u = \frac{q_d - 1}{q_d} v \tag{5.28b}$$

Their intersections with the set of parallel lines

$$u = u_m \quad (m = 1, 2, \ldots) \tag{5.29}$$

are of interest (see Fig. 5.4).

$$v_{1,m} = u_m \frac{q_d}{q_d + 1} \tag{5.30a}$$

$$v_{2,m} = u_m \frac{q_h}{q_h + 1} \tag{5.30b}$$

$$v_{3,m} = u_m \frac{q_h}{q_h - 1} \tag{5.31a}$$

$$v_{4,m} = u_m \frac{q_d}{q_d - 1} \tag{5.31b}$$

---

†*Note:* by definition, all x_i, y_i, z_i, $|X_i|$, and Y_i are positive and therefore both new variables u and v have only positive values; in addition v may be any rational number.

These intersections bound on the real line v $(0 < v < \infty)$ two intervals (sets) of rational numbers v

$$V_{\mathrm{I},m} = (v_{1,m};\ v_{2,m}) \tag{5.32}$$

$$V_{\mathrm{II},m} = (v_{3,m};\ v_{4,m}) \tag{5.33}$$

for which the equation (5.25), and consequently also (5.24), has the solution fulfilling both conditions **A** and **B**.

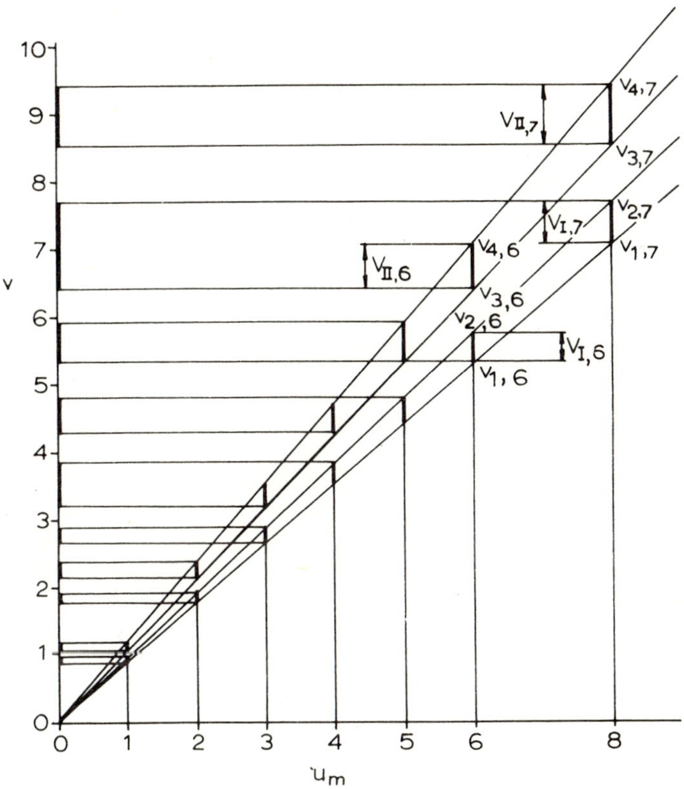

Fig. 5.4 Boundary lines of the family of lines $u = v[q_i/(q_i + 1)]$ for $q_d = 7 \leqslant |q_i + 1| \leqslant 19 = q_h$ with the indicated intervals $V_{I,m}$ and $V_{II,m}$ $(m = 1, 2, \ldots)$. Disjunctions for small m are evident.

Unfortunately the first intervals $V_{\mathrm{I},m}$ and $V_{\mathrm{II},m}$ $(m = 1, 2, \ldots)$ are small and disjointed, as may be seen from Fig. 5.4. Clearly equation (5.25) cannot be solved for all rational numbers. This, however, is not even necessary because it is always possible to put

$$y_1 \doteq Y_1;\ Y_2 = 1 \tag{5.34}$$

and thus to limit v to positive integers X_1'. This case has already been discussed in connexion with relations (5.19) and (5.20), but the upper bound q_h is unreasonably high. This, however, is not prohibitive when crystal filters can be used in the output circuits of mixers. On the other hand synthesizers with LC filters only require that the upper bound q_h should be as small as possible.

The following considerations will reveal $q_{h,\,\min}$. For larger values of m ($m = 1, 2, \ldots$) there are no disjunctions between neighbouring intervals $V_{\mathrm{I},\,m}$ and $V_{\mathrm{II},\,m}$ ($m = r, r+1, \ldots$). If the lower bound of their union is $v_{1,\,r}$ then for

$$q_h > v_{1,r} \tag{5.35}$$

eq. (5.25) may be always solved. But q_h and $v_{1,\,r}$ are mutually dependent through equations (5.27b) and (5.28a), therefore for finding $q_{h,\,\min}$ the iterative graphical solution is used. A chosen value for q_h, e.g. $q_h = 14$, is substituted together with $q_d = 7$ in equations (5.27a,b) and (5.28a,b) and the respective straight lines are drawn. See Fig. 5.5. All disjunctions between the intervals $V_{\mathrm{I},\,m}$ and $V_{\mathrm{II},\,m}$ ($m = 9, 10, \ldots$) in one octave,† including also the suspected ones, have been marked, and the respective upper bounds q_h needed for their removal calculated. It has been found that the smallest q_h which fulfils the relation (5.35) is

$$q_{h,\,\min} = 19 \tag{5.36}$$

Summarizing: if eq. (5.24), and consequently also the general frequency synthesis equation (5.9), has a solution complying with conditions **A**, **B** and **C** for all ratios f_x/f_s, the lower bound of $|q_i + 1|$ ($i = 1, 2, \ldots n-1$) cannot be larger than the maximum q_d in (5.19), and the respective upper bound cannot be smaller than $q_{h,\min}$ in (5.36)

$$7 \leqslant |q_i + 1| \leqslant 19 \tag{5.37}$$

Maintaining at each mixer the ratio of the output frequency to the lower of the input frequencies as $|q_i + 1| \geqslant 7$ ensures that only sidebands of the seventh or higher order will be found in the pass-band of the output filter. In this case the power level of the unwanted signals can be kept below -80 dB, according to the analysis performed in Section 4.4.

It has been pointed out above that, for the upper bound q_h, predominantly economic considerations are decisive. For $q_h = 19 \geqslant |q_i + 1|$ the output filters needed to keep the carrier and image frequencies below the -80 dB level are simple when using balanced mixers (Section 4.3).

† If intervals $V_{\mathrm{I},\,m}$ and $V_{\mathrm{II},\,m}$ ($m = r_1, r+1, \ldots$) form an octave without disjunctions, then there are no disjunctions in the next following octave, etc. The complete proof is given in [5.2].

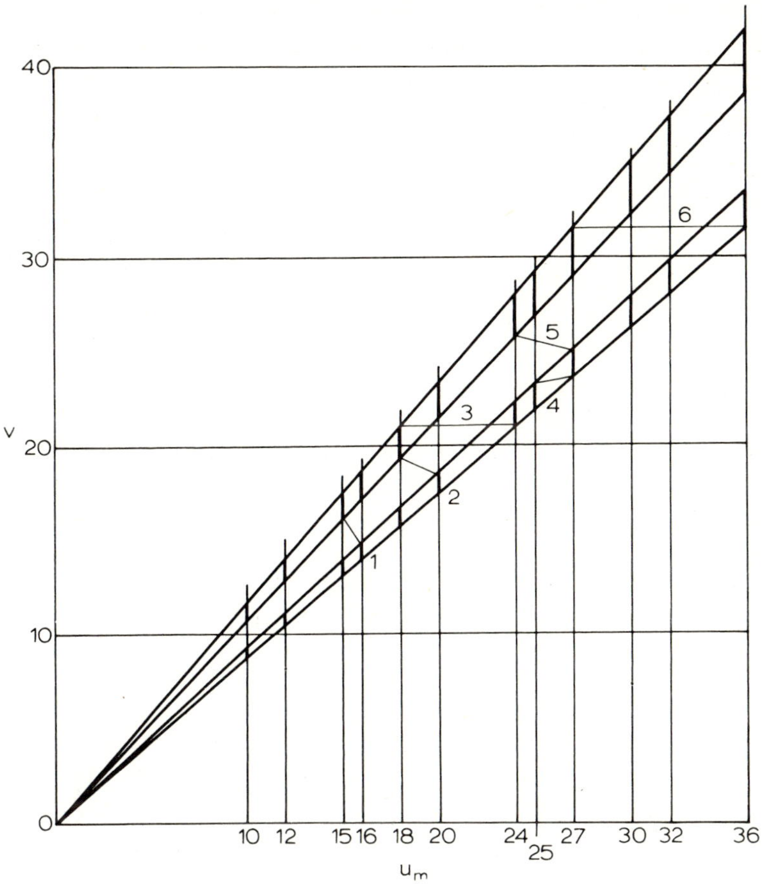

Fig. 5.5 Auxiliary graphical procedure for finding the smallest q_h for which the general frequency synthesis equation (5.9) always has a solution (Kroupa [5.2].)

Sometimes even better suppression of the undesired products at the output of the synthesizer is needed; then the lower bound q_d must be larger. E.g. for $q_d = 9$ the solvability† of (5.9) requires

$$9 \leqslant |q_i + 1| \leqslant 31 \tag{5.38}$$

† There is one singularity $|X_i/Y_i| = 7$ for which eq. (5.9) cannot be solved for **B** fulfilling (5.38).

5.2 SINGLE-FREQUENCY SYNTHESIZERS

First we must find the ratio of the output to the input frequency according to (5.1), and according to (5.2), the product of all small prime numbers $z_0 \in \{u_m\}$ † in the numerator X_1, see eq. (5.10)

$$\frac{f_x}{f_s} = \frac{X_1}{Y} = z_0 \frac{X_1'}{Y} \tag{5.39}$$

Then we proceed to determine the approximate frequency

$$f_s \frac{X_1'}{Y} \tag{5.40}$$

at which the first mixing has to be performed. If this frequency is too high, as happens mainly when

$$\frac{X_1'}{Y} \gg 1, \tag{5.41}$$

multiplication of the denominator Y with a suitable factor $y_0 \in \{u_m\}$ may be the remedy. However, the product of small prime numbers z_0 in the numerator X_1 must also be multiplied by the same factor

$$\frac{f_x}{f_s} = z_0\, y_0 \frac{X_1'}{y_0 Y} = z_1 \frac{X_1'}{Y_1} \tag{5.42}$$

This operation introduces additional multiplication and division by y_0, but this is almost fully compensated by the reduction of the multiplication factor x_1, and possibly also of x_2, and others, in the same proportion as well. What is important is that the number of synthesizing loops $(n-1)$ in the frequency synthesizer remains unchanged because

$$n - 1 < 1{,}2 \log X_1' \quad (q_d \leqslant 7).$$

For the preliminary estimation of the number $(n-1)$ the simplification

$$n - 1 < \log X_1 \tag{5.43}$$

is often sufficient.

Coming to the actual numerical calculation of the first synthesizing loop the opening step is the determination of the smallest y_1', a factor of Y_1, which fulfils the inequality

$$v_1 = \frac{X_1'}{Y_1} y_1' \geqslant v_{1,1} \tag{5.44a}$$

† i.e. z_0 is an element of the set $\{u_m\}$.

According to (5.27a)

$$v_{1,1} = 1 \cdot \frac{q_d}{q_d + 1} = \begin{cases} 0{\cdot}875 & (q_d = 7) \\ 0{\cdot}9 & (q_d = 9) \end{cases} \tag{5.44b}$$

The second step is to find out if, for this v_1, eq. (5.25), with restriction made on $|q_1 + 1|$ according to (5.37) or (5.38), can be solved in such a way that the respective $u_1 = x_1 \epsilon \{u_m\}$; or in other words if this v_1 lies in one of the intervals $V_{I,m}$, $V_{II,m}$ defined by (5.32) and (5.33).

To speed up the calculation of x_1, either the intervals $V_{I,m}$ and $V_{II,m}$, together with the appertaining u_m, must be tabulated (see Table 5.3(a), (b)), or the

Table 5.3(a)

Intervals $V_{I,m}$ and $V_{II,m}$ with the respective u_m solving equation (5.25) when $7 \leqslant |q_i + 1| \leqslant 19$

m	u_m	$V_{I,m}$		$V_{II,m}$	
1	1	0·875	0·950	1·056	1·167
2	2	1·750	1·900	2·111	2·333
3	3	2·625	2·850	3·167	3·500
4	4	3·500	3·800	4·222	4·667
5	5	4·375	4·750	5·278	5·833
6	6	5·250	5·700	6·333	7·000
7	8	7·000	7·600	8·444	9·333
8	9	7·875	8·550	9·500	10·500
9	10	8·750	9·500	10·556	11·667
10	12	10·500	11·400	12·667	14·000
11	15	13·125	14·250	15·833	17·500
12	16	14·000	15·200	16·889	18·667
13	18	15·750	17·100	19·000	21·000
14	20	17·500	19·000	21·111	23·333
15	24	21·000	22·800	25·333	28·000
16	25	21·875	23·750	26·389	29·167
17	27	23·625	25·650	28·500	31·500
18	30	26·250	28·500	31·667	35·000

graphical solution of eq. (5.25) must be applied; the latter procedure is more convenient. In Fig. 5.6, from the family of lines (5.25) only the envelope (5.27a), (5.27b), and (5.28a), (5.28b) is drawn after substituting $q_d = 7$ and $q_h = 19$. To achieve constant resolution logarithmic division on both axes is used. The intervals $V_{I,m}$ and $V_{II,m}$ $(m = 1, 2, \ldots)$ are projected on the v axis; from $v_{1,13} = 15, 75$ onwards there are no further disjunctions.

As soon as the smallest v_1 lying in one of the intervals $V_{I,m}$, $V_{II,m}$ is found, at least one intersection of lines $v = v_1$ and $u = u_m$ lies in the area of one or other branch of the envelope. If there are more intersections in this

Table 5.3(b)

Intervals $V_{I,m}$ and $V_{II,m}$ with the respective u_m solving equation (5.25) when $9 \leqslant |q_i + 1| \leqslant 31$

m	u_m	$V_{I,m}$		$V_{II,m}$	
1	1	0·900 00	0·968 75	1·033 33	1·125 00
2	2	1·800 00	1·937 50	2·066 67	2·250 00
3	3	2·700 00	2·906 25	3·100 00	3·375 00
4	4	3·600 00	3·875 00	4·133 33	4·500 00
5	5	4·500 00	4·843 75	5·166 67	5·625 00
6	6	5·400 00	5·812 50	6·200 00	6·750 00
7	8	7·200 00	7·750 00	8·266 67	9·000 00
8	9	8·100 00	8·718 75	9·300 00	10·125 00
9	10	9·000 00	9·687 50	10·333 33	11·250 00
10	12	10·800 00	11·625 00	12·400 00	13·500 00
11	15	13·500 00	14·531 25	15·500 00	16·875 00
12	16	14·400 00	15·500 00	16·533 33	18·000 00
13	18	16·200 00	17·437 50	18·600 00	20·250 00
14	20	18·000 00	19·375 00	20·666 67	22·500 00
15	24	21·600 00	23·250 00	24·800 00	27·000 00
16	25	22·500 00	24·218 75	25·833 33	28·125 00
17	27	24·300 00	26·156 25	27·900 00	30·375 00
18	30	27·000 00	29·062 50	31·000 00	33·750 00

area the solution of eq. (5.25) is not unique, and for proper choice of x_1 the simplicity of the synthesizer is (using the same multiplier several times and avoiding products with 5, etc.) decisive. Further the position of the intersection indicates the sign of q_1, being in the area of the upper branch, $q_1 > 0$, which means that at the output of the modulator M_{12} (see Fig. 5.3) the sum frequency has to be filtered out. On the other hand, if the intersection is in the area of the lower branch, $q_1 < 0$, and at the output of M_{12}, the difference frequency is used.

After calculation of z_1, y_1, and x_1 the solution of the second synthesizing loop may proceed in the same manner.

Firstly, using (5.3) and (5.4), Y_2 and X_2 must be determined

$$Y_2 = \frac{Y_1}{y_1} \tag{5.45}$$

$$X_2 = \pm(X_1' - x_1\ Y_2) \tag{5.46}$$

using for positive q_1 the (+) sign and for negative q_1 the (−) sign. In the following step the product of all small prime numbers $z_2 \epsilon \{u_m\}$ is separated from X_2

$$X_2 = z_2 \,.\, X_2' \tag{5.47}$$

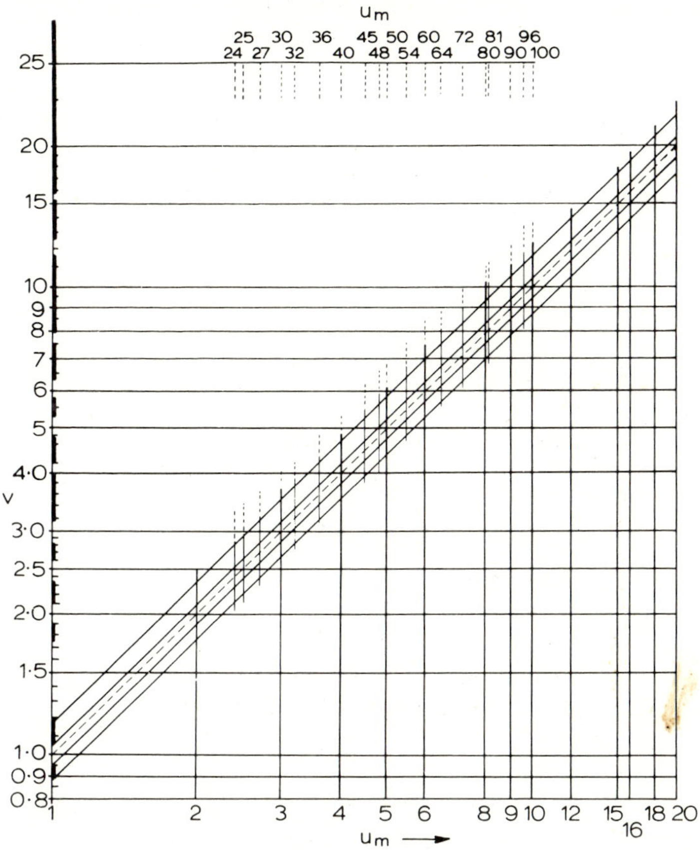

Fig. 5.6 Diagram for the graphical solution of the equation $u = v[q_i/(q_i + 1)]$ with the respective restrictions, $u \epsilon \{u_m\}$ and $7 \leqslant |q_i + 1| \leqslant 19$, made upon u and q_i according to the theory of frequency synthesis. For the thickened parts on the v axis at least one u_m exists. (Kroupa [5.2].)

Finally, the smallest factor y_2 in Y_2 is found for which the x_2 exists and the sign of q_2 is determined.

In exactly the same way the third and all other synthesizing loops are solved.

Example 5.1

Design a single-frequency synthesizer for exciting a television transmitter with the carrier frequency

$$f_x = 59{\cdot}250 \text{ MHz};$$

the standard frequency is

$$f_s = 100 \text{ kHz}$$

Power stages in the transmitter have the total multiplying factor:

$$y_0 = 54$$

Substituting in (5.42) gives:

$$\frac{f_x}{f_s} = 54\,\frac{5925 \times 10^4}{54 \times 10^5} = 54 \times 5 \times \frac{79}{36}$$

Clearly $z_1 = 5 \times 54$

$$X_1' = 79$$

$$Y_1 = 36$$

Preliminary estimation of the number of the synthesizing loops according to (5.23) reveals

$$n - 1 \lesssim 2$$

Choosing the smallest possible y_1

$$y_1 = 1$$

and substituting in (5.44a) we get

$$v_1 \approx 2{\cdot}2$$

From Fig. 5.6 it is readily seen that the intersection of lines $v = 2{\cdot}2$ and $u = 2$ lies in the upper branch of the envelope (5.28), therefore

$$x_1 = 2$$

and at the first mixer M_{12} the addition must be performed. With the help of (5.45) and (5.46), Y_2 and X_2 may be calculated

$$Y_2 = 36$$

$$X_2 = +(79 - 2 \times 36) = 7$$

Clearly $$z_2 = 1$$

The smallest y_2, for which the synthesizer series solution of

$$\frac{X_2}{Y_2} = \frac{7}{36}$$

may be formed, is

$$y_2 = 6$$

$$v_2 \approx 1{\cdot}17$$

$$x_2 = 1$$

Mixing sign '+'

$$Y_3 = 6$$

$$X_3 = +(7 - 1 \times 6) = 1$$

The block diagram shown in Fig. 5.7 illustrates the simplicity of the synthesizer.

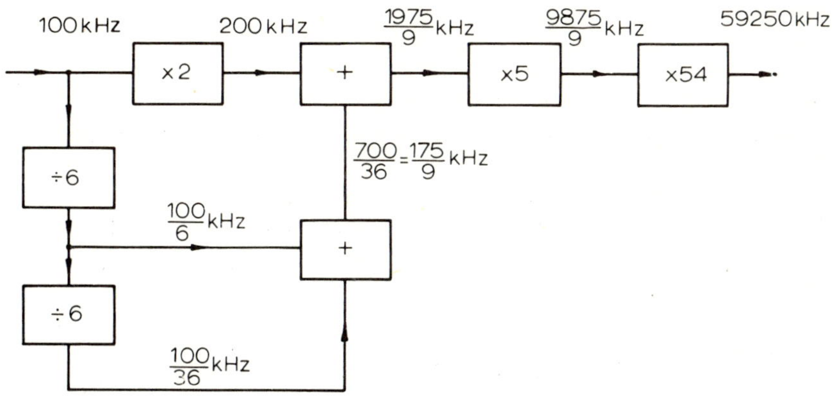

Fig. 5.7 Block diagram of the synthesizer-exciter for a television transmitter having the carrier frequency f_x = 59·250 MHz.

Example 5.2

In the second example we shall deal with a more complicated case, namely with the synthesis of the caesium atomic standard transition frequency [5.4]

$$f_x = 9192\,631\,770; \quad f_s = 5 \times 10^6 \text{ Hz}$$

Substituting in (5.39) gives:

$$\frac{f_x}{f_s} = z_0 \cdot \frac{X_1'}{Y} = 9\,\frac{102\,140\,353}{5 \times 10^5}$$

The approximate frequency (5.40) at which the first mixing should be performed is very high

$$f_s \frac{X_1'}{Y} \approx 1 \text{ GHz}$$

Its reduction by more than one order might be desirable; this is achieved by additional multiplication and division by y_0 and for simplicity a power of two is preferred

$$y_0 = 16$$

Now z_1 and Y_1 defined in eq. (5.42) may be calculated

$$z_1 = 9 \times 16$$

$$Y_1 = 16 \times 5 \times 10^5$$

$$X_1' = 102\ 140\ 353 \text{ remains unchanged.}$$

Preliminary estimation of the number of synthesizing loops according to (5.23) reveals

$$n - 1 \lesssim 9$$

Following the above instructions, especially those of Example 5.1, the circuit elements (dividing and multiplying coefficients) in all loops of the synthesizer will be determined. The numerical calculation is easy if Table 5.4 is completed step by step. A block diagram of the synthesizer is shown in Fig. 5.8. Placing of dividers in the output branches of the synthesizing loops reduces the phase jitter of the preceding stage and renders possible the use of some multipliers several times.†

5.3 VARIABLE FREQUENCY SYNTHESIZERS

In this case a frequency variable over several decimal orders is to be derived from a given standard frequency. The distinguishing feature is that in the numerator in equation (5.1) in every decimal place every decimal digit can occur. Evidently the frequency-synthesizing procedure in all loops of the synthesizer chain must be the same. The decadic operation is guaranteed by placing the divider 1/10 into the output branch of each loop. This arrangement determines of course that all

$$y_r = 10 \quad (r = 2, 3, \ldots, n) \tag{5.48}$$

The coefficients x_i may be found as follows. The condition of variability over ten digits and the condition **B** are fulfilled simultaneously if each x_i is composed of two parts; a large fixed one x_0 and a variable x_{0i} which changes from 0 to 9.

After substitution in the general equation (5.9) we get the mathematical model of the frequency synthesizer with n variable places and with zero in all

---

† The purpose of this example is merely instructional. No magnetic field offset has been considered (as discussed on p. 259). Further, a closer examination of Fig. 5.8 reveals that in the *direct* multiplier chain x_1, z_1 between the input and the output of the synthesizer, additional sources of noise, at a minimum the mixer M_{12} and the divider y_2, have been inserted. Evidently the phase jitter is multiplied by z_1, i.e. by 144, which is undesirable. Therefore in the actual commercial equipment the mixer M_{12} ends the synthesizing process [5.5], [8.13] etc.; see also Example 7.2 and Fig. 7.18.

Table 5.4

Tabulated solution of the single-frequency synthesizer for the caesium atomic frequency standard

i	$Y_i = \frac{Y_{i-1}}{y_{i-1}}$	$X_i = \pm(X'_{i-1} - x_{i-1}Y_i)$	z_i	$X'_i = \frac{X_i}{z_i}$	$\frac{X'_i}{Y_i}$	y_i	$v_i = \frac{X'_i}{Y_i}y_i$	x_i	sign	$\lvert q_i + 1 \rvert = \frac{X'_i}{X_{i+1}}$
1	$16 \times 5 \times 10^5$	$16 \times 919\,263\,177$	9×16	102140353	12·77	1	12·77	12	+	16·64
2	$16 \times 5 \times 10^5$	$+(102\,140\,353 - 96\,000\,000) = 6\,140\,353$	1	6 140353	0·7675	5	3·838	4	–	23·7†
3	16×10^5	$-(6\,140\,353 - 6\,400\,000) = +\,259\,647$	3	86 549	0·054	50	2·7	3	–	9·16
4	32×10^3	$-(86\,549 - 96\,000) = +\,9\,451$	1	9 451	0·295	16	4·74	5	–	17·2
5	2×10^3	$-(9\,451 - 10\,000) = +\,549$	9	61	0·0305	125	3·81	4	–	20·3
6	2^4	$-(61 - 64) = +\,3$								

† Somewhat larger $\lvert q_2 + 1 \rvert$ than the theoretical $q_h = 19$ has been admitted in accordance with (5.38).

UNIV. COLL. N.W. BANGOR LIBRARY

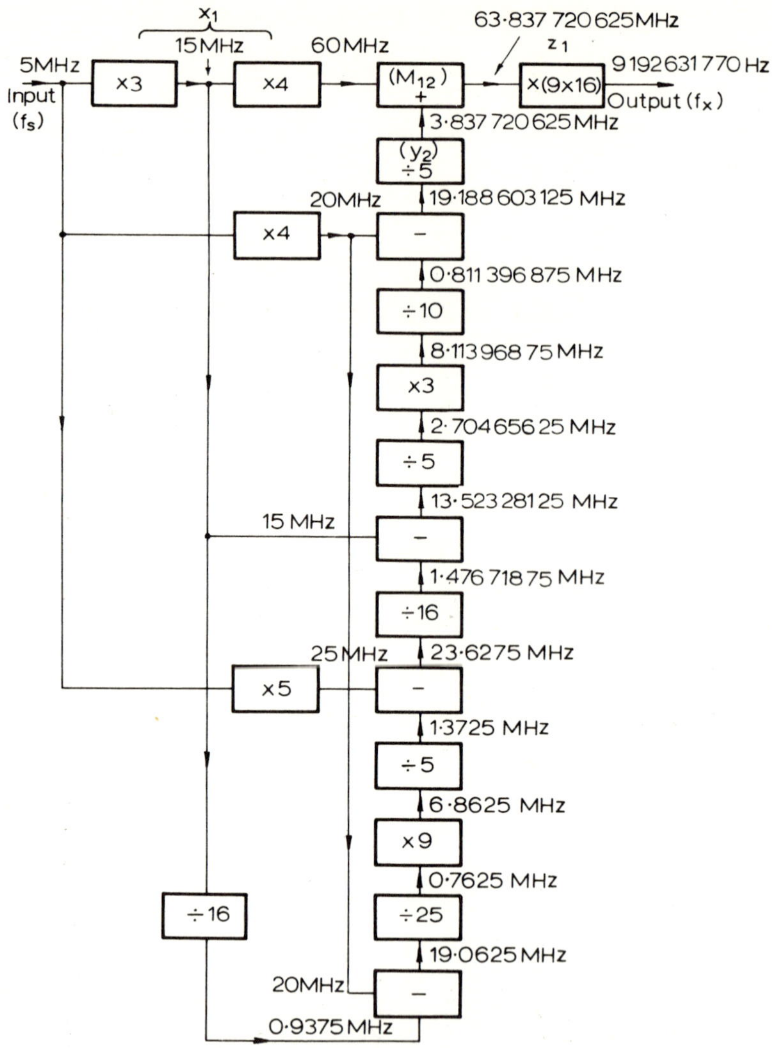

Fig. 5.8 Block diagram of the synthesis of the caesium atomic standard frequency.

other decimal places. The last condition leads to the following infinite series which is in reality a systematic fraction

$$\frac{f_x}{f_s} = \frac{X_1}{Y_1} = \frac{z_1}{y_1}\left[-X_0 + (x_0 + x_{01}) + \frac{x_0 + x_{02}}{10} + \cdots \right.$$
$$\left. + \frac{x_0 + x_{0n}}{10^{n-1}} + \frac{x_0}{10^n} + \frac{x_0}{10^{n+1}} + \cdots \right] \tag{5.49}$$

All terms without variable parts form the infinite geometric series

$$\frac{x_0}{10^n} + \frac{x_0}{10^{n+1}} + \cdots = \frac{x_0}{10^{n-1}} \times \frac{1}{9} \tag{5.50}$$

the sum of which will be added all at once to the last term with the variable part. When we sum all fixed parts of all terms in (5.49) we get the auxiliary term X_0 which must be subtracted at the end of the synthesizing process.

$$X_0 = \tfrac{10}{9} x_0 \tag{5.51}$$

The block diagram of the variable-frequency synthesizer in accordance with the mathematical model (5.49) is shown in Fig. 5.9.

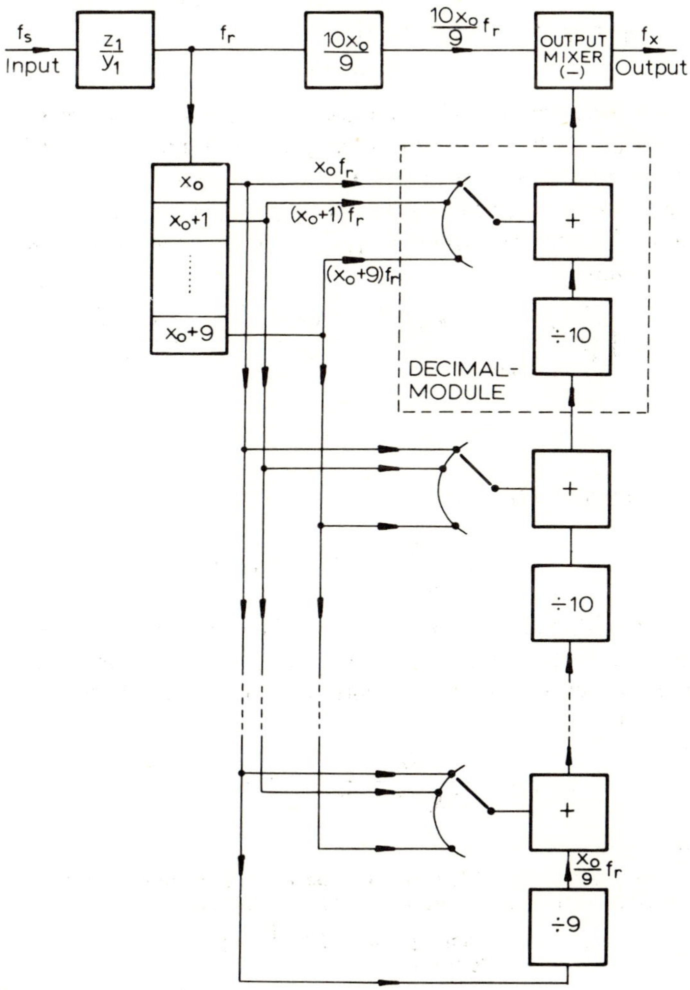

Fig. 5.9 The basic block diagram of the variable-frequency synthesizer.

Solution of eq. (5.49) is simple as there are only three unknowns: z_1, y_1, and x_0. Considering the variability of the frequency f_x, the numerator and the denominator in the ratio of the output to the input frequency could have a common divisor only of the form 10^k ($k = 1, 2, \ldots$). Performing this division the quotient involving the maximum output frequency may be written as

$$\frac{f_{x,\max}}{f_s} = \frac{X_{\max}}{Y} = \frac{z_1 \times 9{\cdot}9 \ldots 9_n}{Y} \tag{5.52}$$

where evidently z_1 is a power of ten

$$z_1 = 10^m \quad (m = \ldots, -2, -1, 0, 1, 2, \ldots) \tag{5.53}$$

and when comparing (5.52) with (5.49)

$$y_1 = Y \tag{5.54}$$

x_0 remains to be determined. However, its auxiliary position suggests at once that the problem is reduced rather to finding an interval of suitable values for x_0. Surely the requirements (5.13) and (5.37) must be maintained, particularly with regard to the lower bound q_d. Therefore $q_{i,\,\min}$ is calculated from (5.49) as the ratio of the smallest possible $x_{i,\,\min} = x_0$ to the largest residue which may be encountered

$$\left(\frac{X_{i+1}}{Y_{i+1}}\right)_{\max} = \frac{x_0 + 9}{10} + \frac{x_0 + 9}{100} + \cdots = \frac{x_0 + 9}{9} \tag{5.55}$$

$$q_{i,\,\min} > 9\frac{x_0}{x_0 + 9} > 0 \tag{5.56}$$

This, substituted in (5.37), gives

$$1 + 9\frac{x_0}{x_0 + 9} > 7 \Rightarrow x_0 > 18^{\dagger} \tag{5.57}$$

Another estimation about x_0 may be made from the ratio, q_1, of frequencies mixed in the first synthesizing loop. If in the output the undesired sidebands of the seventh or higher order only are admitted then it follows from the diagram in Fig. 4.12 that

$$q_{i,\,\max} = \frac{x_0}{10x_0/9 + 10} \geqslant 0{\cdot}8 \Rightarrow x_0 > 36 \tag{5.58}$$

---

† If x_0 is made very large, (5.57) shows that the lowest intermodulation product is always of the 10th order. This follows from (4.62) and the limit

$$\lim_{x_0 \to \infty} \left[1 + 9\frac{x_0}{x_0 + 9}\right] = 10$$

This result can be generalized, for a g-adic fraction expansion of (5.49), the lowest intermodulation product is of the gth order.

When in the output branches of the synthesizing loops no filters with variable selectivity ought to be used, then x_0 must be calculated from the ratio of the highest possible normalized frequency at the input of the modulator to the lowest normalized frequency which has to be filtered from the output signal of the respective modulator (see Fig. 5.9)

$$\frac{x_0 + 9}{10x_0/9} < 1 \Rightarrow x_0 > 81 \tag{5.59}$$

However, the use of the upper bound according to (5.37) gives a more accurate result†

$$\frac{x_0 + 9}{10x_0/9 - x_0 - 9} + 1 \leqslant 19 \Rightarrow x_0 \geqslant 171 \tag{5.60}$$

Evidently the larger x_0 is, the simpler the filters may be and the higher the auxiliary frequency, $f_s(z_1/y_1)X_0$, must be applied. Thus the possibility of physical realization is decisive for the upper bound for the interval of the suitable x_0. Actual values as used by several manufacturers are summarized in Table 5.5.

Table 5.5

Actual values of the transporting carrier, x_0, as used by different manufacturers

ADM – additive decimal module
PLDM – phase locked decimal module

Manufacturer	Type of decimal module	x_0	Reference
Rohde Schwarz	ADM, PLDM	153	[5.8]
Adret-Electronique	PLDM	180	[7.12]
Schlumberger	PLDM	189	[7.13]
Hewlett Packard	ADM	270	[5.7]
General Radio	PLDM	450	[7.11]

Reverting to the modified frequency-synthesis equation (5.49) it can be stated that it is always possible to choose x_0 in such a way that condition **B** holds for all frequency ratios in all mixers. On the other hand condition **A** cannot be maintained because in the set of integers $\{u_m\}$ there is no sequence of elements of the type

$$x_0, x_0 + 1, x_0 + 2, \ldots, x_0 + 9 \tag{5.61}$$

Either the use of ten simple single-frequency synthesizers is needed or filtering of multiples (5.61) directly from a suitable harmonic series must be applied.

† There, a more general application of condition **B** has been made because the normalized 'carrier' frequency $x_0 + 9$ and the first upper sideband $10x_0/9$ do not exist simultaneously.

Considering the block diagram (Fig. 5.9) of the decade frequency synthesizer based on eq. (5.49), hardly anything simpler can be expected because the procedure requires only those algebraic operations needed for representing the respective decimal number with merely one addition and one subtraction of the sum of the auxiliary geometric series. The physical realization consists of a chain of equal stages of 'decimal-modules' (shown dashed in Fig. 5.9); thus the incremental output step can be made as small as one desires. At the same time this arrangement is favourable for the introduction of IC technology [5.6].

Example 5.3

Design of a variable-frequency synthesizer for the frequency band 0–50 MHz with the smallest step

$$0{\cdot}01 \text{ Hz},$$

using the standard frequency

$$f_s = 5 \text{ MHz},$$

and maintaining the level of spurious signals (excluding harmonic multiples of the selected frequency) at least

$$-90 \text{ dB},$$

with the possibility of switching in the search oscillator on all decimal places up to 100 kHz.

Substituting in (5.52) gives

$$\frac{f_{x,\max}}{f_s} = \frac{49{\cdot}9\ldots9_{(9)} \times 10^6}{5 \times 10^6} = \frac{z_1}{y_1} \times 49{\cdot}9\ldots9_{(9)}$$

Evidently:

$$z_1 = 1$$

$$y_1 = 5$$

We have seen that the normalized frequency x_0 according to (5.58) is about one order lower than is given by the estimation based on (5.60). This suggests the idea of subtracting the variable normalized frequency $X_0 - 10x_{01}$ instead of the fixed X_0 particularly if x_{01} changes only over few digits.

The equation (5.49) changes slightly

$$\frac{f_x}{f_s} = \tfrac{1}{5}\,[-(X_0 - 10x_{01}) + x_0 + x_{02} + \cdots]$$

In our case

$$x_{01} = 0, 1, 2, 3, 4.$$

The smallest mixing ratio which may be encountered in the first synthesizing loop is

$$q_{1,\min} > \frac{X_0 - 40}{10x_0/9 + 10} = \frac{10x_0/9 - 40}{10x_0/9 + 10} = \frac{x_0 - 36}{x_0 + 9}$$

Because of the −90 dB spurious signals level, it is reasonable to admit in the output only sidebands of the ninth or higher order i.e. (Section 4.4)

$$\frac{5}{6} < \frac{x_0 - 36}{x_0 + 9}$$

from which

$$x_0 > 261$$

With this value the requirement (5.60) is also fulfilled. Certainly a rounder number is desirable, therefore

$$x_0 = 270$$

However, the real frequencies of the sequences $x_0 + x_{0i}$ are

$$f_s \frac{z_1}{y_1}(x_0 + x_{0i}) = 270, 271, \ldots 279 \text{ MHz}$$

which are too high to be divided with counting circuits (see Section 3.5); therefore the chain of equal decimal modules may start from the term $(x_0 + x_{03})/10$.

Nevertheless the need for filtering frequencies from a dense series together with the necessity to switch rather high frequencies (27·0; 27·1; . . . MHz) to respective mixers still persist. Both disadvantages may be solved if x_0 is split into two parts.

$$x_0 = x_0' + x_0''$$

Consequently in each decimal module an auxiliary mixer must be added (see Fig. 5.10). Evidently to have its output pass-band as free from higher sidebands as possible x_0' should be very high. Consideration of the other mixing process, involving $x_0'' + x_{0i}$, leads to a more definite result; the minimum q_i'' is

$$q''_{i,\min} = \frac{x_0' + x_0/9}{x_0'' + 9} \geqslant q_d - 1$$

After substituting $x_0 = 270$ and $q_d = 7$ we get

$$x_0'' < 35$$

A good choice might be

$$x_0'' = 30; \quad x_0' = 240$$

It is not advisable to go too near to q_h: because of the wide pass-band the output filters following the second (M″) mixer in Fig. 5.10 might be too costly.

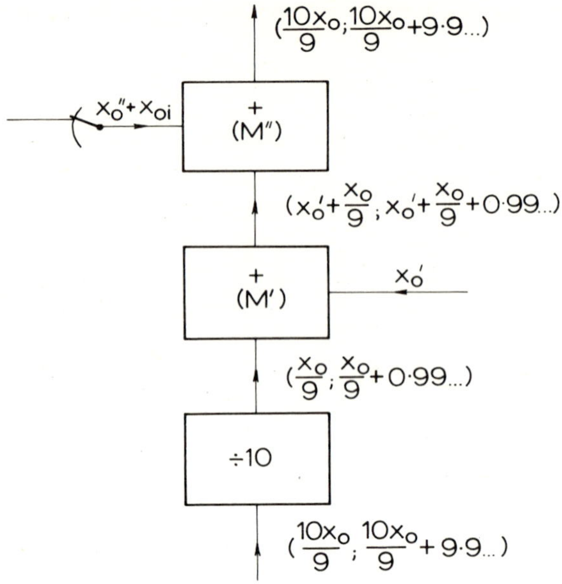

Fig. 5.10 Block diagram of additive decimal module with split transporting carrier $x_0 = x_0' + x_0''$.

The last requirement was the possibility of scanning continuously each of the ranges $(x_0/10^i; (x_0 + 10)/10^i)$, $i = 1, 2, \ldots 7$. This is easily achieved if the input to the respective decimal module is taken from a search oscillator, and the preceding part of the synthesizer is disconnected.

A block diagram of the designed variable-frequency synthesizer is shown in Fig. 5.11. More accurate descriptions of commercial versions employing the system may be found in [5.7] and [5.8].

5.4 DRIFT-CANCELLING SYSTEMS

One of the problems encountered in frequency synthesis is mixing of frequencies having relative ratios outside the interval given by the relation (5.37). Solution may be achieved by adding and subtracting of auxiliary frequencies in *drift-cancelling loops.*†

(i) Drift-Cancelling Loop

There are two possible arrangements: either the auxiliary frequency f_{12} is smaller than the input frequency f_2, i.e.

$$f_2 > f_{12} \quad \text{(Fig. 5.12(a))} \quad \text{or} \quad f_2 < f_{12} \quad \text{(Fig. 5.12(b))}$$

† The general theory has been worked out by Wojciechowski [5.9]. Another solution to this problem uses the phaselock-loop technique (Section 7.3).

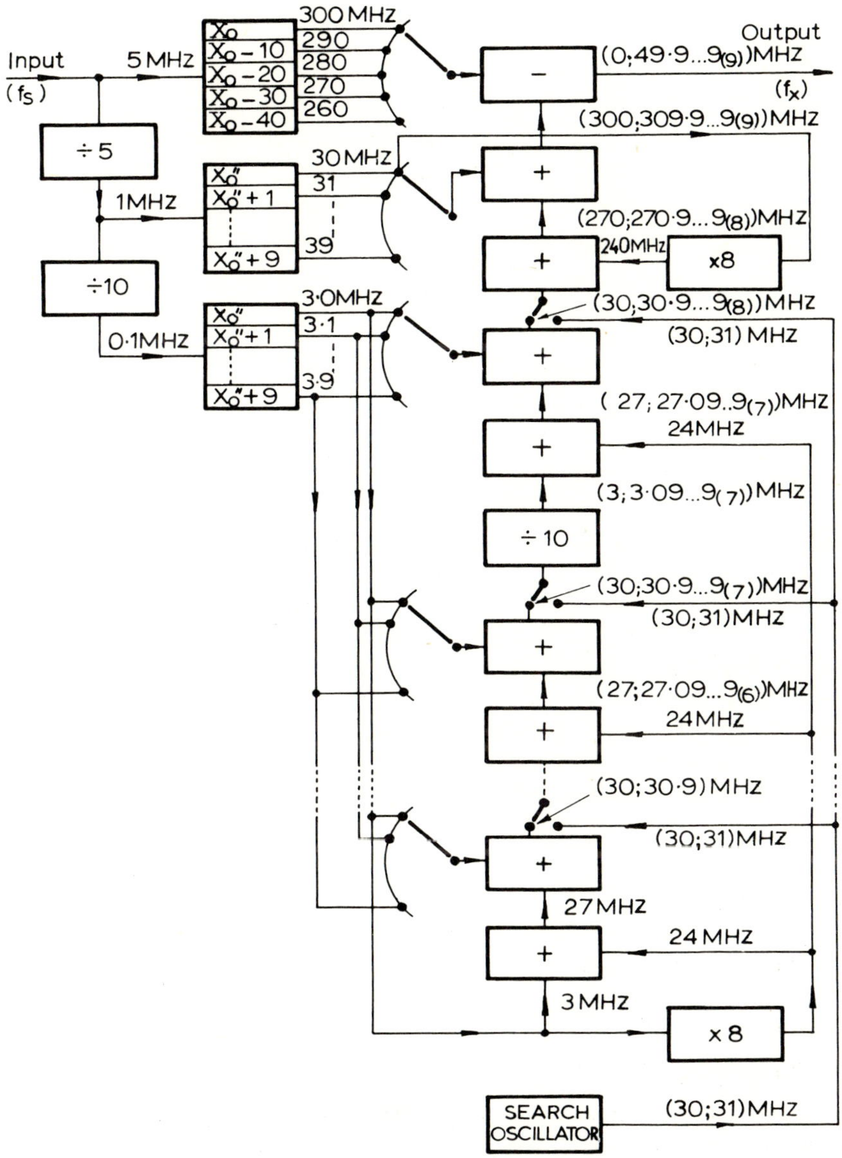

Fig. 5.11 Block diagram of the decade frequency synthesizer, variable in the band (0; 50) MHz in steps as small as 0·01 Hz.

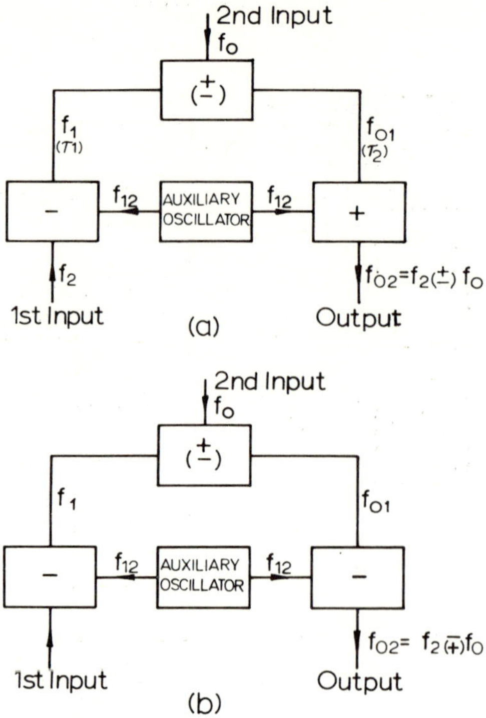

Fig. 5.12 Drift cancelling loops: (a) $f_2 > f_{12}$; (b) $f_2 < f_{12}$.

The loop equations are: ($f_1 > f_o$)

$$f_1 = f_2 - f_{12} \qquad\Big|\qquad f_1 = f_{12} - f_2 \tag{5.62}$$

$$f_{01} = f_1 \,(\pm)\, f_0 = f_2 - f_{12} \,(\pm)\, f_0 \qquad\Big|\qquad f_{01} = f_{12} - f_2 \,(\pm)\, f_0 \tag{5.63}$$

and the output frequency f_{02} is:

$$\begin{aligned} f_{02} &= f_{01} + f_{12} \\ &= f_2 - f_{12} \,(\pm)\, f_0 + f_{12} \\ &= f_2 \,(\pm)\, f_0 \end{aligned} \qquad\Bigg|\qquad \begin{aligned} f_{02} &= f_{12} - f_{01} \\ &= f_{12} - f_{12} + f_2 \,(\mp)\, f_0 \\ &= f_2 \,(\mp)\, f_0 \end{aligned} \tag{5.64}$$

In both cases the auxiliary frequency is cancelled out and its instability has no effect on the output frequency which is equal only to the sum or difference of the input frequencies.

However, the mechanism explained above is true only to the first-order approximation. Both parts of the auxiliary signal propagate in the drift-cancelling loop by a different route and consequently are subjected to different time delays. For evaluating this effect the combination equations (5.62) to (5.64) must be written as phase equations. The result ($f_2 > f_{12}$)

$$\omega_2(t+\tau_1+\tau_2) - \omega_{12}(t+\tau_1+\tau_2) \underset{(-)}{+} \omega_0(t+\tau_2) + \omega_{12}t$$
$$= (\omega_2 \underset{(-)}{+} \omega_0)(t+\tau_2) + \omega_2\tau_1 - \omega_{12}(\tau_1+\tau_2) \qquad (5.65)$$

indicates that changes of the auxiliary frequency (frequency noise) cause changes of the output phase (phase noise), see Section 9.3.

(ii) Multi-Stage Drift-Cancelling Mixing

In a more general case several drift-cancelling loops are cascaded as in Fig. 5.13. All auxiliary frequencies f_{12} to $f_{(k-1)k}$ are eliminated as explained above.

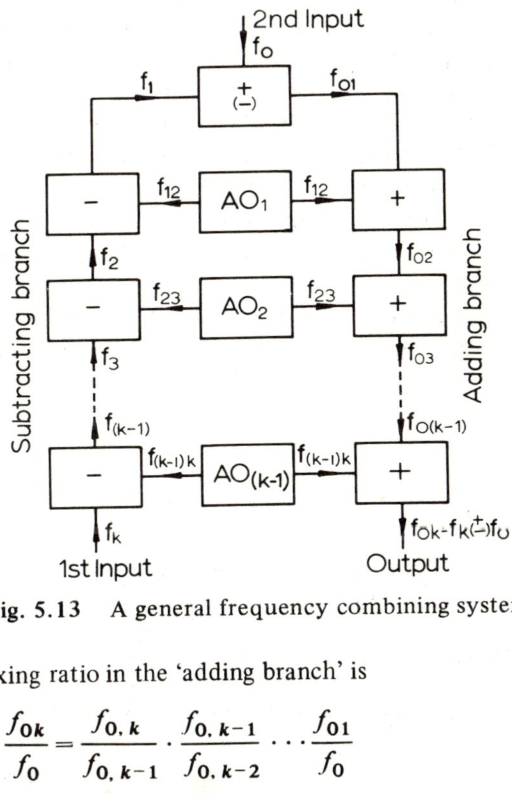

Fig. 5.13 A general frequency combining system.

The overall mixing ratio in the 'adding branch' is

$$\frac{f_{0k}}{f_0} = \frac{f_{0,k}}{f_{0,k-1}} \cdot \frac{f_{0,k-1}}{f_{0,k-2}} \cdots \frac{f_{01}}{f_0}$$
$$= |q_{0k}+1| \cdot |q_{0,k-1}+1| \cdots |q_{01}+1| \qquad (5.66)$$

Introducing the mean mixing ratio $q_{0,\text{ mean}}$ one may write

$$\frac{f_{0k}}{f_0} = |q_{0,\text{ mean}}+1|^k \qquad (5.67)$$

For the lower bound q_d of the $|q_{0,\,\text{mean}} + 1|$ the allowed order of the intermodulation products is decisive, whereas the upper bound q_h must be $q_h \geqslant q_d^2$; this follows from (5.67). Choosing q_d according to (5.19) one gets

$$7 < |q_{0,\,\text{mean}} + 1| < 49 \tag{5.68}$$

The number of mixers in the adding branch is k

$$k \leqslant \frac{\log f_{0k}/f_0}{\log q_d} \approx 1{\cdot}2 \log f_{0k}/f_0 \tag{5.69}$$

whereas the number of the drift cancelling loops is $k - 1$.

(iii) Drift-Cancelling Frequency Synthesizers

In the early generations of direct frequency synthesizers the drift-cancelling technique (which is also designated as the triple-mix system or the Wadley principle) is often encountered [5.10], [5.11], etc. Because the system is still employed, see [5.12], [8.10], etc., we feel obliged to make a brief review.

The feature highlighted in the early history of frequency synthesis was that the cumbersome switching of the highly selective filters was replaced by a much simpler tuning of the auxiliary oscillators. In recent designs the possibility of lowering the output phase noise of the synthesizer is the characteristic that is emphasized, see Section 9.3.

The fundamental equation (5.49) holds also for decade drift-cancelling frequency synthesizers. The only change is adding and simultaneous subtracting of the auxiliary terms x_{ai} ($i = 0, 1, 2, \ldots$) which are the normalized auxiliary-oscillator frequencies.

$$\frac{f_x}{f_s} = \frac{z_1}{y_1}\left[-X_0 + x_{a0} - x_{a0} + (x_0 + x_{a1} - x_{a1} + x_{01}) + \frac{x_0 + x_{a2} - x_{a2} + x_{02}}{10} + \cdots \right] \tag{5.70}$$

The service which is provided by the drift-cancelling loops is the filtration of the individual harmonic frequencies $f_r(x_0 + x_{0i})$ from the input spectra. Thus the decimal modules combine the function of filters for $x_0, x_0 + 1, \ldots$ and switches. The operation will be explained in conjunction with Fig. 5.14 which is based upon an actual device [5.12].

The input 100-kHz spectrum is peaked between 2·3 to 3·2 MHz. Via a low-pass filter it is applied to the first mixer and heterodyned with a frequency from the auxiliary oscillator which is switched in 100-kHz steps from 6·95 through to 6·05 MHz. The following narrow-band filter-amplifier is centred on 3·75 MHz with a pass-band of ± 10 kHz; thus only one harmonic of the reference can

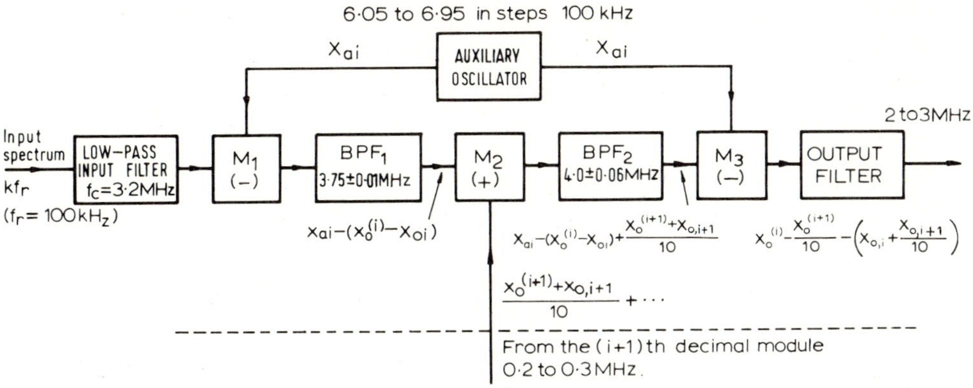

Fig. 5.14 Drift cancelling decimal module.

beat with the selected auxiliary-oscillator frequency to produce the required output, all others falling outside the pass-band of the filter.

Example 5.4

If 7 is selected on the 100-kHz dial the auxiliary frequency will be 6·25 MHz, which, to produce an output of 3·75 MHz, heterodynes with the 25th harmonic of the 100-kHz reference. The nearest frequencies, 3·85 and 3·65 MHz, provided from the 24th and 26th harmonics respectively, lie outside the 3·75 ± 0·01 MHz pass-band and are suppressed.

If half of the pass-band is reserved for the oscillator drift (the other for the filter drifts) the required stability will be of the order 1×10^{-3}.

Next in the synthesis process is adding of the output of the preceding decimal module, 0·2 to 0·3 MHz, to the 3·75 MHz signal, the upper sideband being in the range 3·95 to 4·05 MHz, and the band-pass of the BPF_2 being 4·0 ± 0·06 MHz.

Finally in the third mixer the 4-MHz intermediate frequency is heterodyned, once more, with the frequency of the auxiliary oscillator and in this way its influence is cancelled out.

Returning to the example:

$$6{\cdot}25 - (3{\cdot}95 \text{ to } 4{\cdot}05) = 2{\cdot}3 \text{ to } 2{\cdot}2 \text{ MHz}$$

which is derived solely from the reference and the output of the preceding decimal module.

Examination of the mixing ratios in the decimal module reveals that at M_1 and M_3 the condition (5.37) is severely violated. And actually one may easily find, using the diagram in Fig. 4.12, that the intermodulation product of the 5th order falls into the pass-band of the BPF_1. This need not be disastrous when using balanced modulators. Another remedy might be provided by increasing the auxiliary frequency; however, the stability problem would soon become prohibitive when employing a switched LC oscillator. A further solution, i.e., lowering of the transporting carrier $f_r x_0$, would lead to a very high mixing ratio in the M_2 stage. This difficulty can be removed if the subtracting of the auxiliary frequency, $f_r x_{ai}$, is postponed and performed in the next decimal module, see Fig. 5.15.

Several block diagrams of drift cancelling frequency synthesizers, which were marketed in Europe in the 50s, may be found in [5.11].

5.5 CONCLUSIONS

It has been shown that the frequency synthesis process may be reduced to the well-known algebraic operation of the approximation of real numbers. On this basis a sound mathematical theory has been developed and a general frequency synthesis equation derived, valid for both single-frequency and variable-frequency synthesizers. Application of the theory has been demonstrated on practical examples.

Development of frequency synthesizers progressed for a quarter of a century before their use became widespread. During this period many, sometimes quite sophisticated, schemes were used, the value of which is now merely historical. In this connexion we feel that due tribute should be paid to the pioneering work of Finden [5.1] as well as others [5.13], [5.14], etc. who first made decade frequency generators.

Finally, the advantages and disadvantages of direct frequency synthesis in comparison with other methods used for derivation of a new frequency f_x from a standard frequency f_s will be given:

Advantages

(1) Without any difficulty, frequencies to fractions of one Hz may be generated.

(2) Times needed for switching the variable-frequency synthesizer from one frequency to another are very short – below one millisecond without any difficulty, this being due to the absence of phaselocked oscillators.

(3) Because of the absence of the latter there is practically no risk of the output frequency being incorrect.

(4) The power level of discrete spurious frequencies in the output signal can be kept below the –80 to –90 dB level.

(5) The lowest phase-noise that has been reported.

Disadvantages

(1) Larger dimensions and weight; because of the filters, integrated circuits cannot be used throughout.

(2) For the same reasons a higher price may be expected.

The final inference is that the place of a direct-frequency synthesizer should be in laboratories, in fixed communications centres, and wherever the ultimate in switching times and phase noise is required.

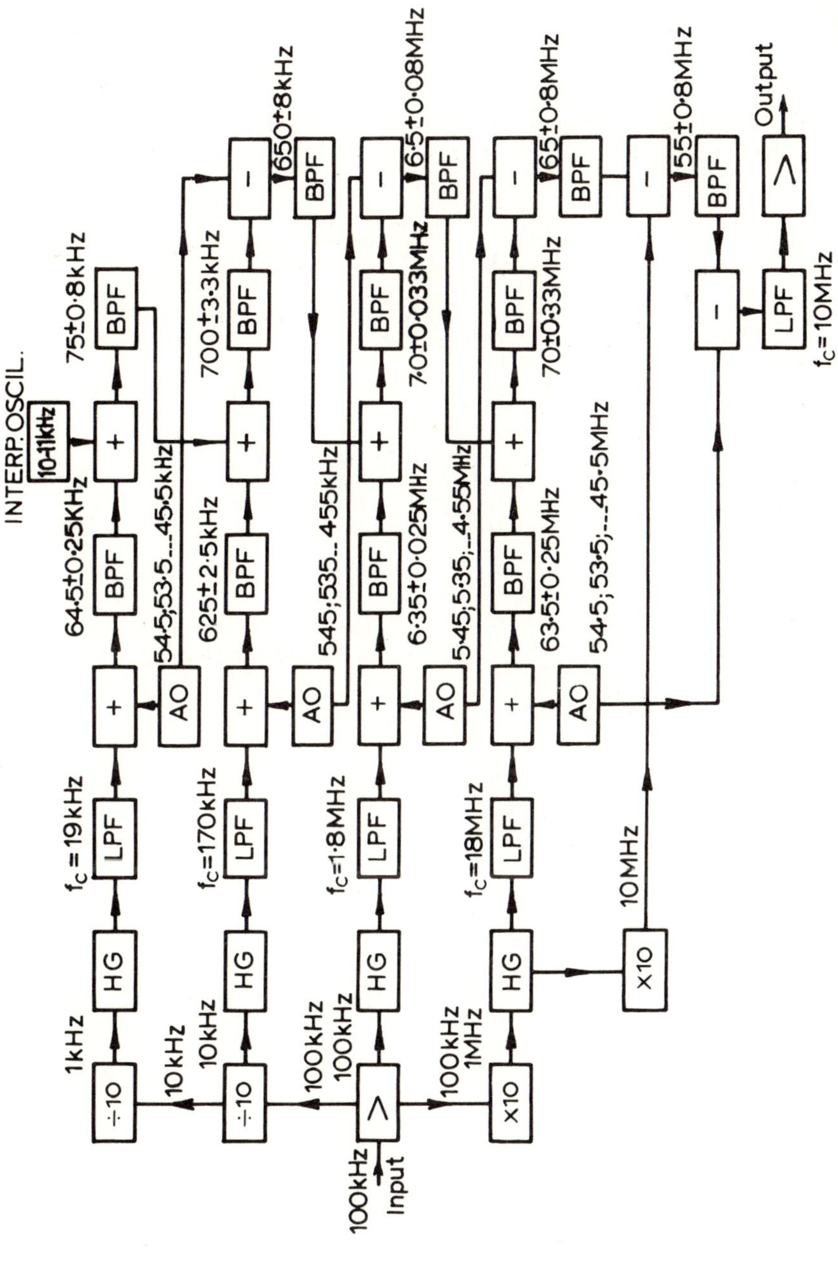

Fig. 5.15 Drift cancelling decimal variable frequency synthesizer with delayed subtraction of auxiliary frequencies.

REFERENCES

[5.1] FINDEN, H. J. The frequency synthesizer. *Journal I.E.E.*, Part III, **90**, 165-80, 1943.

[5.2] KROUPA, V. Theory of frequency synthesis. *I.E.E.E. Transactions on Instrumentation and Measurement*, IM-17, 56-68, 1968.

[5.3] PERRON, O. *Irrationalzahlen*, 3rd ed. Berlin: Walter de Gruyter, 1947.

[5.4] ESSEN, L., HOPE, E. G. and PARRY, J. V. L. Circuits employed in the N.P.L. caesium standard. *Proc. I.E.E.*, **106**, Part B, 240-4, 1959.

[5.5] HOLLOWAY, J., *et al.* Comparison and evaluation of cesium atomic beam frequency standards. *Proc. I.R.E.*, **47**, 1730-6, 1959.

[5.6] MONTGOMERY, W. E. Application of integrated electronics to military communications and radar systems. *Proc. I.E.E.E.*, **52**, 1721-31, 1964.

[5.7] VAN DUZER, V. E. A 0–50 Mc frequency synthesizer with excellent stability, fast switching, and fine resolution. *Hewlett-Packard Journal*, **15**, 1-6, May 1964.

[5.8] GERHOLD, J. Dekadischer HF-Messender SMDH. *Neues von Rohde und Schwarz*, **8**, 5-12, Jan. 1968.

[5.9] WOJCIECHOWSKI, B. M. Theory of a frequency-synthesizing network. *Bell System Technical Journal*, **39**, 649-73, May 1960.

[5.10] COLAS, M. Le stabilidyne. *L'Onde Électrique*, **36**, 83-93, Feb. 1956.

[5.11] FLICKER, H. Stand der Frequenzmesstechnik nach dem Überlagerungsverfahren. *Handbuch für Hochfrequenz- und Elektro-Techniker*, VI Band, pp. 349-92. Berlin: Verlag für Radio-Foto-Kinotechnik GMBH, 1960.

[5.12] The Marconi Company. Hydrus HF receiver. *Telecommunications*, **2**, 40-1, July 1968.

[5.13] BOELLA, M. Generatore di frequenze campione per misure di alta precisione. *Alta Frequenza*, **14**, 183-94, Sept.–Dec. 1945.

[5.14] LEONHARDT, R. and FLICKER, H. Eine Neuentwicklung: Dekadische Frequenzmessanlage 10 Hz bis 30 MHz mit Absolutkontrolle. *Rohde u. Schwarz-Mitt.*, p. 69, Aug. 1952.

6 Phaselock Loop

Another approach to the solution of frequency synthesis problems makes use of oscillators, the output frequency of which is maintained in an exact rational-number relation to the standard reference frequency.

The feedback system involved is generally called the phaselock loop. The basic mathematical theory, as described in many textbooks, e.g. [6.1], forms a solid background for understanding the problem. But bearing in mind the importance which the phaselock loop plays in frequency synthesis we shall briefly recall the essentials of the theory. But discussion of the properties will be limited to second-order phaselock loops only. We feel this narrowing justified because loops of this type enable an independent choice of three circuit parameters to be made and this is generally sufficient to cope with the requirements met in frequency synthesis.

Because of limited space, in many places we shall simply refer to published results and so save lengthy derivations. In this connexion it should be noted that there is some confusion in respect of symbols and definitions as between different authors. We have accordingly decided to follow the terminology of Gardner [6.2] and to a smaller extent that of Byrne [6.3].

6.1 FUNDAMENTAL EQUATIONS

The basic components of a phaselock loop, shown schematically in Fig. 6.1(a), are the voltage controlled oscillator (VCO) and phase detector (PD) usually with a low-pass filter (F) between them.

The system works as follows: in the PD the phases of the input signal and of the VCO are compared; the output voltage of the PD passes through the loop filter F, where eventually only the h.f. components are suppressed, to the control element of the VCO, and changes its frequency in such a way that the phase difference between the input signal and the local oscillator is reduced.

When the loop is locked the average output frequency of the VCO is exactly equal to the average frequency of the input signal, i.e. for each output cycle there is one, and only one, cycle at the oscillator input.

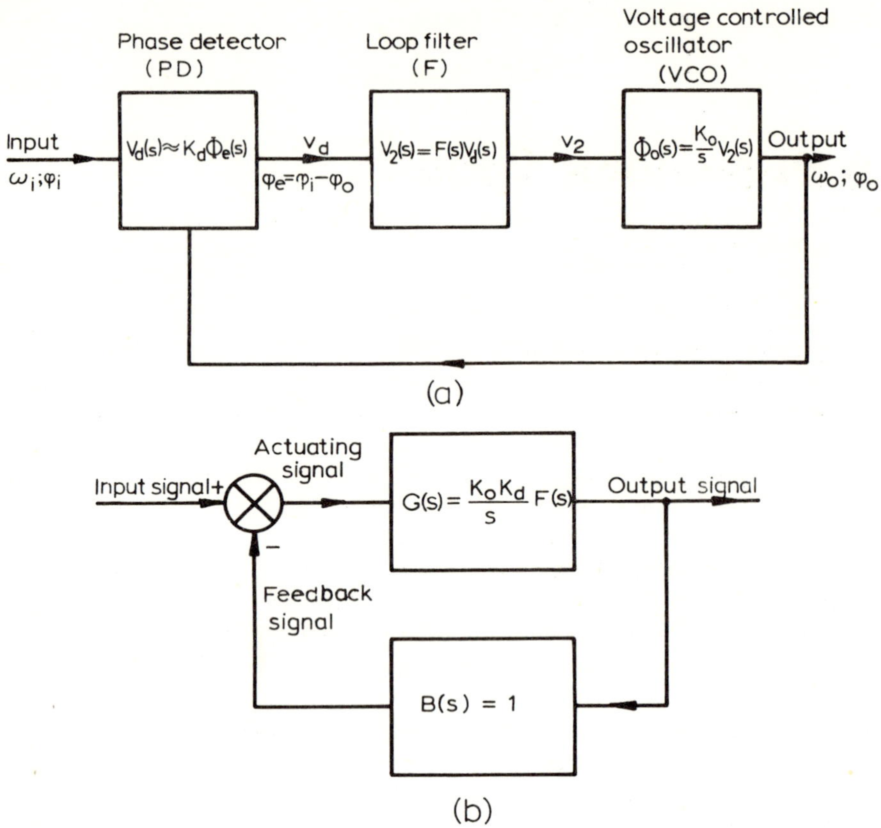

Fig. 6.1 The phaselock loop: (a) basic block diagram; (b) equivalent feedback network.

The mathematical description of the phaselock loop is in a nonlinear differential equation. Assuming that the oscillator (VCO) has a linear control-characteristic of a slope K_0 (2π Hz/volt) and that its free-running frequency is ω_c, the actual oscillator frequency ω_0 is

$$\omega_0 = \omega_c + K_0 v_2 \tag{6.1}$$

where v_2 is the applied control voltage.

Integration of (6.1) gives the output phase

$$\varphi_0 = \omega_c t + \int_0^t K_0 v_2 \, dt \tag{6.2}$$

The phase error is

$$\varphi_e = \varphi_i - \varphi_0 = \omega_i t - \omega_c t - \int_0^t K_0 v_2 \, dt \tag{6.3}$$

Substituting $\omega_i - \omega_c = \Delta\omega$ and differentiating gives

$$\frac{d\varphi_e}{dt} = \Delta\omega - K_0 v_2 \tag{6.4}$$

Very often the output of the phase detector has the form (see Section 6.4)

$$v_d = K_d \sin \varphi_e \tag{6.5}$$

If the filter passes the beat (6.5) without modification while rejecting all other (i.e. high frequency) components in the detector output, then

$$v_2 = v_d = K_d \sin \varphi_e \tag{6.6}$$

and the differential equation of the automatic phase control (APC) loop may be written

$$\frac{d\varphi_e}{dt} = \Delta\omega - K_d K_0 \sin \varphi_e \tag{6.7}$$

Generally the phase error is small; this entails

$$\sin \varphi_e \approx \varphi_e \tag{6.8}$$

from which the linearization of (6.7) results

$$\frac{d\varphi_e}{dt} = \Delta\omega - K\varphi_e; \quad K = K_d K_0 \tag{6.9}$$

Solution in the time domain of both (6.7) and (6.9) is performed in Appendix 6.1. However, the more usual approach is the application of the Laplace transform. When the phase angles are measured in a coordinate system which moves at the free-running speed ω_c of the controlled oscillator we get from (6.2)

$$\Phi_0(s) = \frac{1}{s} K_0 V_2(s) = \frac{1}{s} K_0 F(s) V_d(s) = \frac{1}{s} K_0 K_d F(s) \mathscr{L}[\sin(\varphi_i - \varphi_0)] \tag{6.10}$$

where $F(s)$ is the transfer function of the low-pass filter F. For small phase errors, which is generally the case when the loop is locked, eq. (6.10) simplifies to

$$\Phi_0(s) = \frac{1}{s} K F(s)[\Phi_i(s) - \Phi_0(s)] \tag{6.11}$$

Rearrangement results in the basic loop equations:

$$\frac{\Phi_0(s)}{\Phi_i(s)} = H(s) = \frac{K F(s)}{s + K F(s)} \tag{6.12}$$

$$\frac{\Phi_i(s) - \Phi_0(s)}{\Phi_i(s)} = \frac{\Phi_e(s)}{\Phi_i(s)} = 1 - H(s) = \frac{s}{s + K F(s)} \tag{6.13}$$

which are the familiar feedback-network equations (see Fig. 6.1(b)) with the forward transfer function (gain)

$$G(s) = \frac{K F(s)}{s} \tag{6.14}$$

and with the feedback transfer function

$$B(s) = 1 \tag{6.15}$$

Before proceeding further it is necessary to specify the loop filter $F(s)$. Its task is to attenuate fast changes in phase error due to noise in the input signal; it also helps to smooth out the high-frequency components of the phase detector output. Usually a simple RC filter or a phase-lag-lead filter, as shown in Fig. 6.2, is used. The transfer function of both filter types may be written in the same form

$$F(s) = \frac{1 + sT_2}{1 + sT_1} \tag{6.16}$$

When substituting in (6.12) the loop transfer function is

$$H(s) = \frac{(K/T_1)(1 + sT_2)}{s^2 + s[(1 + KT_2)/T_1] + K/T_1} \tag{6.17}$$

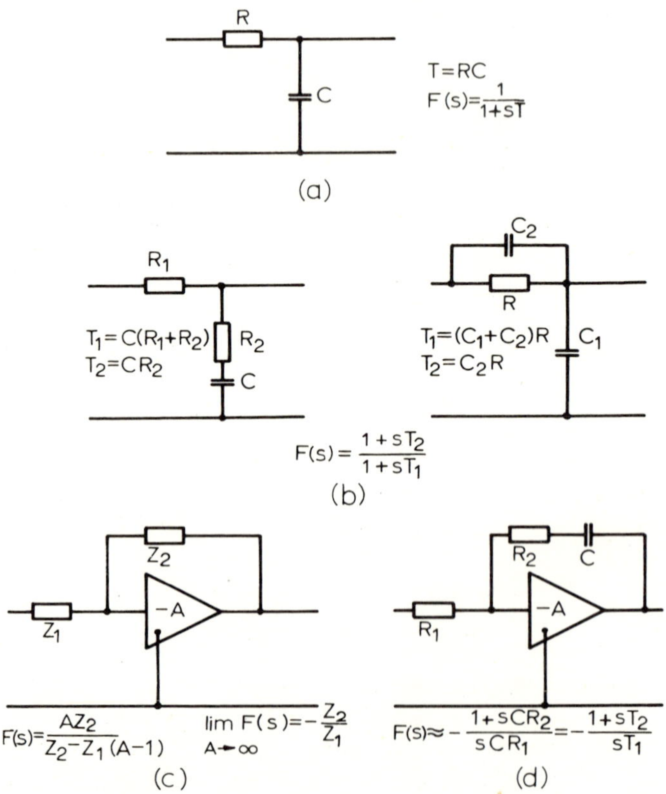

Fig. 6.2 Second-order phaselock loop filters: (a) simple RC filter (lag network); (b) phase lag-lead or proportional plus integral networks; (c) general active filter configuration; (d) active phase lag-lead network.

Because the highest power of s in the denominator of (6.17) is two the loop is known as a 'second-order loop'. Because of its simplicity and good performance it is widely applied in practice. Often, in agreement with servo terminology, the 'natural frequency' ω_n and the damping factor ζ are used as defining parameters

$$\omega_n = \sqrt{\frac{K}{T_1}}; \quad \zeta = \frac{1}{2}\sqrt{\frac{K}{T_1}}\left(T_2 + \frac{1}{K}\right) = \frac{\omega_n}{2}\left(T_2 + \frac{1}{K}\right) \tag{6.18}$$

and the transfer function $H(s)$ may be written as

$$H(s) = \frac{s\omega_n(2\zeta - \omega_n/K) + \omega_n^2}{s^2 + 2\zeta\omega_n s + \omega_n^2} \tag{6.19}$$

When an active filter, as in Fig. 6.2(d), is used

$$H_a(s) = \frac{(K/T_1)(1 + sT_2)}{s^2 + s(KT_2/T_1) + K/T_1} \quad \dagger \tag{6.20}$$

or

$$H_a(s) = \frac{2\zeta\omega_n s + \omega_n^2}{s^2 + 2\zeta\omega_n s + \omega_n^2} \tag{6.21}$$

where

$$\omega_n = \sqrt{\frac{K}{T_1}}; \quad \zeta = \frac{1}{2}\sqrt{\frac{K}{T_1}}T_2 = \frac{\omega_n}{2}T_2 \tag{6.22}$$

Comparing (6.20) with (6.17) it is readily seen that for $KT_2 \gg 1$ (high gain loop conditions) both equations are the same; consequently this is true for pairs (6.19)–(6.21) and (6.18)–(6.22) as well.

There are other-order loops too. If the filter is omitted entirely

$$H(s) = \frac{K}{s + K} \tag{6.23}$$

and the first-order loop results. Whereas the third order is reached when the filter contains two cascaded integrators. However, the stability must be examined; this is not so with first- and second-order loops (as commonly built) which are unconditionally stable.

6.2 SMALL-SIGNAL PROPERTIES

For frequency synthesis applications knowledge of the interference level at the output of the phase-controlled oscillator is of great importance. There are three main sources of spurious signals which may influence the output phase jitter: (1) undesired components superimposed on the reference signal, (2) remainder of

---

† The minus sign of $F(s)$ is removed by $(-K_d)$, i.e. by referring to the input signal shifted by 180°.

the 'carrier' at the reference frequency and of its multiples at the phase detector output, and (3) the instability of the voltage-controlled oscillator itself.

(A) Response to Spurious Sine-Wave Input Signals

Consider the input voltage composed of the reference and a small spurious signal

$$v_i(t) = V_i \cos(\omega_i t + \varphi_i) + V_{sp} \cos(\omega_{sp} t + \varphi_{sp}) \tag{6.24}$$

Using (9.11) we may write

$$v_i(t) = V_i[1 + m_a \cos(\Omega t + \Delta\varphi_{sp})] \cos[\omega_i t + \varphi_i + m_p \sin(\Omega t + \Delta\varphi_{sp})] \tag{6.25}$$

where

$$m_a = m_p = \frac{V_{sp}}{V_i} \ll 1 \quad \text{and} \quad \Omega = (\omega_{sp} - \omega_i)$$

$$\Delta\varphi_{sp} = (\varphi_{sp} - \varphi_i)$$

Expansion of (6.25) leads finally to

$$v_i(t) \approx V_i \cos(\omega_i t + \varphi_i) + m_a V_i \cos(\Omega t + \Delta\varphi_{sp}) \times \cos(\omega_i t + \varphi_i) - m_p V_i \sin(\Omega t + \Delta\varphi_{sp}) \sin(\omega_i t + \varphi_i) \tag{6.26}$$

The VCO output voltage is (in the locked state)

$$v_0(t) = -V_0 \sin(\omega_i t + \varphi_0)^{\dagger} \tag{6.27}$$

Assuming the phase detector to be a perfect multiplier, its output (neglecting double-frequency terms that will be removed by the loop filter) is

$$v_d = \frac{V_i V_0 K_m}{2}[\sin(\varphi_i - \varphi_0) + m_a \cos(\Omega t + \Delta\varphi_{sp}) \sin(\varphi_i - \varphi_0) - m_p \sin(\Omega t + \Delta\varphi_{sp}) \cos(\varphi_i - \varphi_0)] \tag{6.28}$$

When $(\varphi_i - \varphi_0)$ is small, which is generally true, the term caused by the spurious amplitude modulation is of the second order and (6.28) simplifies to

$$v_d = K_d[\varphi_i - \varphi_0 - m_p \sin(\Omega t + \Delta\varphi_{sp})] \tag{6.29}$$

where the phase detector gain $K_d = \frac{1}{2} V_i V_0 K_m$ has been introduced.

Evidently, only that component in (6.26) which is in quadrature to the reference signal and consequently responsible for the input phase modulation needs to be considered. The same inference may be readily deduced from the vector diagram in Fig. 6.3.

The amplitude of the respective output phase variations follows from (6.12)

$$\Phi_0(\Omega) = m_p |H(j\Omega)| \tag{6.30}$$

† Note that v_i and v_0 are nearly 90 degrees out of phase with one another which is due to the integrating properties of the voltage-controlled oscillator.

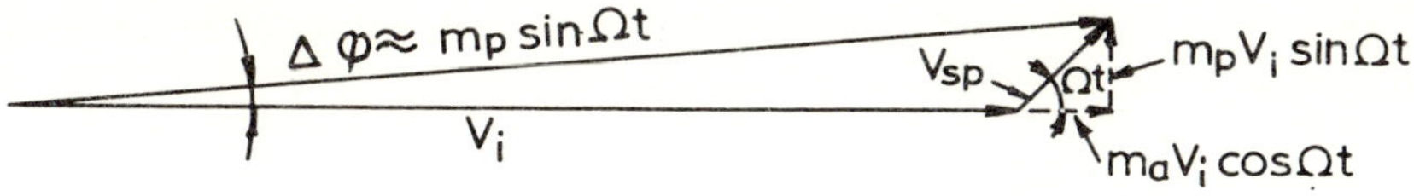

Fig. 6.3 Vector diagram of reference voltage with a low-index spurious amplitude and phase modulation due to the superposition of a small sinusoidal signal.

The normalized phase jitter attenuation curves for different loop filter configurations are plotted in Fig. 6.4 and 6.5. The low-pass function of the phaselock loop is evident. Examination reveals for high values of normalized frequencies (ω/K) or (ω/ω_n) a 6 dB per octave (20 dB per decade) slope whereas in the case of a simple RC filter the slope is 12 dB per octave (40 dB per decade).

If a spurious voltage v_{sp} is added to the error signal v_d (e.g. remainder of detected carriers) the respective output phase response is

$$\Phi_{0,\,sp}(s) = \frac{1}{K_d} H(s)\, V_{sp}(s) \tag{6.31}$$

In the case of a sinusoidal disturbance with the amplitude V_{sp} the index of the output phase modulation is

$$m_{p0} = \frac{V_{sp}}{K_d} |H(j\Omega)| \tag{6.32}$$

When $m_{po} \ll 1$ the ratio between the output signal and the spurious sideband (S/SpSB) is

$$(\text{S/SpSB}) \approx \frac{2}{m_{p0}} \tag{6.33}$$

(B) Noise Performance

If the noise power is concentrated in a relatively narrow band around the frequency ω_i the noise voltage $n(t)$ may be expressed by (9.12); the input phase noise is

$$\varphi_{ni}(t) = \frac{n_s(t)}{V_i} \tag{6.34}$$

and its variance

$$\overline{\varphi_{ni}^2} = \frac{\overline{n^2(t)}}{V_i^2} = \frac{P_N}{2P_S} = \frac{1}{2(\text{SNR})_i} \tag{6.35}$$

where P_S/P_N is the input signal-to-noise ratio.

Assuming that the spectrum of $n(t)$ is flat within the input bandwidth B_i then the input spectral density is

$$N_i = \frac{\overline{n^2(t)}}{B_i} \quad (\text{volts}^2/\text{Hz}) \tag{6.36}$$

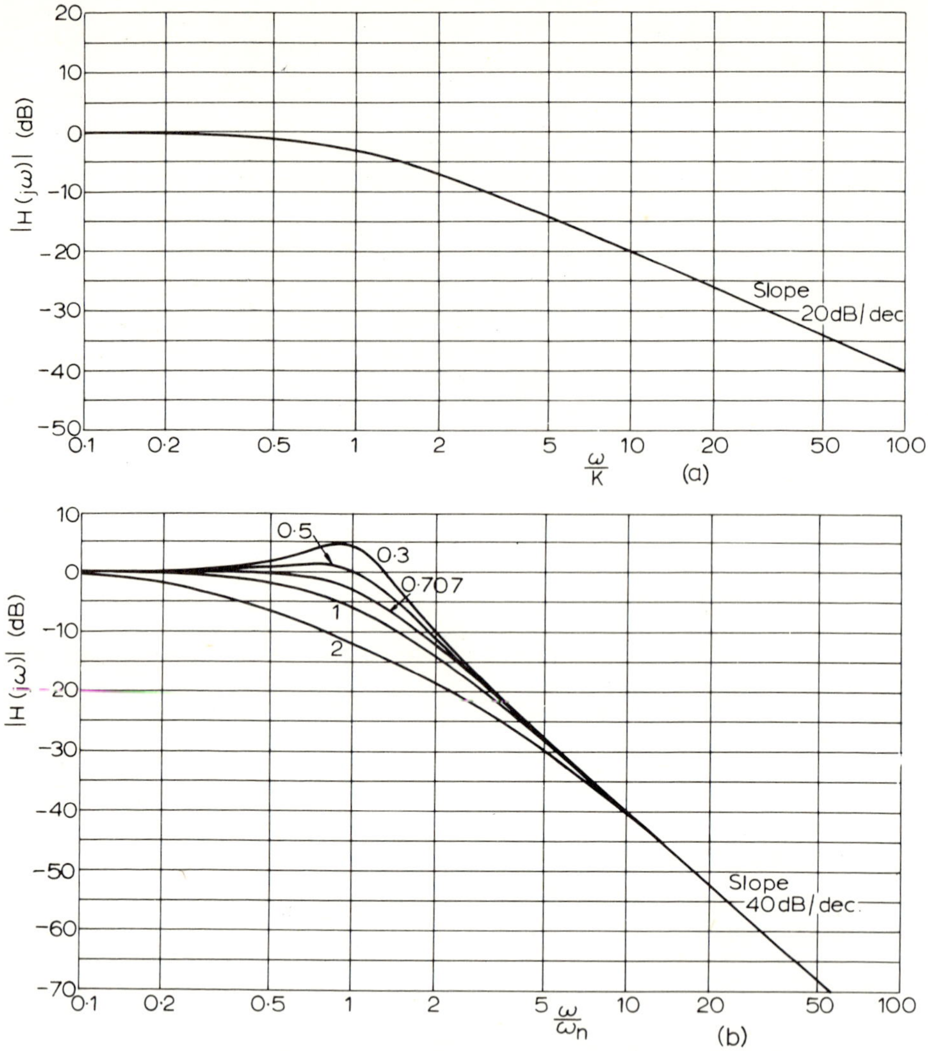

Fig. 6.4 Jitter attenuation in phaselock loops: (a) with no filter (first-order loop); (b) with a simple RC filter (for different damping parameters $\zeta = 0{\cdot}3$; $0{\cdot}5$; . . . 2).

and the equivalent low-pass spectrum ('one-sided', for definition see Section 9.2) of the phase noise φ_{ni} has a bandwidth $B_i/2$ and a density

$$\left.\begin{aligned} S'_\varphi(f) &= \frac{\overline{\varphi_{ni}^2}}{B_i/2} = \frac{2N_i}{V_i^2} \quad && (0 < f < B_i/2) \\ S'_\varphi(f) &= 0 && (B_i/2 < f) \end{aligned}\right\} \tag{6.37}$$

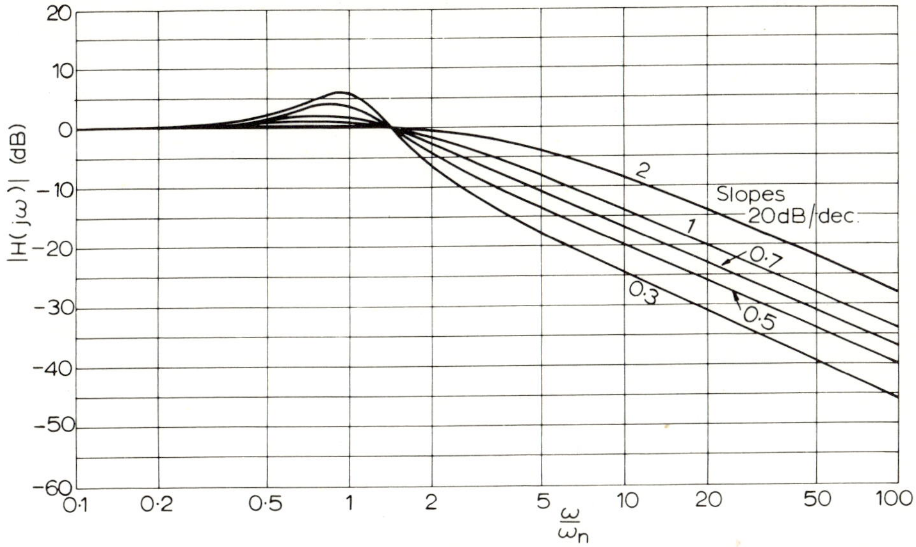

Fig. 6.5 Jitter attenuation in high-gain second-order phaselock loop with lag-lead filter (for different damping parameters $\zeta = 0{\cdot}3; 0{\cdot}5; \ldots 2$).

Using the above relation together with (6.30) we may calculate the variance of the output phase jitter (fluctuation of the VCO phase)

$$\overline{\varphi_{n0}^2} = \int_0^{B_i/2} \frac{2N_i}{V_i^2} |H(j\omega)|^2 \, df \approx \frac{2N_i}{V_i^2} \int_0^{\infty} H(j\omega)^2 \, df^{\,\dagger} \tag{6.38}$$

Introducing the 'loop-noise bandwidth' B_L

$$B_L = \int_0^{\infty} |H(j\omega)|^2 \, df \quad (\text{Hz}) \tag{6.39}$$

The mean square output phase jitter is

$$\overline{\varphi_{n0}^2} = \frac{P_N}{P_S} \frac{B_L}{B_i} = \frac{1}{2\,(\text{SNR})_L} \tag{6.40}$$

where in analogy to (6.35) the 'loop signal-to-noise ratio' $(\text{SNR})_L$ has been introduced.

By substituting in (6.39) the transfer function $H(j\omega)$ according (6.17) we get

$$B_L = \frac{K}{4} \frac{1 + (KT_2)^2/KT_1}{1 + KT_2} \quad (\text{Hz}) \tag{6.41}$$

---

† This approximation is valid if $B_i \gg B_L$, see (6.39).

For no filter or any RC filter $T_2 = 0$ and consequently

$$B_L = \frac{K}{4} \quad \text{(Hz)} \tag{6.42}$$

In the high-gain second-order loop, $KT_2 \gg 1$ and the loop noise bandwidth is approximately

$$B_L = \frac{1}{4}\,\frac{1 + (KT_2)^2/KT_1}{T_2} = \frac{\omega_n}{2}\left(\zeta + \frac{1}{4\zeta}\right) \quad \text{(Hz)} \tag{6.43}$$

The normalized loop-noise bandwidth $B_L/(\omega_n/2)$ has the minimum for $\zeta = 0{\cdot}5$ with a rather slow increase toward higher damping factors (see Fig. 6.6). In the range $0{\cdot}25 < \zeta < 1{\cdot}0$ the respective noise power variations are below 1 dB. Therefore the approximation

$$B_L \approx \frac{\omega_n}{2} \quad \text{(Hz)} \tag{6.44}$$

as a rule of thumb is justified. (Note that B_L is in Hz but ω_n in radians/s.) When $(KT_2)^2/KT_1$ is much greater than one

$$B_L \approx \frac{KT_2}{4T_1} \quad \text{(Hz)} \tag{6.45}$$

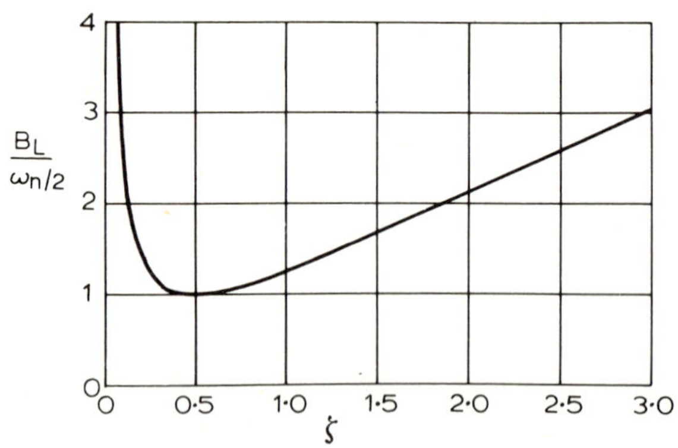

Fig. 6.6 Normalized loop-noise bandwidth of the high-gain second-order loop. (From Gardner [6.2].)

(C) Responses to Step Changes in Phase and Frequency

Sudden changes in phase or frequency may be of different origins: VCO instabilities, or changes in counting rates if counting circuits are associated with the phaselock loop, etc. The phase errors φ_e will be found when evaluating (6.13).

(i) Phase Step

Consider a step change of input phase of magnitude $\Delta\varphi_i$. The respective Laplace transform is

$$\Phi_i(s) = \frac{\Delta\varphi_i}{s} \tag{6.46}$$

The steady-state error follows readily from the final value theorem of Laplace transforms

$$\lim_{t\to\infty} y(t) = \lim_{s\to 0} sY(s) \tag{6.47}$$

Substituting (6.46) we get

$$\lim_{t\to\infty} \varphi_{e1} = \lim_{s\to 0} \frac{s\Delta\varphi_i}{s + K\,F(s)} = 0 \tag{6.48}$$

The transient time response of the phase error in the second-order loop is found by the inverse Laplace transform of (6.13) after substituting (6.19)

$$\varphi_{e1}(t) = \Delta\varphi_i \,.\, \exp(-\zeta\omega_n t)\left[\cos\sqrt{1-\zeta^2}\,\omega_n t - \frac{\zeta - \omega_n/K}{\sqrt{1-\zeta^2}} \times \sin\sqrt{1-\zeta^2}\,\omega_n t\right]^{\dagger} \tag{6.49}$$

For the critical damping ($\zeta = 1$) and the high-gain loop the limiting of (6.49) reveals:

$$\varphi_{e1}(t) = \Delta\varphi_i\;(1 - \omega_n t)\exp(-\omega_n t) \tag{6.50}$$

(ii) Frequency Step

Consider a step change of frequency of magnitude $\Delta\omega$. The input phase is a ramp and therefore $\Phi_i(s) = \Delta\omega/s^2$. The steady-state value is, according to (6.47)

$$\lim_{t\to\infty} \varphi_{e2}(t) = \lim_{s\to 0} \frac{\Delta\omega}{s + K\,F(s)} = \frac{\Delta\omega}{K\,F(0)} \tag{6.51}$$

The product $K\,F(0)$ is often called the 'velocity constant' or 'd.c. loop gain' and is denoted by K_v. The 'static phase error' or 'velocity error' necessary to keep the phaselock when the free-running frequency of the VCO and the reference frequency differ one from another by $\Delta\omega$, is also given by (6.51)

$$\varphi_v = \frac{\Delta\omega}{K_v} \tag{6.52}$$

† For the over-damped case ($\zeta > 1$) the arguments of trigonometric functions are imaginary, and according to identities, $\cos jx = \cosh x$ and $\sin jx = j \sinh x$, hyperbolic functions are inserted; accordingly for ($\zeta > 1$) $\cos\sqrt{(1-\zeta^2)} = \cosh\sqrt{(\zeta^2 - 1)}$ and $[\sin\sqrt{(1-\zeta^2)}]/\sqrt{(1-\zeta^2)} = [\sinh\sqrt{(\zeta^2-1)}]/\sqrt{(\zeta^2-1)}$ this is also true for (6.53), (6.54), and (6.60).

In the second-order loop with a passive filter (Fig. 6.2(a) and (b)) $K_v = K$, but when an operational amplifier, with d.c. gain A, is used, $K_v = K \cdot A$, and consequently much smaller φ_v may be attained (values of $A = 10^2$ to 10^7 are typical).

The transient-time phase error $\varphi_{e2}(t)$ is

$$\varphi_{e2}(t) = \frac{\Delta\omega}{K}\left\{1 - \exp(-\zeta\omega_n t)\left[\cos\sqrt{1-\zeta^2}\,\omega_n t - \frac{(K/\omega_n) - \zeta}{\sqrt{1-\zeta^2}}\sin\sqrt{1-\zeta^2}\,\omega_n t\right]\right\} \tag{6.53}$$

which simplifies for the high-gain second order loop to

$$\varphi_{e2}(t) = \frac{\Delta\omega}{K_v} + \frac{\Delta\omega}{\omega_n}\left(\frac{1}{\sqrt{1-\zeta^2}}\sin\sqrt{1-\zeta^2}\,\omega_n t\right)\exp(-\zeta\omega_n t) \tag{6.54}$$

and in the particular case of critical damping to

$$\varphi_{e2}(t) = \frac{\Delta\omega}{K_v} + \frac{\Delta\omega}{\omega_n}(\omega_n t)\exp(-\omega_n t) \tag{6.55}$$

(iii) Frequency Ramp

If the input frequency is linearly changing with time at a rate $\Delta\dot{\omega}$ (radians per second^2) the Laplace transform of the input phase is $\Phi_i(s) = \Delta\dot{\omega}/s^3$. Applying the final value theorem to the derivative of the phase error we get

$$\lim_{t\to 0}\frac{d\varphi_{e3}(t)}{dt} = \lim_{s\to 0} s\,\frac{s^2(\Delta\dot{\omega}/s^3)}{s + K\,F(s)} = \frac{\Delta\dot{\omega}}{K_v} \tag{6.56}$$

Evidently when K_v is finite a steady-state frequency error results.

Assuming that the operational amplifier in Fig. 6.2(d) has an infinite gain then $F(s) = (1 + sT_2)/sT_1$ and a steady-state phase error, called 'acceleration error' or 'dynamic tracking error', is

$$\varphi_a = \lim_{t\to\infty}\varphi_{e3}(t) = \lim_{s\to 0}\frac{s^3(\Delta\dot{\omega}/s^3)}{s^2 + K[(1+sT_2)/T_1]} = \frac{\Delta\dot{\omega}}{K/T_1} = \frac{\Delta\dot{\omega}}{\omega_n^2}\ \text{(radians)} \tag{6.57}$$

For the sinusoidal detector, however, the correct expression should be

$$\sin\varphi_a = \frac{\Delta\dot{\omega}}{\omega_n^2} \tag{6.58}$$

and consequently the maximum rate of change

$$\Delta\dot{\omega} = \omega_n^2 \tag{6.59}$$

In the third-order loop with $F(s)$ of the type $G(s)/s^2$ where $G(0) \neq 0$ (two cascaded integrators) the acceleration error may be reduced to zero.

The time response of $\varphi_{e3}(t)$ in the high-gain second-order loop is

$$\varphi_{e3}(t) = \frac{\Delta\dot{\omega}t}{K_v} + \frac{\Delta\dot{\omega}}{\omega_n^2}\left[1 - \exp(-\zeta\omega_n t)\left(\cos\sqrt{1-\zeta^2}\,\omega_n t + \frac{\zeta}{\sqrt{1-\zeta^2}}\sin\sqrt{1-\zeta^2}\,\omega_n t\right)\right] \tag{6.60}$$

When $\zeta = 1$ (critical damping)

$$\varphi_{e3}(t) = \frac{\Delta\dot{\omega}t}{K_v} + \frac{\Delta\dot{\omega}}{\omega_n^2}\left[1 - (1 + \omega_n t)\exp(-\omega_n t)\right] \tag{6.61}$$

Normalized values $\varphi_{e1}(t)/\Delta\varphi_i$, $\varphi_{e2}(t)/(\Delta\omega/\omega_n)$, and $\varphi_{e3}(t)/(\Delta\dot{\omega}/\omega_n^2)$ (without steady-state parts) for the important case of the damping factor $\zeta = 0{\cdot}707$ are drawn in Fig. 6.7. Values for other ζ may be found in [6.2].

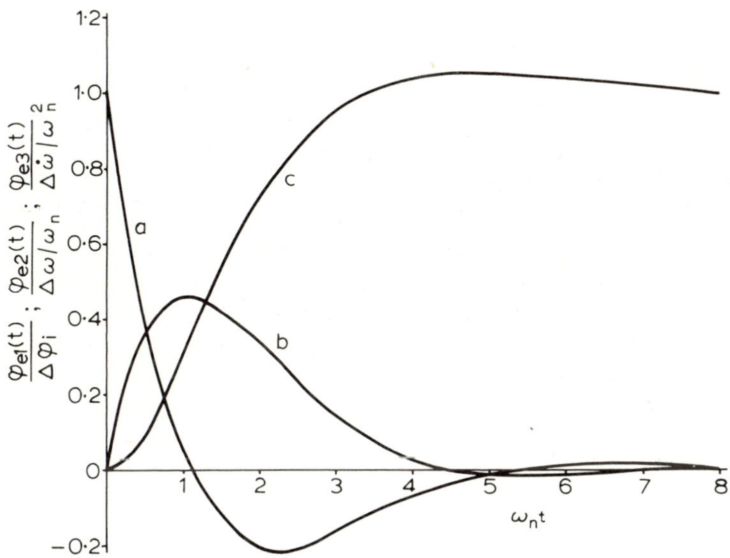

Fig. 6.7 Normalized transient phase errors (without steady-state parts) in high-gain second-order loop ($\zeta = 0{\cdot}707$):

(a) $\dfrac{\varphi_{e1}(t)}{\Delta\varphi_i}$; (b) $\dfrac{\varphi_{e2}(t)}{\Delta\omega/\omega_n}$; (c) $\dfrac{\varphi_{e3}(t)}{\Delta\dot{\omega}/\omega_n^2}$

6.3 LARGE-SIGNAL PROPERTIES

The linear approach used for solving phaselock problems in previous sections must be discarded when limiting conditions, such as lock-in or lock-out performances are examined.

(A) Hold-In Range

The largest frequency difference $\Delta\omega_H$, between the free-running frequency of the VCO and the reference frequency, for which the loop may be kept locked, is called the 'hold-in range'. $\Delta\omega_H$ is determined by the maximum output of the phase comparator for which the steady-state solution of (6.4), i.e. $d\varphi_e/dt = 0$, holds. When a detector with the sinusoidal characteristic is used

$$\Delta\omega = K_v \sin\varphi_e \tag{6.62}$$

and

$$\Delta\omega_H = \pm K_v \tag{6.63}$$

whereas for the sawtooth detector characteristic

$$\Delta\omega_H = \pm\pi K_v \tag{6.64}$$

According to (6.63) or (6.64) the hold-in range could be made arbitrarily large simply by increasing K_v. However, this is not entirely possible because in this case some other loop component would become saturated (e.g. the operational amplifier, or the tuning range of the VCO, etc.).

(B) Pull-Out Frequency

Evaluation of (6.53) will show that for a sudden frequency step the transient phase error may exceed by many times the final static value (6.52). Therefore the possibility of disturbing the locked state of the loop, even if the hold-in conditions are otherwise fulfilled, cannot be excluded. Gardner [6.2], having recourse to published results, concludes that the answer to this problem is not simple, that it depends on circumstances and is perhaps not yet fully established in the literature. Nevertheless, there is some frequency-step limit $\Delta\omega_{PO}$ called 'pull-out frequency' below which the loop does not skip cycles, and remains in lock. For the first-order loop

$$\Delta\omega_{PO} = \Delta\omega_H = \pm K_v \tag{6.65}$$

and the high-gain second-order loop

$$\Delta\omega_{PO} = 1{\cdot}8\ \omega_n(\zeta + 1) \tag{6.66}$$

(at least for the range $(0{\cdot}4 < \zeta < 1{\cdot}5)$.

Performance of moderate-gain second-order loops might be expected to be degraded with respect to (6.66). According to [6.4] the lock is lost permanently when $\Delta\omega > 0{\cdot}72\ K_v$.

If noise is present, as always happens, there is a finite probability, though in the extreme very small, of cycle-slipping in the phaselocked loop. Sanneman and Rowbotham investigated the behaviour of the high-gain second-order loop ($\zeta = 0{\cdot}707$) by computer simulation [6.5]. The result is a simple exponential law between the average time (T_{av}) to unlock and the $(\mathrm{SNR})_L$

$$T_{av} = \frac{2}{\omega_n} \exp[\pi(\mathrm{SNR})_L] \tag{6.67}$$

This relation was verified experimentally by Keblavi [6.6], see Fig. 6.8.

For $(SNR)_L = 1$, T_{av} is of the order of seconds and the loop performance is certainly unsatisfactory. However, the behaviour of the phaselock loop improves rapidly with $(SNR)_L$. If the extrapolation in Fig. 6.8 to larger values is justified, then for $(SNR)_L = 10$ to 20 the expected T_{av} will be of the order of 10^8 seconds or even more, which is compatible with reliability requirements of solid-state devices.

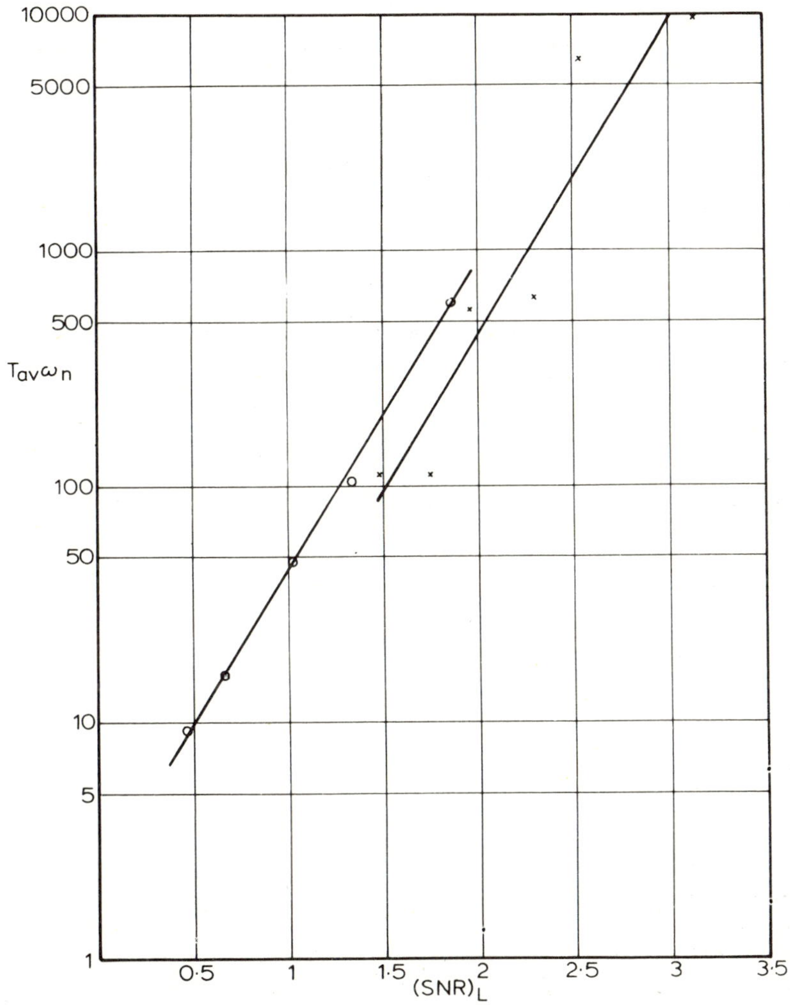

Fig. 6.8 Normalized mean time to unlock ($T_{av}\omega_n$) vs. $(SNR)_L$ in a high-gain second-order loop ($\zeta = 0{\cdot}707$): (○) theoretical results (Sanneman and Rowbotham); (X) experimental verification (Keblawi).

(C) Pull-In Range

The pull-in range $\Delta\omega_P$ of a phaselock loop is generally defined by the condition that for all input frequencies within this range the circuit will reach the synchronized state independently of the initial conditions of the system.

(i) First-Order Loop

In the first-order loop the pull-in range is equal to the hold-in range, i.e.

$$\Delta\omega_P = \Delta\omega_H \tag{6.68}$$

This is readily proved when dividing eq. (6.7) by K_v and plotting $\dot{\varphi}_e/K_v$ versus φ_e for different values of $\Delta\omega$ as in Fig. 6.9. It is clear that, if $\Delta\omega/K_v < 1, \dot{\varphi}_e/K_v$ goes

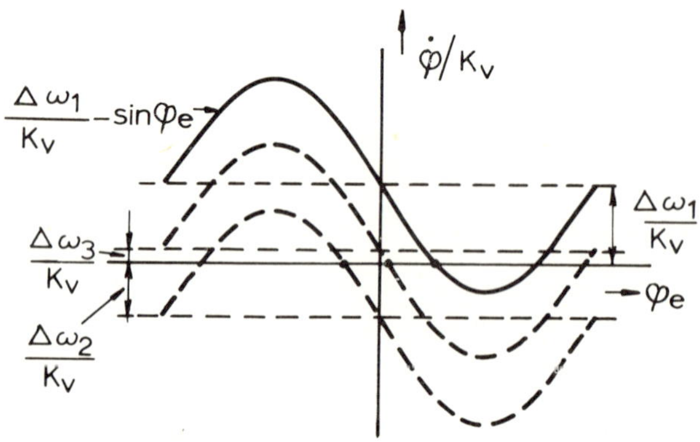

Fig. 6.9 Phase plane plot of first-order loop for different normalized detunings ($\Delta\omega/K_v$); stable nulls are indicated.

twice through zero in each interval of 2π. However, only one of the nulls is stable: for a small displacement $\Delta\varphi_e$ in the neighbourhood of $\dot{\varphi}_e = 0$ the application of the Taylor expansion on (6.7) gives

$$\frac{d\Delta\varphi_e}{dt} = -K_v\,\Delta\varphi_e \cos\varphi_{e0} \tag{6.69}$$

Integration of the above equation reveals

$$\Delta\varphi_e(t) \sim \exp(-K_v\, t \cos\varphi_{e0}) \tag{6.70}$$

It is clear that $\Delta\varphi_e(t)$ will increase with time if cos φ_{e0} is negative and converge to zero if cos φ_{e0} is positive, i.e. $-\pi/2 < \varphi_{e0} < \pi/2$; consequently only this null is stable.

Every cycle has a stable null, therefore locking takes place without cycle-skipping; at the same time φ_e does not change more than a half cycle and the conclusion is that the pull-in and hold-in ranges are the same.

The behaviour of the first-order loop outside the hold-in range provides a deeper insight into the pulling mechanism of higher-order loops.

Considering a sawtooth characteristic of the phase detector, the locked state cannot be reached when $|\Delta\omega/K_v| > \pi$, and the phase error must vary periodically from $-\pi$ to $+\pi$. The beating waveform computed from (6.118) is shown in Fig. 6.10. What should be noted is its unsymmetrical shape which is responsible for the presence of the d.c. component and consequently for the reduction of the original difference frequency. Evidently, the first-order loop will tend to pull towards the lock even when it cannot be reached. The same result will be found for detectors with other characteristics [6.4].

An alternative explanation for the pulling effect may be based on the frequency modulation of the VCO through the phase-detector beat-note; the first upper or lower sideband ($\omega_0 \pm \Delta\omega$) has the same frequency as the input signal; their combination produces at the phase-detector output a 'pulling' d.c. component.

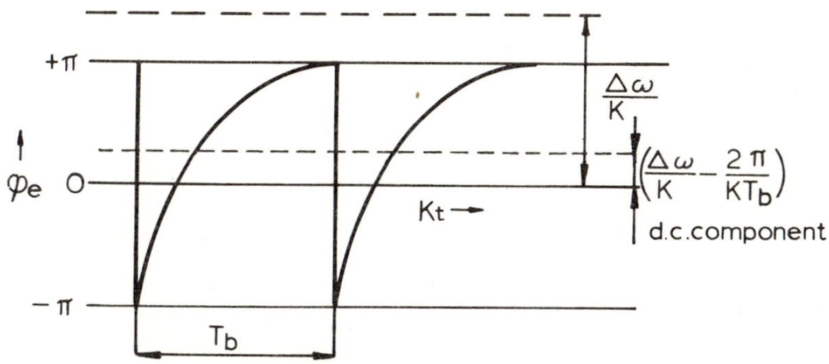

Fig. 6.10 Beating waveform of phase error φ_e in first-order loop with sawtooth comparator ($\Delta\omega/K = 1{\cdot}4\pi$).

(ii) Second-Order Loop

Understanding of the behaviour of the second-order loop will be assisted by realizing that at sufficiently high frequencies, the loop filter is merely a voltage divider: according to (6.16)

$$\lim_{s\to\infty} F(s) = T_2/T_1 \tag{6.71}$$

and the loop behaves as a first-order loop with the reduced gain $K(T_2/T_1)$. This applies also to the shape of the phase-detector beat-note when $\Delta\omega > K(T_2/T_1)$; however, due to the loop integrator, the effect of the d.c. component is cumulative until phaselock is eventually reached. The scope trace in Fig. 6.11 demonstrates the effect in the second-order loop with the sawtooth phase comparator.

Exact solution of the pull-in problem leads to complicated relations between $\Delta\omega_P$ and other circuit parameters; one approach based on the above ideas is described in [6.7]. To get workable results, suitable for engineering designs, approximate formulae with limited validity have been derived.

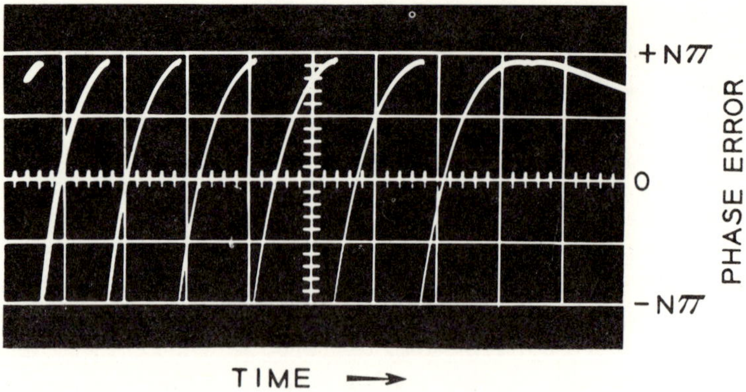

Fig. 6.11 Scope trace of phase error beat-note in second-order loop with a sawtooth phase comparator after the mistuning $\Delta\omega$ is brought just below the pull-in frequency. (Copyright, 1962, The American Telephone and Telegraph Co.; reprinted by permission.)

For second-order loops containing sawtooth comparators

$$\Delta\omega_P \approx \pi K \frac{2}{\sqrt{3}} \sqrt{T_2/T_1} \quad (T_2/T_1 < 0{\cdot}5;\ T_2 K \gg 1) \tag{6.72}$$

For the case of the sinusoidal phase detector the generally accepted results may be summarized as follows [6.8].

Simple RC filter:

$$\Delta\omega_P \approx 1{\cdot}25\sqrt{K/T_1} = 1{\cdot}25\,\omega_n \quad (KT_1 > 5) \tag{6.73}$$

Phase lag-lead filter

$$\Delta\omega_P \approx K\sqrt{2T_2/T_1}^{\dagger} \quad (KT_1 > 5) \tag{6.74}$$

or

$$\Delta\omega_P \approx \sqrt{2(2\zeta\omega_n K_v - \omega_n^2)} \tag{6.75}$$

which may be simplified for a high-gain loop to

$$\Delta\omega_P \approx 2\sqrt{\zeta\omega_n K_v} \tag{6.76}$$

(D) Lock-In Range

As long as the initial frequency difference is in the pull-in range, we can be sure that the circuit will lock; but in the extreme this may take a very long time. Therefore the frequency range $\Delta\omega_L$‡ in which the steady state is reached 'immediately' is of considerable practical importance.

---

† If the loop filter contains an operational amplifier, as in Fig. 6.2(c) or (d), then, in equations (6.72) to (6.74), K must be replaced by $\sqrt{KK_v}$.

‡ The seize frequency according to Byrne [6.3].

(i) First-Order Loop

When discussing the pull-in problem we have seen that in the first-order loop the locking takes place without cycle-skipping as soon as $|\Delta\omega| \leqslant K_v$, therefore

$$\Delta\omega_L = \Delta\omega_H = \pm K_v \tag{6.77}$$

(ii) Second-Order Loop

Because the second-order loop behaves at higher frequencies as a first-order loop, but with reduced gain, eq. (6.77) may be applied and

$$\Delta\omega_L = K\frac{T_2}{T_1} \approx 2\zeta\omega_n \tag{6.78}$$

(E) Settling Time

The settling time is the time required for the phase error to almost reach its steady-state value.

(i) Lock-in Time

In the lock-in range, where neither discontinuity nor cycle-skipping is expected, the decay of the damping term to the value, say 0·1, may supply the desired information.

For the first-order loop, with sawtooth or triangular phase detector, the use of (6.118) gives

$$T_{L,1} \approx 2{\cdot}3/K \tag{6.79}$$

and similarly (6.113) reveals for the sinusoidal comparator

$$T_{L,1} \approx \frac{2{\cdot}3}{K\sqrt{1-(\Delta\omega/K)^2}} \tag{6.80}$$

which for all values $|\Delta\omega/K| < 0{\cdot}9$ may be simplified to

$$T_{L,1} \lesssim 5/K \tag{6.81}$$

For the second-order loop (sawtooth or triangular PD) the time constant in (6.49) and (6.53) supplies the locking time

$$T_{L,2} \approx \frac{2{\cdot}3}{\zeta\omega_n} \approx \frac{4{\cdot}6}{K}\frac{T_1}{T_2} \tag{6.82}$$

Using the probability approach, Keblawi [6.9] calculated the average locking time

$$(T_{L,2})_{av} = 2{\cdot}5/\omega_n \approx 1{\cdot}3/B_L \quad (\zeta = 0{\cdot}707) \tag{6.83}$$

For the sinusoidal phase detector the upper bound is according to [6.8]

$$T_{L,2} \lesssim \frac{10}{K}\frac{T_1}{T_2} \tag{6.84}$$

(ii) Pull-in Time

The pull-in time T_p may be orders larger; Richman's result is [6.10]

$$T_P \approx \frac{(\Delta\omega)^2}{2\zeta\omega_n^3} \tag{6.85}$$

where $\Delta\omega$ is the initial frequency difference between the VCO and the reference signal. This relation gives the best results when $\Delta\omega$ is neither near $\Delta\omega_L$ nor near $\Delta\omega_P$. In the special case of a high-gain loop with $\zeta = 0{\cdot}707$

$$T_P \approx \frac{4{\cdot}15(\Delta f)^2}{B_L^3} \quad \text{(sec; Hz)} \tag{6.86}$$

For comparison the mean time to phaselock (6.83) and the pull-in times (6.86) are plotted in Fig. 6.12.

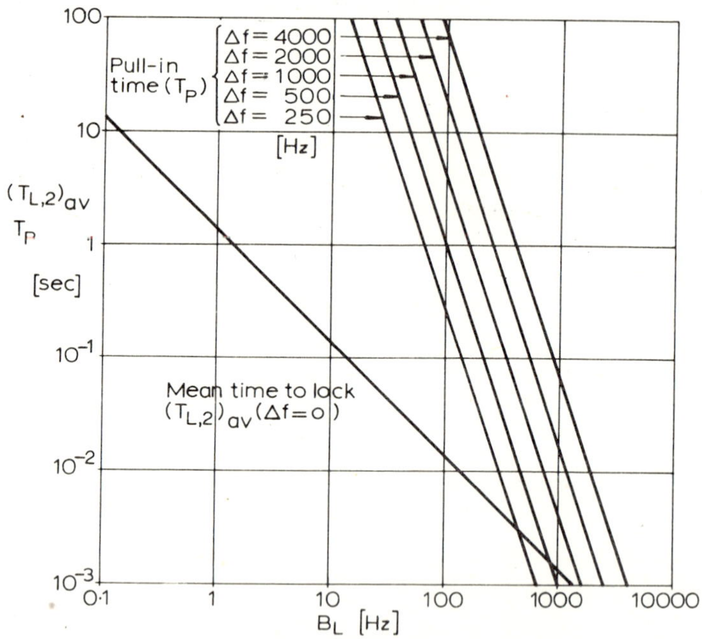

Fig. 6.12 Mean time to lock $(T_{L,2})_{av}$ and pull-in times (T_P) as functions of the loop-noise bandwidth B_L; high-gain second-order loop ($\zeta = 0{\cdot}707$). (From Keblawi [6.9].)

(F) Phaselock Acquisition

If the original mistuning between the VCO and the reference signal is smaller than $\Delta\omega_H$, and outside the lock-in range $\Delta\omega_L$, establishment of the phaselock state in the loop may require too long a time or even not take place at all (outside the pull-in range). There are several methods of dealing with this [6.8].

(i) Sweeping the VCO

Frequency sweep of the VCO may be obtained by an independent 'hunting' oscillator or simply by exploiting the charging and discharging phenomena of the loop capacitor caused by step voltages applied to the input of the filter. As soon as phaselock takes place the feedback action reduces the impedance presented by the loop to the sweeping generator. This feature is often exploited to switch the hunting oscillator on and off automatically [6.11].

In accordance with (6.59) the rate of change of the VCO frequency $\Delta\dot{\omega}$ cannot exceed ω_n^2 (considering the sinusoidal characteristic of the VCO). However, it is recommended that $\Delta\dot{\omega}$ should be kept below $\omega_n^2/2$; in this case the probability of acquiring phaselock is very high (≈ 1) irrespective of initial phase and frequency conditions.

It seems that a larger damping factor ζ will assist the acquisition of phaselock; Gardner [6.2] suggests that for the best performance ζ should be between 0·5 and 2.

(ii) Changing the Loop Bandwidth

A forced increase of ω_n during the acquisition period widens, according to (6.73) and (6.74), the pull-in range and drastically reduces the pull-in time (6.85). Examination of (6.18) reveals that either T_1 must be lowered or the gain K increased.

In the first case part of the filter resistance is short-circuited (advisable because there are no memory charges), for example by shunting R_1 with two diodes as in Fig. 6.13 and as suggested by Rey [6.4]. Note that at the beginning of the acquisition period the PD output voltage is large and one or the other diode is conducting. As soon as phaselock is established the voltage across R_1 diminishes and the diodes are put out of action.

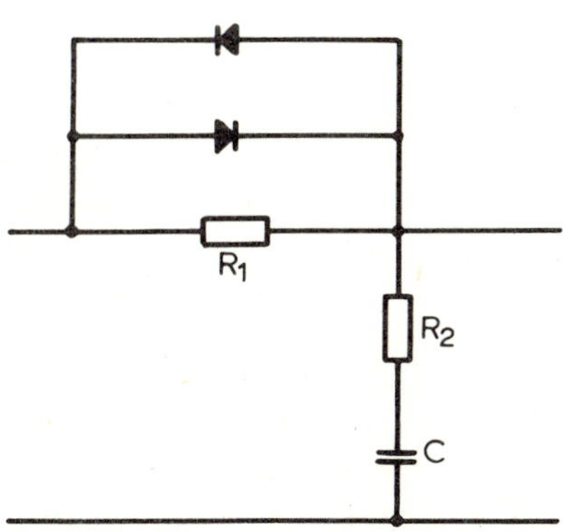

Fig. 6.13 A method for widening the pull-in range at the beginning of the phaselock acquisition period.

In the second case the coherent automatic gain control (AGC) may be exploited when no limiters are involved. As soon as the loop locks, AGC voltage appears and forces the input signal, the phase detector input, and consequently the loop gain K down.

(iii) Addition of Frequency Discrimination

Simultaneous application of automatic-frequency and phase-control circuits is another way of assisting acquisition of phaselock. A common arrangement often used in frequency synthesis applications is shown in Fig. 6.14. When the difference frequency between 'fed-back' and reference frequencies is large, the output of the frequency discriminator diminishes the original 'error' until the lock-in range is approached; this situation established, the phase detector takes-over and finishes the locking procedure. An actual digital frequency and phase comparator is described, for example, in [3.12].

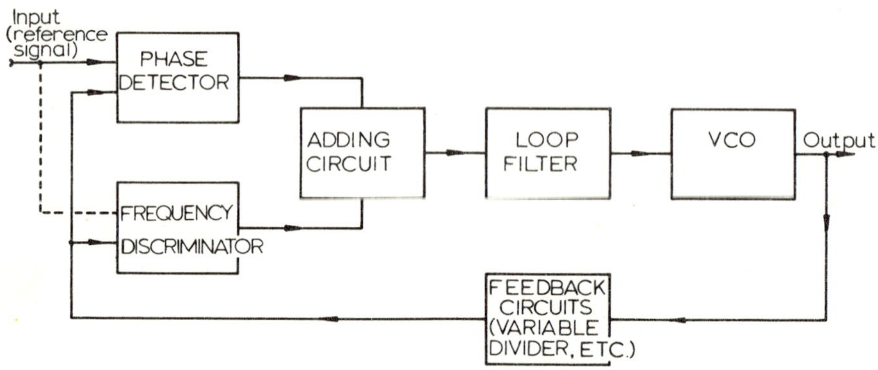

Fig. 6.14 Phaselock acquisition assisted by frequency discriminator.

(G) Phaselock Indication

The most convenient method of indication of the locked state in the loop is the use of another phase detector with one input voltage shifted in phase by 90° as in the block diagram in Fig. 6.15. Generally the phase error φ_e is small (6.8) and the output voltage of the loop phase detector is $v_d = K_d \sin \varphi_e \approx K_d \varphi_e$. In this case the auxiliary phase detector, which is driven in quadrature, has the output $v_d' = K_d' \cos \varphi_e \approx K_d'$, which is practically the maximum d.c. output; when the lock is lost a beat-note at the difference frequency appears with a rather small d.c. component – the d.c. detector output is almost zero.

Monitoring the output of the hunting oscillator, which is automatically switched on and off (see Section 6.3(F)), is another method of lock indication, where used.

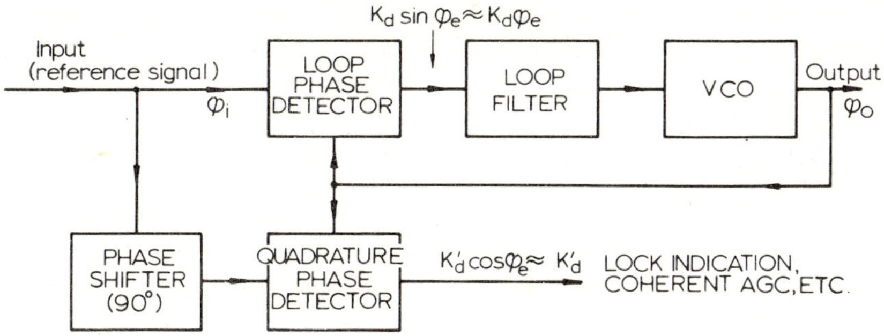

Fig. 6.15 Phaselock indication based on the 'quadrature phase detector'. (From Gardner [6.2].)

6.4 PHASE DETECTORS

Phase detection, in the same way as frequency mixing, may be based on exploitation of the nonlinear voltage versus current characteristics.

Examination of the general intermodulation product derived in Appendix 4.1 reveals that, if two signals with the same frequency, but differing in phase (by φ), are applied to a nonlinear element, the d.c. component contains terms which are proportional to $\cos n\varphi$ ($n = 1, 2, \ldots$). The task of the phase detector is to emphasize the fundamental phase-dependent part and suppress all others, particularly the phase-independent part in the d.c. output signal. This is generally achieved when using balanced circuits.

(A) Switching Phase Detectors

First we shall consider the bridge arrangement shown in Fig. 6.16. To simplify discussion and clarify the basic properties, ideal switches will be substituted for the diodes. The instantaneous voltages in circuits 1 and 2 are $e_1 + e_2$ and $e_1 - e_2$ respectively. Assume

$$e_1 = E_1 \cos \omega t \quad \text{and} \quad e_2 = E_2 \cos(\omega t + \varphi') \tag{6.87}$$

The amplitudes of $e_1 \pm e_2$ normalized to E_1 are, according to (9.5)

$$\frac{|e_1 \pm e_2|}{E_1} = \alpha_\pm = \sqrt{1 \pm 2\alpha_2 \cos \varphi' + \alpha_2^2} \tag{6.88}$$

For small values of $\alpha_2 = E_2/E_1$, i.e. $\alpha_2 \ll 1$, eq. (6.88) simplifies to

$$\alpha_\pm \approx 1 \pm \alpha_2 \cos \varphi' \tag{6.89}$$

If no biasing voltage is present the current flow angle is π and the d.c. output (v_d) across R_{z1} and R_{z2} is proportional to

$$\frac{1}{\pi} E_1(\alpha_+ - \alpha_-) = \frac{2}{\pi} E_2 \cos \varphi' \tag{6.90}$$

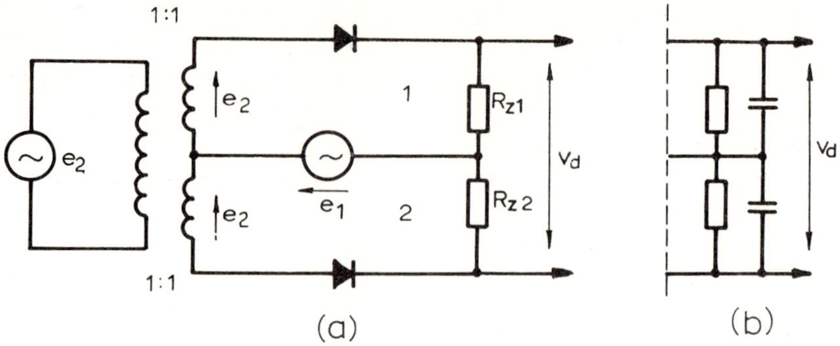

Fig. 6.16 Diode phase detectors: (a) mean value detection; (b) peak value detection.

The highest sensitivity to small phase changes requires that both signals e_1 and e_2 are in quadrature, i.e. $\varphi' = (\pi/2) - \varphi$, then

$$\frac{1}{\pi}E_1(\alpha_+ - \alpha_-) \approx \frac{2}{\pi}E_2 \sin\varphi \tag{6.91}$$

The other extreme is that both amplitudes E_1 and E_2 are nearly the same. In this case $\alpha_2 \approx 1$ and

$$\frac{1}{\pi}E_1(\alpha_+ - \alpha_-) \approx \frac{E_2}{\pi}\sqrt{2}[\sqrt{1+\sin\varphi} - \sqrt{1-\sin\varphi}] = \frac{E_2}{\pi}2\sqrt{2}\sin\frac{\varphi}{2} \tag{6.92}$$

Comparing this with (6.91) the linear range for small φ is larger and the characteristic of the phase discriminator tends towards a triangular waveform (see Fig. 6.17).

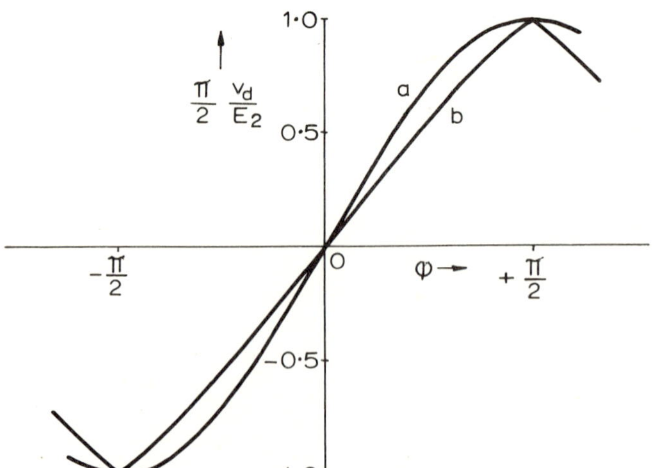

Fig. 6.17 Characteristics of diode phase detectors when: (a) $E_1 \gg E_2$ the normalized output is $(\pi/2)(v_d/E_2) \approx \sin\varphi$; (b) $E_1 \approx E_2$ the normalized output is $(\pi/2)(v_d/E_2) \approx \sqrt{2}\sin\varphi/2$.

When the diodes are working as peak detectors the output circuit is changed as in Fig. 6.16(b), and in the above results only the factor $1/\pi$ in eq. (6.90) to (6.92) disappears, i.e. for the same input voltages the output is π-times larger. An important advantage is a much lower level of undesired a.c. components (see the sampling phase detectors below).

Another phase detector of this type is a mere switch, e.g. the series or shunt modulator described in Section 4.3. The operation is evident from the waveforms in Fig. 6.18. In the interval $0 < \omega t < \pi$ the switch is closed and $e_0 \approx e_i$ whereas for $\pi < \omega t < 2\pi$, $e_0 = 0$. The d.c. component is

$$E_{\mathrm{d.c.}} \approx \frac{E_i}{2\pi} \int_0^{\pi} \sin(\omega t + \varphi')\, d\omega t = \frac{E_i}{\pi} \cos\varphi' = \frac{E_i}{\pi} \sin\varphi \qquad (6.93)$$

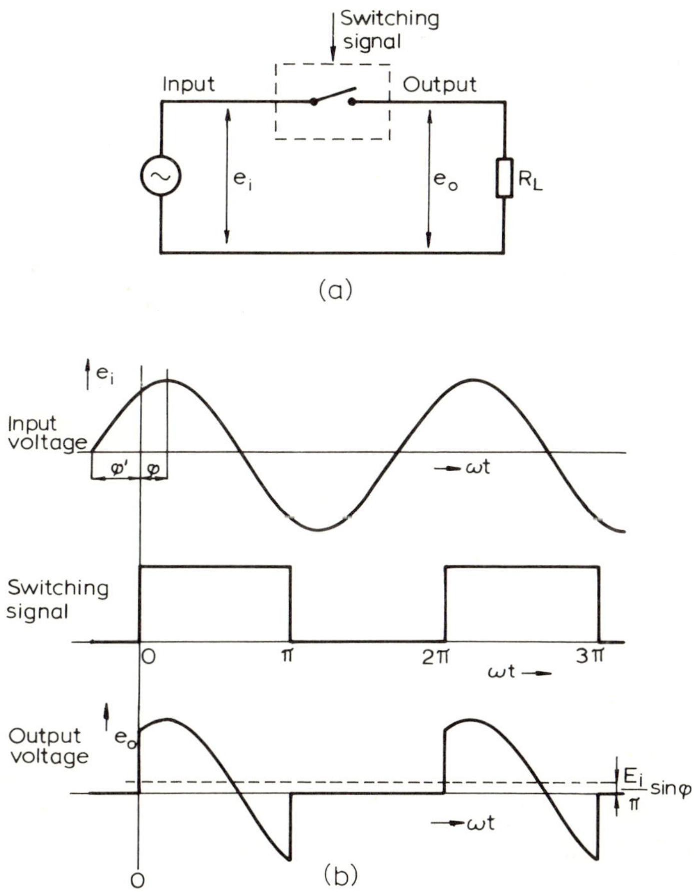

Fig. 6.18 Switch as phase detector: (a) principle; (b) typical waveforms.

We have found that both types of phase detecting circuits described above have sinusoidal characteristics. However, this is not the inherent property of the network arrangement but merely the consequence of the sinusoidal inputs. It may be easily shown, using (6.93), that transition to the rectangular waveforms will result in a triangular relationship between the phase and the output voltage. The advantage is a significant improvement in the linearity which extends through the whole interval $(-\pi/2 < \varphi < \pi/2)$.

The sawtooth phase-detector characteristic pushes the linear range even beyond $\pm\pi/2$ to $(-\pi < \varphi < \pi)$. Practical realization can be achieved by several different circuits; the simple flip-flop, described in Section 3.5(B), may serve as an example. Both input signals must be available in the form of pulse trains which are applied to the 'Set' and 'Reset' terminals respectively (see Fig. 3.7). The 'space-mark' ratio at either output, Q or $\bar{Q}$, is proportional to the phase difference between the input signals as shown in Fig. 6.19. When $\varphi \to n\pi$ ($n = 1, 3, \ldots$) the pulses will eventually pass each other; there will be a sudden discontinuity and the level of the d.c. component will change quickly from one extreme to the other.

In frequency synthesis applications of phaselock systems the phase jitter of the VCO caused by the a.c. components in the phase detector output may be a serious problem. Full-wave operation (Fig. 6.16) is a partial remedy, but the improvements are comparatively small. More effective reduction of the residual phase-detector jitter can be achieved by the sample-and-hold technique.

(B) Sampling Phase Detectors

The principle is shown in Fig. 6.20. In each period the storage (hold) capacitor is connected for a very short time to the source of the reference signal and retains the gained information about the instantaneous reference voltage until the next sampling takes place

$$v_d \approx E_i \sin(\omega_i kT_i + \varphi) = E_i \sin\varphi \quad (k = 0, 1, 2, \ldots) \tag{6.94}$$

The residual phase jitter is due to the finite sampling time ΔT on the one hand and to the finite reverse impedance of the holding circuit on the other hand. In actual circuits the leakage of the sampling signal must also be considered. An improvement in reference frequency rejection of 110 dB relative to a conventional flip-flop phase detector was reported [3.11] with the circuit reproduced in Fig. 6.21. The first switch S_1 (a MOSFET – metal-oxide-surface field-effect-transistor) is normally closed and C_2 forms part of the derivative circuit $R(C_1 + C_2)$ which changes the input rectangular waveform into an approximately triangular one. At the sample time, S_1 is opened, causing C_2 to hold its value. This provides a constant level to S_2, a second MOSFET, which samples the voltage to the holding capacitor C_3.

Another type of the sample-and-hold phase detector has been suggested by Ulicki [6.12] for extracting the information from the pulse-width modulated

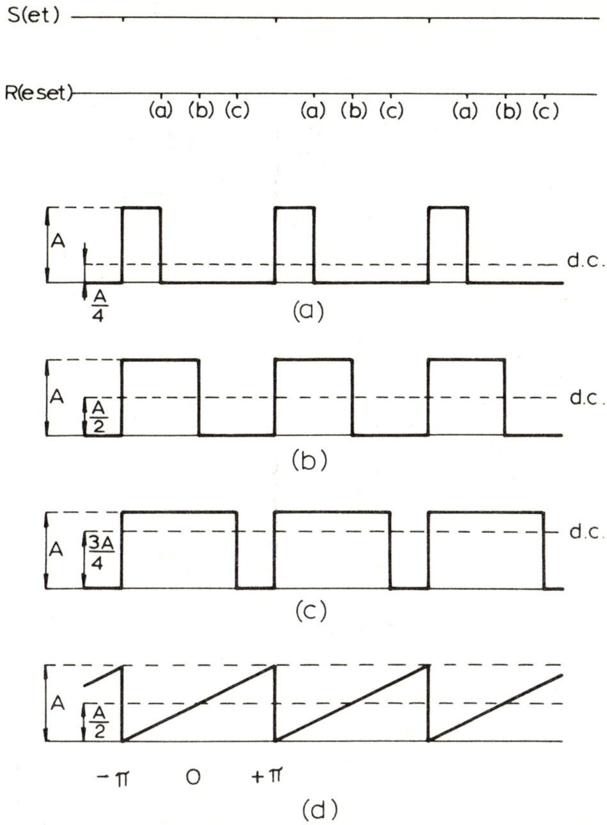

Fig. 6.19 Waveforms in the sawtooth (flip-flop) phase detector: (a) for the phase difference $-\pi/2$; (b) for the phase difference 0; (c) for the phase difference $+\pi/2$; (d) the smoothed d.c. components.

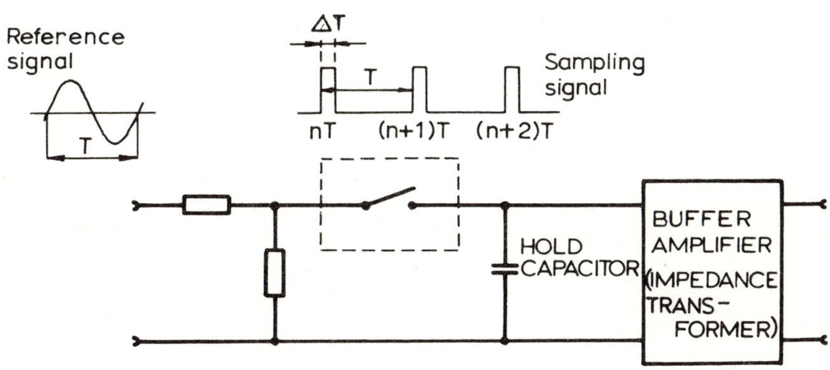

Fig. 6.20 Sampling phase detector.

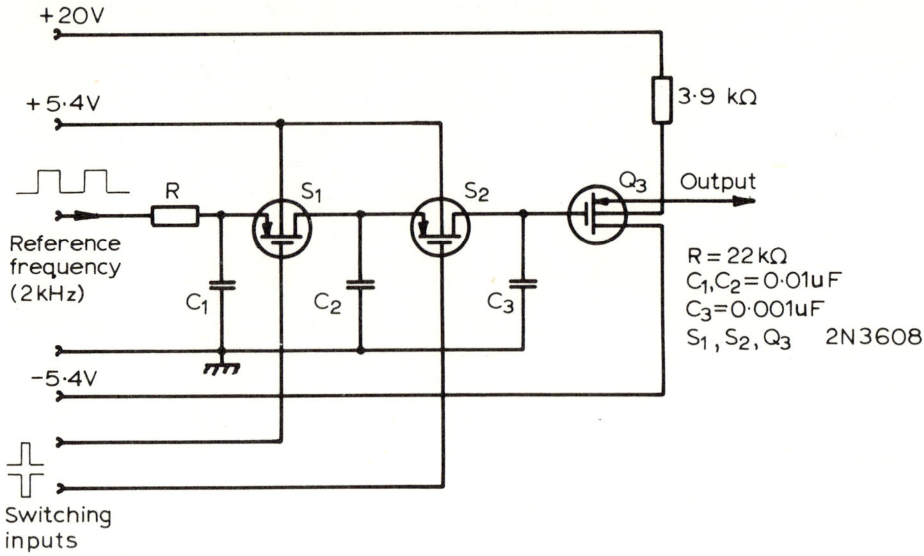

Fig. 6.21 A double sampling phase detector. (Adopted from *Electronics*, May 2, 1966).

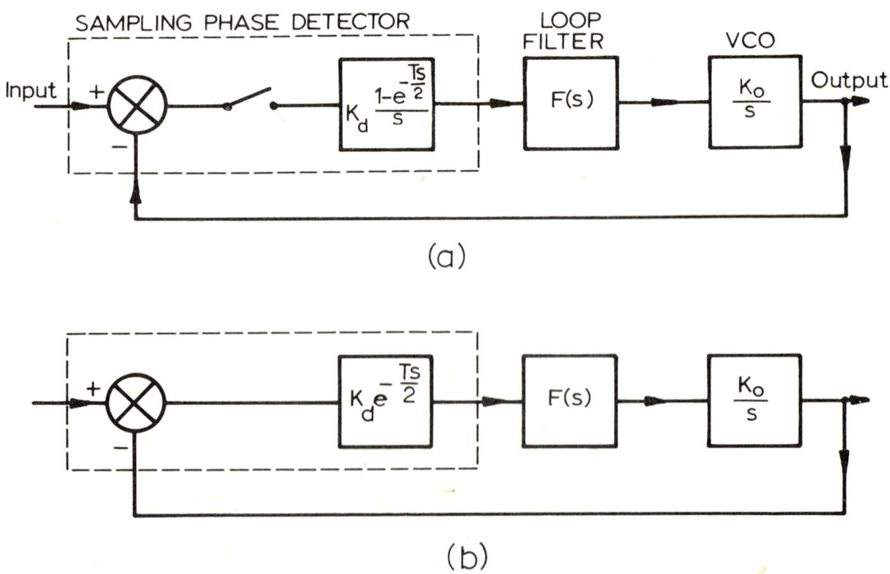

Fig. 6.22 (a) Phaselock loop with sampling phase detector; (b) approximating continuous system.

output of the set-reset flip-flop (mentioned above). A linear integrating circuit is operated over the pulse duration, and in this way a d.c. voltage proportional to the pulse width is obtained; when this d.c. voltage is sampled and stored, at the end of the period, in a memory (hold) capacitor, the integrator may be reset and the cycle repeated. (See also [3.12] for an actual circuit of this type.)

A refinement of the sample-and-hold technique is the sample-hold-delay-sample-hold configuration. The phase detector of this type contains two sample-and-hold circuits (shown schematically in Fig. 6.20) in cascade; but the activation of the second circuit, in contradistinction to the system shown in Fig. 6.21, is delayed until the transient on the first hold capacitor ends. If both hold capacitors are properly chosen this phase detector behaves as a simple phase detector followed by an RC filter. An actual phase detector of this type (IC version) is described in [6.17].

Finally we must consider changes in the transfer function of the phaselock loop caused by the transition to the sampling technique.

The sampling phase detector is a 'zero-order data hold' circuit, the transfer function of which is indicated in Fig. 6.22(a). In the approximating continuous system (Fig. 6.22(b)) the delay factor exp $(-Ts/2)$ must be introduced, T being the duration of the sampling period. For greater detail relevant textbooks should be consulted, e.g. [6.13].

6.5 VOLTAGE-CONTROLLED OSCILLATORS

Choice of type of VCO depends predominantly on the desired properties of the device incorporating the phaselock loop. A common condition, encountered almost always, is to keep the noise and spurious levels as low as possible. Other requirements, large tuning range, and large and eventually frequency-independent gain factor K_0, are in direct opposition to the former condition.

Considering the noise properties of the output signal generated by a phaselock system one is obliged to take into account the phase fluctuations of the reference as well as of the VCO itself. In this case the basic block diagram in Fig. 6.1(a) must be modified to that shown in Fig. 6.23. Assuming that the system is linear, the loop equation is

$$[\Phi_i(s) - \Phi_0(s)]\, K\, F(s)\frac{1}{s} + \Phi_{osc}(s) = \Phi_0(s) \tag{6.95}$$

Rearrangement gives the output phase

$$\Phi_0(s) = \Phi_i(s)\frac{K\, F(s)/s}{1 + K\, F(s)/s} + \Phi_{osc}(s)\frac{1}{1 + K\, F(s)/s} \tag{6.96}$$

or, by substituting (6.12) and (6.13)

$$\Phi_0(s) = \Phi_i(s)\, H(s) + \Phi_{osc}(s)\,[1 - H(s)] \tag{6.97}$$

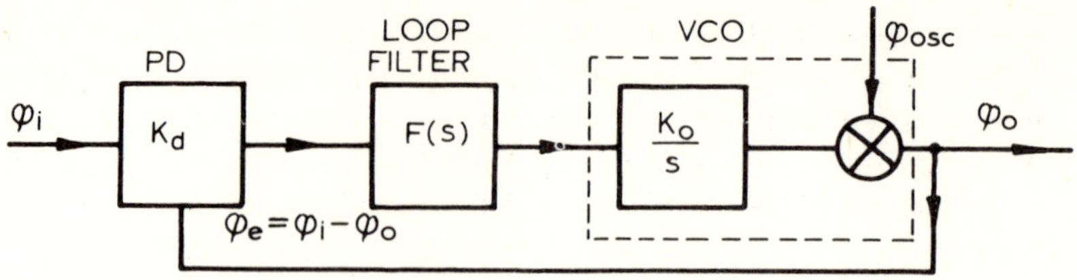

Fig. 6.23 Modification of the phaselock loop diagram taking account of the phase noise of the voltage controlled oscillator (VCO).

If the input phase jitter and the oscillator phase jitter are uncorrelated the spectral density (one-sided) of the output phase noise is (see Section 9.2)

$$S_\varphi'(\omega)_0 = S_\varphi'(\omega)_i |H(j\omega)|^2 + S_\varphi'(\omega)_{osc} |1 - H(j\omega)|^2 \tag{6.98}$$

Making an approximate evaluation of the loop transfer function $H(j\omega)$ we get (see Fig. 6.4 and 6.5)

$$\left.\begin{aligned} H(j\omega) &\approx 1 \quad \omega \ll \omega_n \\ |H(j\omega)| &\ll 1 \quad \omega \gg \omega_n \end{aligned}\right\} \tag{6.99}$$

and consequently for low-offset frequencies

$$S_\varphi'(\omega)_0 \approx S_\varphi'(\omega)_i \quad \omega < \omega_n \tag{6.100}$$

whereas for high-offset frequencies

$$S_\varphi'(\omega)_0 \approx S_\varphi'(\omega)_{osc} \quad \omega > \omega_n \tag{6.101}$$

The final inference from the above discussion is that the phase noise of the VCO is reduced by the phaselock action only in a relatively small range but otherwise retains its importance. This the designer must have in mind when low ouput noise is one of the chief requirements, as in frequency synthesis applications.

(A) Crystal Oscillators

When a simple frequency translation from one range to another is required, phaselock loops with voltage-controlled crystal oscillators (VCXO) are often used. Atomic-frequency standards or single-frequency synthesizers for transmitters are examples.

Oscillator circuits which are commonly used are transistor versions of the well-known Pierce configuration. The principle is shown in Fig. 6.24. The crystal oscillates near its series resonance. The voltage tuning is accomplished by varactors which form part of the total circuit capacity. Considering the equivalent circuit in Fig. 6.25 and assuming C_1 large compared with C_L the series resonance of the combination is approximately

$$\omega = \frac{1}{\sqrt{L_e[C_e(C_0 + C_L)/(C_e + C_0 + C_L)]}} = \omega_r \sqrt{1 + \frac{C_e}{C_0 + C_L}} \tag{6.102}$$

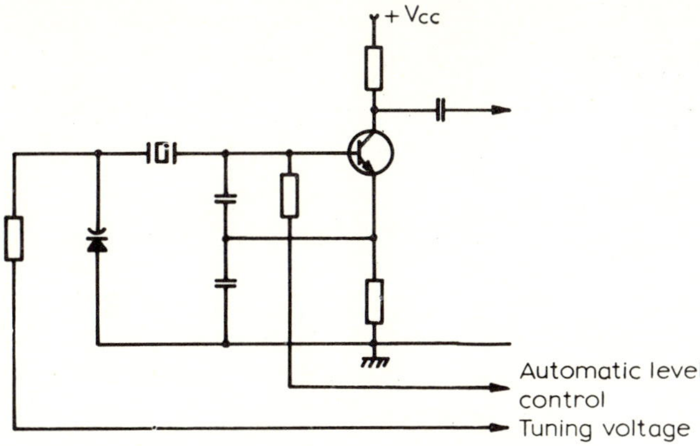

Fig. 6.24 Varactor tuned crystal oscillator.

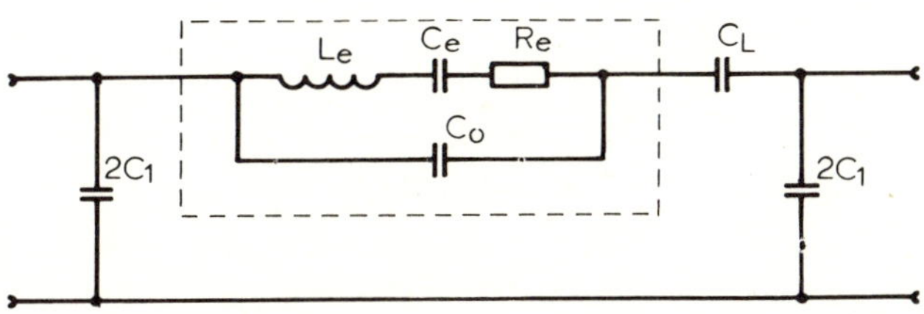

Fig. 6.25 Crystal equivalent circuit; C_0 is the static capacity between electrodes, C_L the 'load capacity', and $2C_1$ are the coupling capacitors. (From Marker [6.14].)

By computing the relative incremental change we get, after neglecting higher-order terms,

$$\frac{d\omega}{\omega_r} \approx -\frac{1}{2}\frac{C_e}{C_0 + C_L}\frac{dC_L}{C_0 + C_L} \tag{6.103}$$

If the load capacity C_L is voltage dependent, i.e. $C_L \approx C_v$, the incremental VCO gain $K_0(v)$ is

$$K_0(v) \approx \frac{d\omega}{dC_v}\frac{dC_v}{dv} \tag{6.104}$$

Substituting from (6.103) and performing the derivation of dC_v/dv, having recourse to (2.9), leads to the result

$$K_0(v) \approx \omega_r \frac{\gamma}{2}\frac{C_e}{C_0 + C_v}\frac{C_v}{C_0 + C_v}\frac{1}{v + \Phi} \tag{6.105}$$

No varactors are needed in cases where crystals such as GT or ring-type are used, the resonance frequencies of which can be changed by d.c. voltage applied to their electrodes. The advantage of this arrangement is excellent isolation between the oscillator and the automatic phase control (APC) circuit. The principle is shown in Fig. 6.26. The relative gain K_0/ω_r for a GT resonator (in the 100 kHz range) is of the order 10^{-8}/volt and practically voltage independent.

Actual design of crystal oscillators is beyond the scope of the present book; some details with further references are summarized in [6.14] and in [6.2, Section 5.3]. But above all, recommendations of manufacturers should be followed as closely as possible. Crystal oscillators employing IC concepts are described in [6.15].

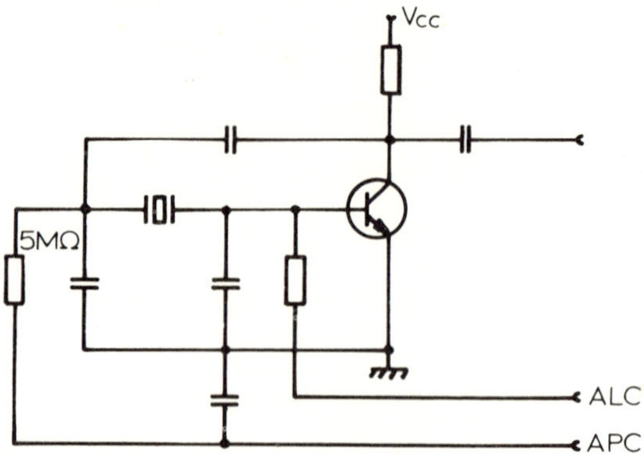

Fig. 6.26 Crystal oscillator tuned by the voltage applied to crystal electrodes.

(B) LC Oscillators

If a wider range must be covered the varactor voltage-tuned LC oscillator is the common solution.

The incremental gain $K_0(v)$ is given, as in the previous case, by (6.104). For the highly general resonant circuit configuration shown in Fig. 6.27 we shall find (assuming a voltage-independent doping profile constant γ)

$$K_0(v) \approx \omega \frac{\gamma}{2} \frac{C_1}{C_1 + C_2 + C_v} \frac{C_v}{C_2 + C_v} \frac{1}{v + \Phi} \tag{6.106}$$

The maximum tuning range (f_H/f_L) is reduced by the inevitable resonant-circuit stray capacities and the varactor-case capacity, to far below the theoretical value, i.e.

$$\left(\frac{f_H}{f_L}\right)_{\max} < \sqrt{\frac{C_{v,\,\max}}{C_{v,\,\min}}} \approx \left(\frac{V_B}{v_{\min}}\right)^{\gamma/2} \tag{6.107}$$

The common practice is to use a bank of LC oscillators whenever a larger frequency band must be covered. This approach is also good from the stability point of view. At lower frequencies Clapp or Colpitts oscillator circuits are used;

at higher frequencies the inductive feedback is preferred because the amount of the fixed capacitance may be smaller.†

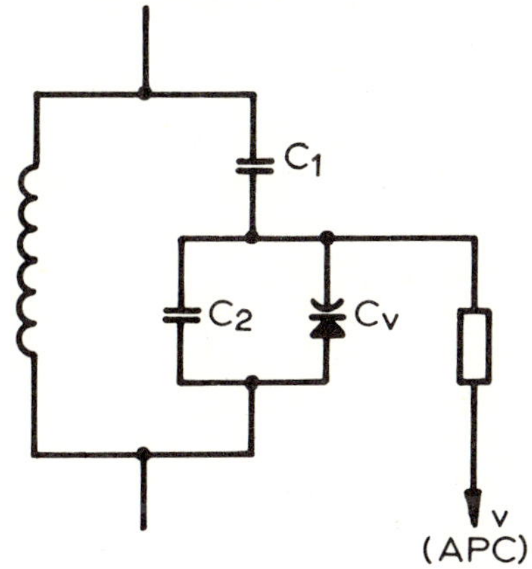

Fig. 6.27 LC resonant circuit tuned by the voltage-dependent capacitor C_v; C_1 is the separating capacitor; stray and varactor case capacities are lumped in C_2.

(C) RC Oscillators

The frequency is generally inversely proportional to a time constant RC. Therefore the same ratio of variable capacity results in a larger frequency range when compared with LC oscillators. In addition voltage-dependent resistors (diodes etc.) may be used. However, the phase stability and noise characteristics of RC oscillators (astable multivibrators, phase-shift oscillators, Wien bridge oscillators, etc.) are insufficient to compete with requirements usually encountered in frequency synthesis applications. An attractive advantage is the possible compatibility with thin-film and IC technology [6.16].

6.6 CONCLUSIONS

A phaselock loop behaves as a narrow filter, with extremely small phase slope in the pass-band, which may be realized at any frequency with RC elements only. When exploited in frequency synthesis, a reduction in size, weight, and power-consumption may be expected. The importance of the phase-locking technique is further enhanced by the fact that it is compatible with IC technology.

Performance properties of phaselocked loops which are generally investigated by designers and looked for by users are:

(i) loop stability,
(ii) level of spurious signals,
(iii) tuning- or switching-time

† Some problems of varactor-tuned LC oscillators are discussed in more detail in Section 7.5(D).

It is common practice to employ second-order loops, which are unconditionally stable.

In frequency synthesis applications, leakage of signals with the reference frequency and its harmonics is a serious problem which simple phase detectors and second-order loops cannot cope with. For this reason sample-and-hold comparators are generally used, with additional RC sections attached to the basic lag-lead network. In these cases the stability of the modified loop must be tested; the usual tool is the Bode plot (see Appendix 6.2).

The tuning time, which is composed of the pull-in and lock-in times, should be of the order of milliseconds, to save precious communications time, etc. This requirement can be met by sufficiently large loop-noise-bandwidth (B_L) but this is in opposition to the previous condition. If no satisfactory compromise can be reached, means for speeding-up acquisition of phaselock (sweeping the VCO, addition of a frequency discriminator, etc.) may solve the problem.

APPENDIX 6.1 SOLUTION OF FIRST-ORDER LOOP EQUATIONS

(i) Sinusoidal Phase Detector

$$\frac{d\varphi_e}{dt} = \Delta\omega - K\sin\varphi_e \tag{6.108}$$

Separation of variables leads to

$$\frac{d\varphi_e}{\Delta\omega - K\sin\varphi_e} = dt \tag{6.109}$$

The integral of the left-hand side is readily found by referring to published tables, e.g. [1.15, p. 804].

$$t - t_0 = -\frac{2}{\sqrt{(\Delta\omega)^2 - K^2}} \arctan\left[\sqrt{\frac{\Delta\omega + K}{\Delta\omega - K}} \tan\left(\frac{\pi}{4} - \frac{\varphi_e}{2}\right)\right] \tag{6.110}$$

where t_0 is an arbitrary constant; in the hold-in range $K > \Delta\omega$, and both $\sqrt{(\Delta\omega)^2 - K^2}$ and $\sqrt{\Delta\omega - K}$ are imaginary; application of the relation

$$\tan(-jx) = -j\tanh(x) \tag{6.111}$$

changes (6.110) into

$$t - t_0 = \frac{2}{\sqrt{K^2 - (\Delta\omega)^2}} \operatorname{arctanh}\left[\sqrt{\frac{K + \Delta\omega}{K - \Delta\omega}} \tan\left(\frac{\pi}{4} - \frac{\varphi_e}{2}\right)\right]$$

$$= \frac{1}{\sqrt{K^2 - (\Delta\omega)^2}} \operatorname{ig} \frac{1 + \sqrt{(K + \Delta\omega)/(K - \Delta\omega)}\tan(\pi/4 - \varphi_e/2)}{1 - \sqrt{(K + \Delta\omega)/(K - \Delta\omega)}\tan(\pi/4 - \varphi_e/2)} \tag{6.112}$$

Evaluation of tan $(\pi/4 - \varphi_e/2)$ gives finally

$$\varphi_e = 2\arctan\left[\sqrt{\frac{K-\Delta\omega}{K+\Delta\omega}}\;\frac{1-\exp[-\sqrt{K^2-(\Delta\omega)^2}\,(t-t_0)]}{1+\exp[-\sqrt{K^2-(\Delta\omega)^2}\,(t-t_0)]}\right]+\frac{\pi}{2} \qquad (6.113)$$

For $t \to \infty$ the steady state is reached and

$$\varphi_e = \arcsin\frac{\Delta\omega}{K} \qquad (6.114)$$

(ii) Linear Phase Detector

$$\frac{d\varphi_e}{dt} = \Delta\omega - K\varphi_e \qquad (6.115)$$

$$\frac{d\varphi_e}{\Delta\omega - K\varphi_e} = dt \qquad (6.116)$$

$$K\varphi_e - \Delta\omega = C\exp(-Kt) \qquad (6.117)$$

The integration constant C depends on initial conditions: $t = 0, \varphi_e = \varphi_{e0}$ (where φ_{e0} is an arbitrary phase constant) thus

$$\varphi_e = \exp(-Kt)\left(\varphi_{e0} - \frac{\Delta\omega}{K}\right) + \frac{\Delta\omega}{K} \qquad (6.118)$$

APPENDIX 6.2 STABILITY OF PHASELOCK LOOPS

The phaselock loop, as any feedback system, can oscillate if the denominator in (6.12) equals zero, i.e.

$$1 + \frac{1}{s}K\,F(s) = 0 \qquad (6.119)$$

Writing the open loop gain in the form $|\,(1/j\omega)\,K\,F(j\omega)\,|\exp(j\psi)$ the above equation is fulfilled for

$$\left|\frac{1}{j\omega}K\,F(j\omega)\right| = 1 \quad \text{and} \quad \psi = (2n+1)\pi; \quad (n = 0, \pm 1, \pm 2, \ldots) \qquad (6.120)$$

When plotting the locus of roots of (6.119) for different gains K in the s-plane a considerable degree of insight into the behaviour of the respective phaselock loop can be obtained. Particularly, the locus entering the right half-plane signifies the instability of the system for that condition.† This never happens with the second-order loop with simple lag or lag-lead filter. If, however, additional filtering is used, or time delay is present (sampling phase detectors, etc.) and the

† More about the root-locus method may be found in [6.1] or [6.2], or any other textbook dealing with feedback control systems.

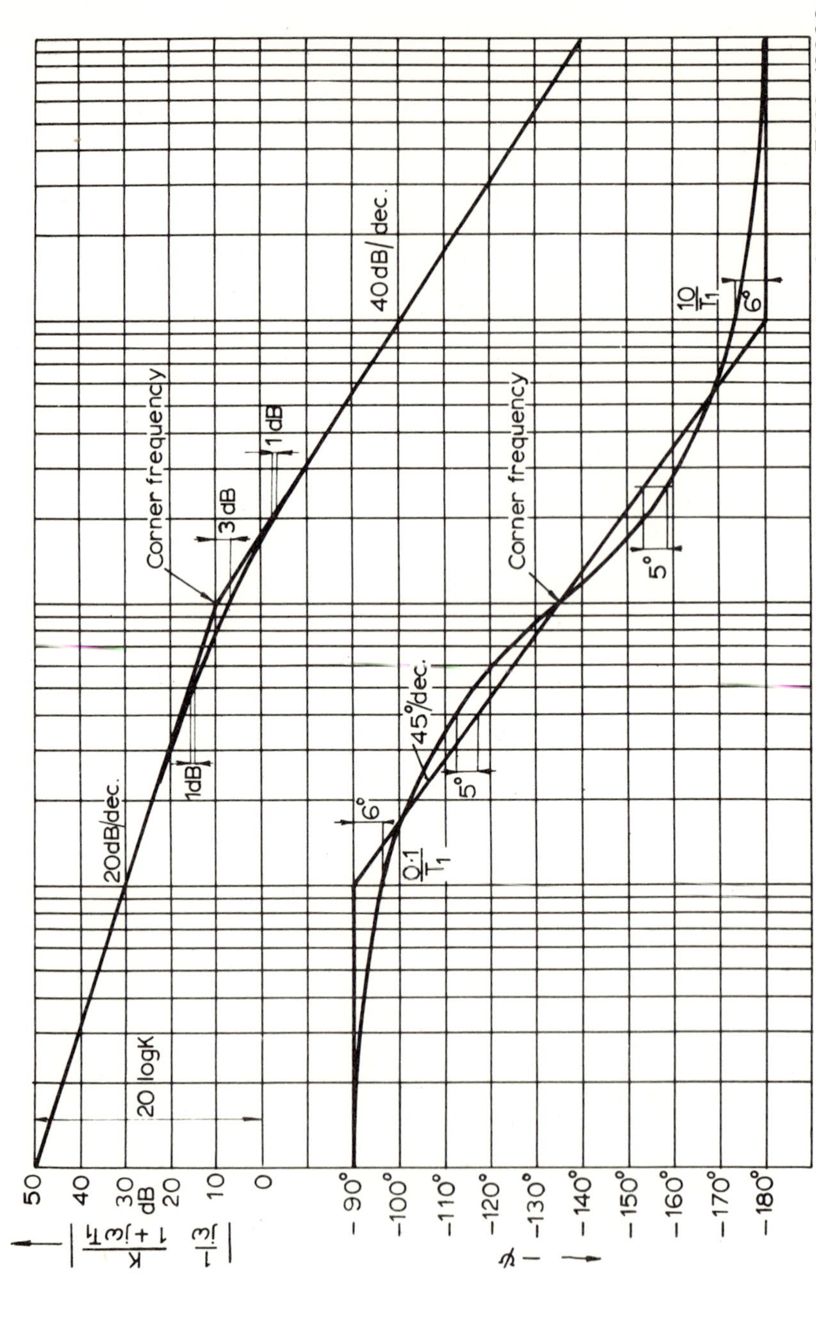

Fig. 6.28 Asymptotic approximation of the Bode plot for the phaselock loop with simple RC filter; the maximum errors are indicated ('corner frequency' = $1/T_1 = 1/RC$).

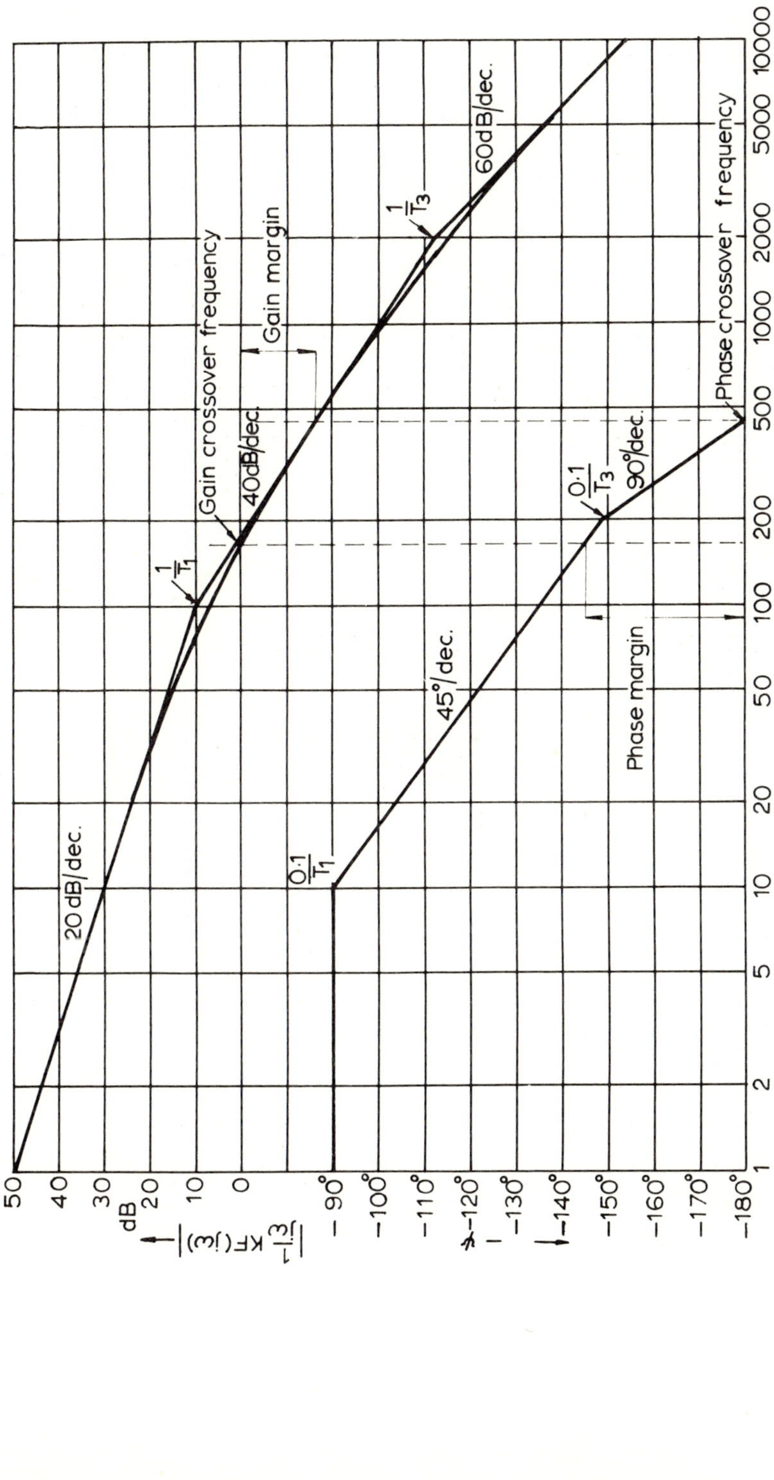

Fig. 6.29 Bode plot (asymptotic approximation) of a stable loop.

Fig. 6.30 Bode plot (asymptotic approximation) of an unstable loop.

gain varies (variable dividers in the feedback path, etc.), stability conditions must be checked. Bode plots of the open-loop gain, i.e. $|(1/j\omega) K F(j\omega)| \exp(j\psi)$ versus log f or log ω is a simple and generally used method. The criterion for unconditional stability is that the gain must fall below unity ($\log |(1/j\omega) K \times F(j\omega)| = 0$) before the phase shift reaches $-180°$.

The simple asymptotic solution will generally supply sufficient information to decide about loop stability. In Fig. 6.28 the construction of the respective asymptotes is shown and the errors are indicated when the filter is formed by a simple RC section, i.e. $F(j\omega) = 1/(1 + j\omega T_1)$.† The loop is unconditionally stable because the $-180°$ line is an asymptote of the phase response.

However, the situation may change radically when adding another independent RC section, i.e. $F(j\omega) = [1/(1 + j\omega T_1)] \times [1/(1 + j\omega T_3)]$. In Fig. 6.29 is shown the Bode plot of a stable loop exhibiting a 'phase margin' of 35°(±)11° and a 'gain margin' of about 16 dB. On the other hand, the 'phase cross-over frequency' being smaller than the 'gain cross-over frequency', as illustrated in Fig. 6.30, indicates instability, and the feedback system must be redesigned.

REFERENCES

[6.1] SAVANT, C. J., Jr. *Basic Feedback Control System Design,* New York, Toronto, London: McGraw-Hill, 1958.

[6.2] GARDNER, F. M. *Phaselock Techniques,* New York, London, Sydney: Wiley 1966.

[6.3] BYRNE, C. J. Properties and design of the phase-controlled oscillator with a sawtooth comparator. *Bell System Technical Journal,* **41,** 559-602, March 1962.

[6.4] REY, T. J. Automatic phase control: theory and design, *Proc. I.R.E.,* **48,** 1760-71, 1960; corrections in *Proc. I.R.E.,* p. 590, March 1961.

[6.5] SANNEMAN, R. W. and ROWBOTHAM, J. R. Unlock characteristics of the optimum Type II phase locked loop, *I.E.E.E. Trans. on Aerospace and Navigational Electronics,* **ANE-11,** 15-24, 1964.

[6.6] KEBLAWI, F. S. Unlock behavior of the second order phase-locked loop with and without interfering carriers, *RCA Review,* **28,** 277-96, June 1967.

[6.7] McALEER, H. T. A new look at the phase-locked oscillator, *Proc. I.R.E.,* **47,** 1137-43, 1959; correction in *Proc. I.R.E.,* p. 1771, Oct. 1960.

[6.8(a)] ROBSON, P. G. The pull-in range of a phase-locked loop. *Conference on Frequency Generation and Control for Radio Systems: London, Conference Publication* No. 31, pp. 139-43, May 1967.

[6.8(b)] MEER, S. A. A generalized analysis for the acquisition time and pull-in range of phase-locked loops. *Op. cit.* pp. 144-8.

[6.9] KEBLAWI, F. S. Probability distribution of time to phase lock for a second-order phase-locked loop. *RCA Review,* **29,** 106-21, March 1968.

[6.10] RICHMAN, D. Color-carrier reference phase synchronization accuracy in NTSC color television. *Proc. I.R.E.,* **42,** 106-33, 1954.

[6.11] BRUNK, M. and SCHRÖDER, J. Automatische Fangbereicherweiterung einer Frequenzsynchronisationsschaltung mit Hilfe eines Suchoszillators. *Frequenz,* **16,** 511-5, December 1962.

[6.12] ULICKI, E. M. Cubic inch frequency synthesizers, *Proc. of the 19th Annual Frequency Control Symposium,* pp. 580-616, Fort Monmouth, N.Y., 1965.

† For more about the Bode plot construction see, for example, [6.1, chapter 5].

[6.13] RAGAZZINI, J. R. and FRANKLIN, G. F. *Sampled-data control systems,* New York, Toronto, London: McGraw-Hill, 1958.

[6.14] MARKER, T. F. Crystal oscillator design notes. *Frequency,* **6**, 12-6, April 1968.

[6.15] DOYLE, N. P. A comparison of solid-state subcarrier oscillators for color TV receivers. *I.E.E.E. Transactions on Broadcasting and Television receivers,* **BTR-16**, 37-43, April 1970.

[6.16] GREBENE, A. B. and CAMENZIND, H. R. Frequency-selective integrated circuits using phase-lock techniques. *I.E.E.E. Journal of Solid-State Circuits,* **SC-4**, 216-25, August 1969.

[6.17] RENSCHLER, E. and WELLING, B. Frequency synthesizing with the phase-locked loop. *The Electronic Engineer,* **29**, 84-90, June 1970. Or *Motorola Application Note* AN 463.

7 Phaselocked Frequency Synthesizers

As a general concept all devices generating output frequencies which are rational multiples (or nearly rational multiples, in the case of interpolation etc.) of a standard input frequency are called frequency synthesizers, irrespective of the actual frequency-changing process involved.

From the very beginning we see competition between two techniques: real or 'passive-element' synthesis (as described in chapter 5), and the application of phaselock loops involving 'active' elements – i.e. VCOs. The latter has the advantage of simpler and cheaper selective circuits which can be reduced to mere RC networks only.

Before the solid state age there were objections to the use of 'active' elements, particularly, in field applications. But this changed with the introduction of microminiaturization and IC technology: increased reliability, small size, inexpensive large scale production, and reduced power consumption paved the way for widespread use of the phaselock technique in frequency synthesis.

7.1 PHASELOCKED FREQUENCY MULTIPLIERS

It was pointed out in chapter 2 that multiplication factors in conventional circuits, based on simple nonlinear elements, should be restricted to small numbers (2, 3, and 5); otherwise effective suppression of undesired harmonic components is difficult. Another difficulty, particularly with long chains of multiplier stages, is the large number of tuned circuits, which impairs output phase stability. The benefits of the phaselock technique, which make possible, by simple means, the realization of a very narrow pass-band with extremely small phase slope, are evident.

(A) The VCO and Loop Selectivity Determine the Desired Multiple

The principle is indicated in Fig. 7.1. Taking into account results found in Section 6.4(B) the output of the sampling phase detector in Fig. 7.1 is

$$v_d \approx E_0 \sin(\omega_0 t_k - \varphi_0) \tag{7.1}$$

The sampling takes place at instants

$$t_k = kT_i + \Delta t_i \quad (k = 0, 1, 2, \ldots) \tag{7.2}$$

where

$$T_i = \frac{2\pi}{\omega_i} \quad \text{and} \quad \Delta t_i = \frac{\varphi_i(t_k)}{2\pi} T_i = \frac{\varphi_i(t_k)}{\omega_i} \tag{7.3}$$

By substituting in (7.1) we get

$$v_d \approx E_0 \sin\left[2\pi k \frac{\omega_0}{\omega_i} + \varphi_i(t_k) \frac{\omega_0}{\omega_i} - \varphi_0\right] \tag{7.4}$$

When ω_0 is an exact multiple of ω_i, i.e.

$$\frac{\omega_0}{\omega_i} = N \quad (N \text{ integer}) \tag{7.5}$$

and the sampling frequency is large compared with the natural loop frequency ω_n then

$$v_d \approx E_0 \sin(N\varphi_i - \varphi_0) \tag{7.6}$$

Considering the last equation, we conclude that the frequency multiplier in Fig. 7.1 may be replaced by the equivalent circuit in Fig. 7.2.

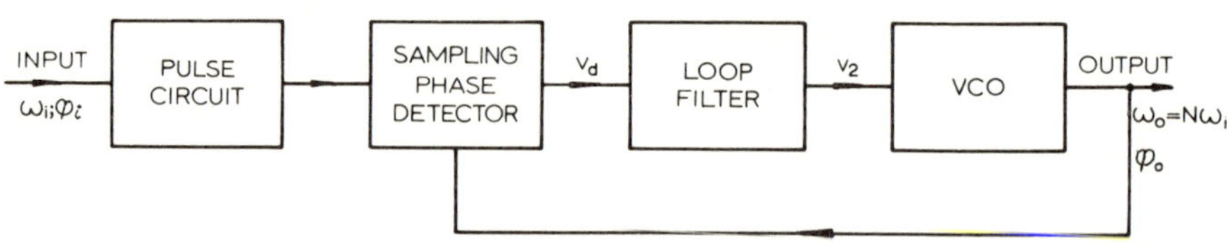

Fig. 7.1 Phaselocked frequency multiplier; the VCO and the loop selectivity determine the desired multiple.

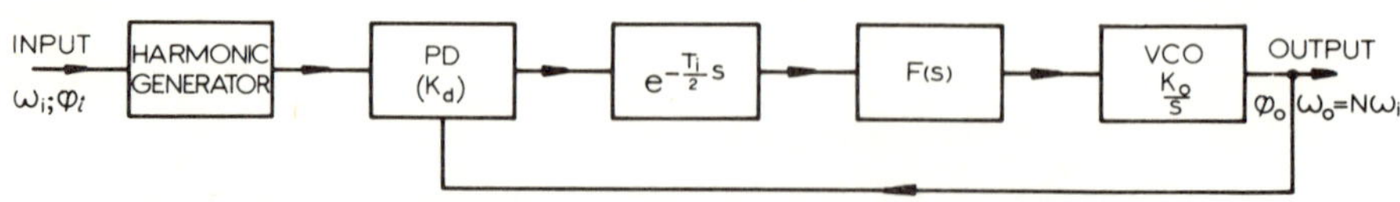

Fig. 7.2 Equivalent circuit of the phaselocked frequency multiplier in Fig. 7.1.

Choice of the maximum harmonic multiple N depends upon the instability of the VCO and on the pull-in (or phaselock acquisition, if used) range $\Delta\omega_P$. When

$$|\Delta\omega_0| < \Delta\omega_P < \frac{\omega_i}{2} \tag{7.7}$$

where $|\Delta\omega_0|$ is the peak value of the possible VCO instability and tuning inaccuracy, then there is no danger of a false lock to $(N-1)\,\omega_i$ or $(N+1)\,\omega_i$. From (7.5) and the inequality (7.7) follows the upper bound for the multiplication factor

$$N_{max} < \frac{1}{2}\left|\frac{\omega_0}{\Delta\omega_0}\right| \tag{7.8}$$

The spurious 'sidebands' ($\omega_0 \pm M\omega_i; M = 1, 2 \ldots$) at the output may be traced back to imperfections in the sampling phase detector. If additional attenuation is desired the natural frequency ω_n of the loop must be much smaller than ω_i. Evidently the simple phaselocked frequency multiplier as in Fig. 7.1 is particularly suitable when the input and output frequencies are fixed (standard frequency and its multiples etc.) because in this case the VCO may be a crystal oscillator, and the loop fulfilling $\omega_n \ll \omega_i$ is easily realized [7.1].

(B) Auxiliary Frequency Multiplier Determines the Desired Multiple

When multiplication factors are large the choice of the correct multiple by the phaselocked VCO must be assisted; application of an auxiliary frequency multiplier, as shown in Fig. 7.3 solves the problem [7.2].

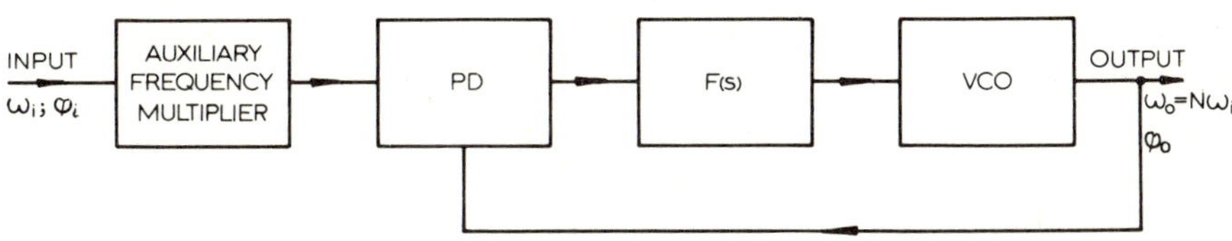

Fig. 7.3 Phaselocked frequency multiplier with resolution enhanced by an auxiliary multiplier.

Multiplication of low standard frequencies into the microwave region is performed, often, in this way (see atomic frequency standards, microwave spectroscopy, etc.). Besides the filtering properties of the loop, the higher power supplied by the VCO is appreciated. To simplify the phase detection problem at high frequencies and raise the gain K_d, IF amplification is introduced as in Fig. 7.4. With its help the loop equation is easily derived

$$[(N\Phi_i(s) - \Phi_0(s))\,F_M(s) - M\Phi_i(s)]\,F(s)\,K_d\,K_0/s = \Phi_0(s) \tag{7.9}$$

where $F_M(s)$ is the IF-filter modulation transfer function, which is discussed in more detail in Appendix 7.1. By introducing the loop gain $K = K_d K_0$ and measuring the input phase in the output frequency coordinate system, i.e. $\Phi_i'(s) = (N+M)\Phi_i(s)$, eq. (7.9) may be written as

$$\frac{\Phi_0(s)}{\Phi_i'(s)} = \frac{N}{N+M} \cdot \frac{[F_M(s) - M/N]\,K\,F(s)/s}{1 + K\,F(s)\,F_M(s)/s} \tag{7.10}$$

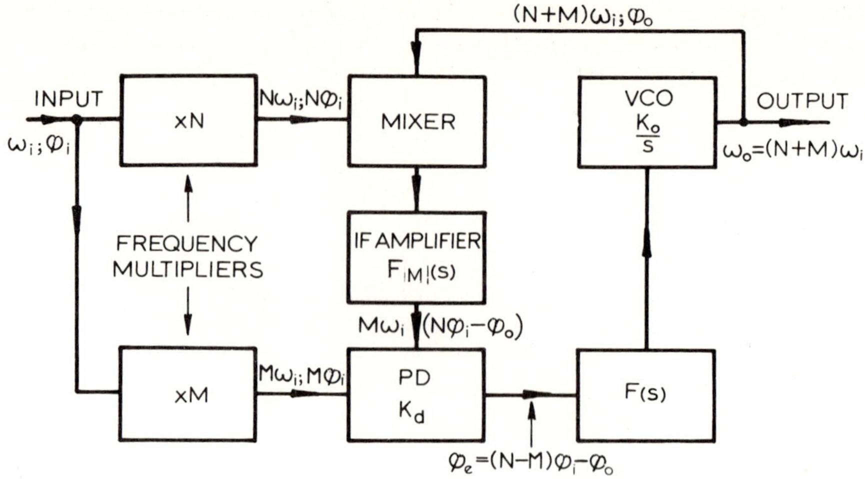

Fig. 7.4 Phaselocked frequency multiplier with auxiliary frequency multiplication (by N) and IF amplification.

In frequency multipliers of this type, as generally built, the IF frequency $M\omega_i$ is small compared with $\omega_0 = (N + M)\omega_i$, and $N\omega_i$; (7.10) may therefore be simplified to

$$\frac{\Phi_0(s)}{\Phi_i'(s)} \approx \frac{K\,F(s)\,F_M(s)}{s + K\,F(s)\,F_M(s)} \tag{7.11}$$

(C) Frequency Divider Determines the Desired Multiple

Fail-safe choice of large multiplication factors may be secured when a frequency divider is put in the feedback path of the phaselock loop as shown in Fig. 7.5.

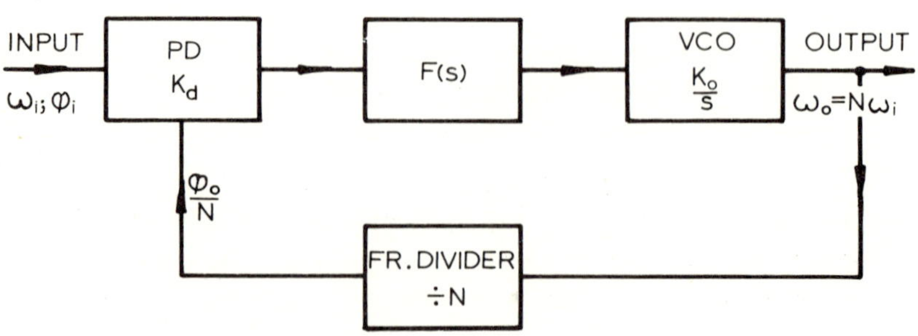

Fig. 7.5 Phaselocked frequency multiplier in which the desired multiple is determined by the frequency divider in the feedback path.

When dividing the actual feedback system equation (in which time delays due to the frequency divider and the sampling phase detector, Section 6.4(B), have been neglected)

$$\frac{\Phi_0(s)}{\Phi_i(s)} = \frac{K\,F(s)/s}{1 + K\,F(s)/sN} \tag{7.12}$$

by N, the familiar phaselock loop, but with reduced gain ($K' = K/N$), output frequency ($\omega_0' = \omega_0/N$), and phase ($\Phi_0'(s) = \Phi_0(s)/N$), results. Accordingly the phase-locked frequency multiplier with the feedback path divider may be replaced by the equivalent circuit in Fig. 7.6 which has the loop equation

$$\frac{\Phi_0'(s)}{\Phi_i(s)} = \frac{K'\,F(s)}{s + K'\,F(s)} \tag{7.13}$$

the natural frequency

$$\omega_n = \sqrt{K'/T_1} = \sqrt{K/NT_1} \tag{7.14}$$

and the loop-noise bandwidth

$$B_L = \frac{1}{4}\left[\frac{K}{N}\cdot\frac{T_2}{T_1} + \frac{1}{T_2}\right] \approx KT_2/4NT_1 \tag{7.15}$$

The pull-in range of the primed system is, according to (6.74),

$$\Delta\omega_P' \approx \frac{K}{N}\sqrt{2\frac{T_2}{T_1}} \tag{7.16}$$

or in terms of the output frequency ω_0

$$\Delta\omega_P \approx K\sqrt{2T_2/T_1} \tag{7.17}$$

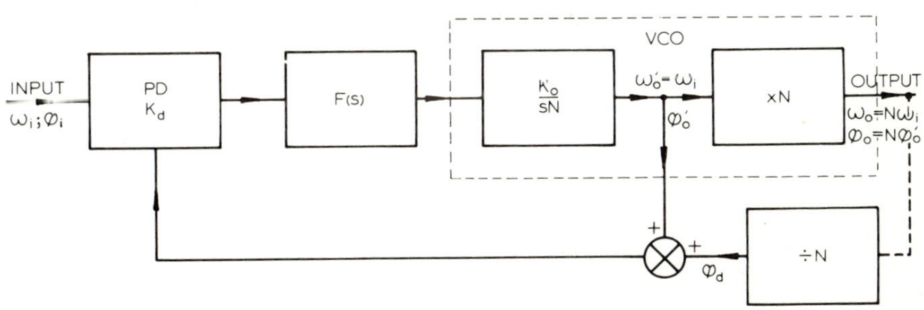

Fig. 7.6 Equivalent circuit of the phaselocked frequency multiplier in Fig. 7.5.

It is interesting to note that the insertion of the divider reduces the loop-noise bandwidth but does not affect the pull-in range. Byrne [6.3] defines the 'figure of merit', which is the ratio

$$\Delta\omega_P/B_L \approx 4N\sqrt{2T_1/T_2} \tag{7.18}$$

Making a qualitative appreciation of the effect of the phase jitter originating in the divider and the oscillator itself we shall apply the same procedure as used for deriving (6.97). The result is

$$\Phi_0(s) = N[\Phi_i(s) - \Phi_d(s)]\,H'(s) + \Phi_{osc}(s)[1 - H'(s)] \qquad (7.19)$$

7.2 PHASELOCKED FREQUENCY DIVIDERS

The main application field for phaselocked frequency dividers is at very high frequencies which are beyond the range of counting dividers. Reduction of atomic standard frequencies (e.g. of the ammonia or hydrogen masers) into the range of precision crystal oscillators is a practical example [7.3], [7.4]. Measurement of microwave frequencies is another example [8.1]. In frequency synthesis, division in decimal modules is sometimes also made in this way [7.11], [7.13].

Phaselocked frequency dividing results when in the feedback path either a harmonic generator (e.g. sampling phase detector switched at the rate of the output VCO frequency) or a frequency multiplier is placed. In both cases the block diagram in Fig. 7.7 may be used for derivation of the main properties of the dividing loop.

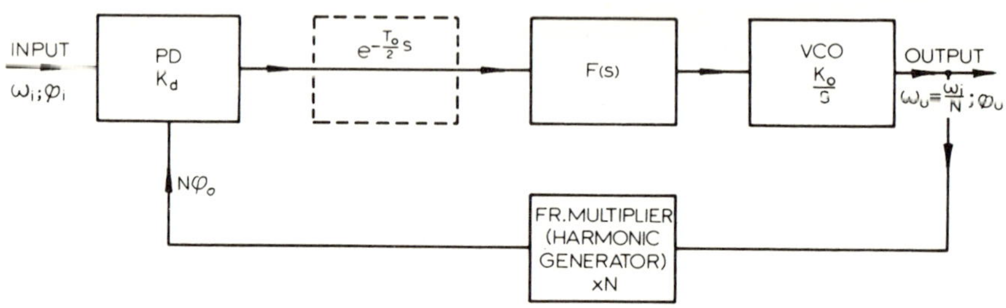

Fig. 7.7 Basic diagram of the phaselocked frequency divider.

The feedback system equation is

$$\frac{\Phi_0(s)}{\Phi_i(s)} = \frac{K\,F(s)/s}{1 + NK\,F(s)/s} \qquad K = K_d\,K_0 \qquad (7.20)$$

When multiplied by the dividing factor N the primed system results: $\Phi'_0(s) = N\Phi_0(s)$; $\omega'_0 = \omega_i$, and $K' = NK$; the equivalent circuit is redrawn in Fig. 7.8.

The natural frequency of the loop is

$$\omega_n = \sqrt{KN/T_1} \qquad (7.21)$$

The pull-in range is

$$\Delta\omega'_P = KN\sqrt{2T_2/T_1} \qquad (7.22a)$$

and in terms of the output frequency ω_0

$$\Delta\omega_P = K\sqrt{2T_2/T_1} \tag{7.22b}$$

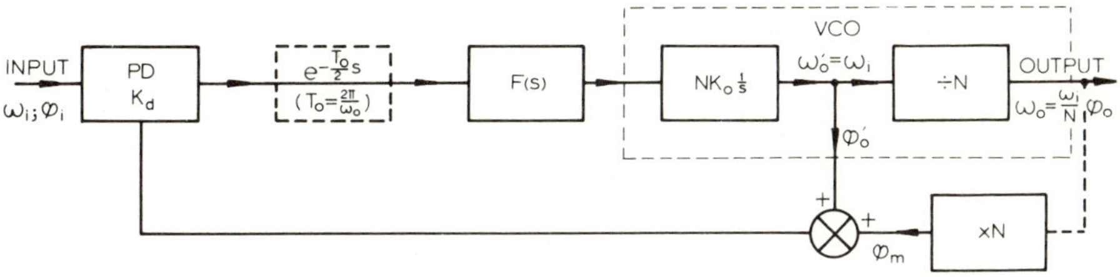

Fig. 7.8 Equivalent circuit of the phaselocked frequency divider.

The multiplier phase jitter φ_m adds to the input phase φ_i; proceeding in the same way as earlier with eq. (7.19) we get

$$\Phi_0(s) = \frac{1}{N}[\Phi_i(s) - \Phi_m(s)]\,H'(s) + \Phi_{osc}(s)\,[1 - H'(s)] \tag{7.23}$$

When the feedback frequency multiplier is a simple harmonic generator, locking on the neighbouring spectral lines will be prevented if

$$|\Delta\omega_0| < \Delta\omega_P < \frac{\omega_i}{2N(N+1)} \approx \frac{\omega_0}{2N} \tag{7.24}$$

The upper bound for the dividing factor follows from the above relation and is effectively the same as found earlier for frequency multipliers

$$N_{max} < \frac{1}{2}\left|\frac{\omega_0}{\Delta\omega_0}\right| \tag{7.25}$$

We have pointed out that the field of applications for phaselocked dividers is at extremely high frequencies. Evidently there will be the same reasons for introduction of IF amplification as mentioned in the previous section. As an example, a block diagram of an actual hydrogen maser phaselocked frequency divider is shown in Fig. 7.9.

7.3 PHASELOCKED MIXERS

Actual mixing is performed in a conventional mixer, as described in chapter 4, but for the selection of the desired product the filtering properties of the phase-lock loop are employed. The obvious advantage is that large, cumbersome, and sometimes even critical LC circuits can be replaced by simple and uncritical RC circuits. Consequently much larger mixing ratios can be realized without difficulty.

(A) Addition or Subtraction of Two Frequencies

A block diagram of the phaselocked mixing arrangement is shown in Fig. 7.10. In the feedback path of the loop there is a mixer and an IF filter; the former

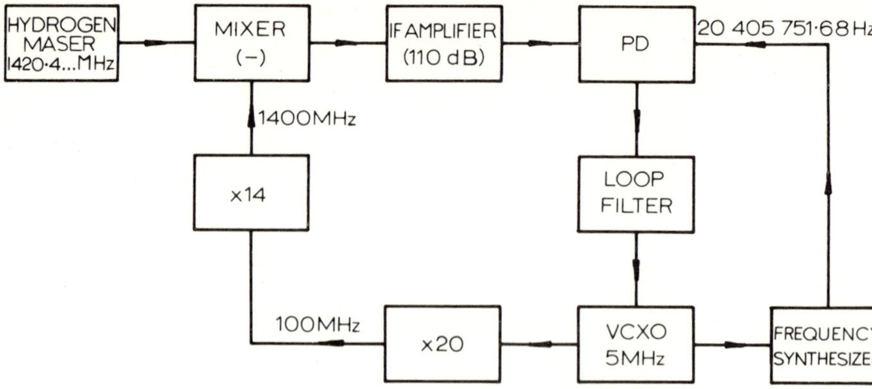

Fig. 7.9 Hydrogen maser phaselocked frequency divider (From Vessot et al. [7.4].)

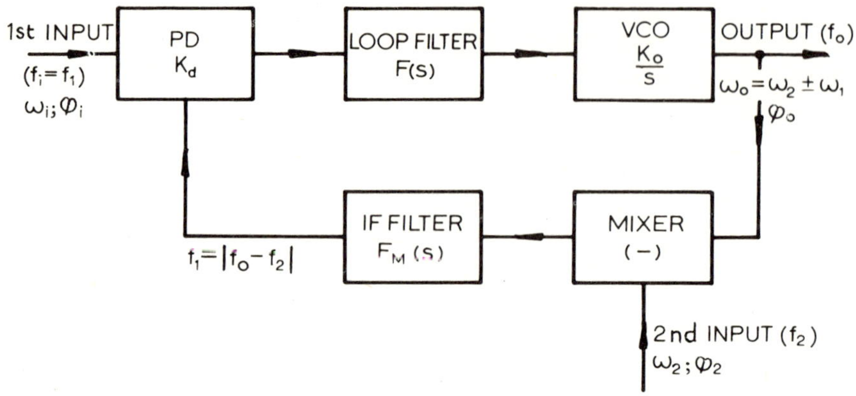

Fig. 7.10 Block diagram of the phaselock mixing arrangement.

behaves with respect to the phase as a simple adder with the transfer function either (+1) or (−1); the latter may be replaced by an equivalent 'modulation transfer function' $F_M(s)$ as shown in Appendix 7.1. Referring to Fig. 7.10 the loop equation may be written

$$\frac{\Phi_0(s)}{\Phi_i(s)} = \frac{K F(s)}{s + K F(s) F_M(s)} \tag{7.26}$$

The VCO output frequency, f_0, is that of the desired sideband, and the reference frequency, f_i, is equal to the lower of the input frequencies, f_1 and f_2, to be mixed ($f_1 < f_2$). In the locked state

$$f_0 = f_2 \pm f_1{}^{\dagger} \tag{7.27}$$

† Further, only the positive sign will be considered, and in agreement with the symbolism introduced in chapters 4 and 5 the mixing ratio $q = f_2/f_1$ will be positive in the case of adding and negative in the case of subtracting.

and it is the subtraction, $f_0 - f_2$ or $f_2 - f_0$, which is always performed in the mixer. The advantages of this working mode are evident from examination of the new mixing ratio. By dividing (7.27) by f_2 we get

$$q_m = \frac{f_0}{f_2} = 1 + \frac{f_1}{f_2} = \frac{q+1}{q}; \quad q = f_2/f_1 \tag{7.28}$$

Considering the case where q is an integer, and using (4.66), one finds for the order, N, of intermodulation products

$$N \geqslant |q+1| + |q| - 2B = 2[|q| - B] \pm 1 \tag{7.29}$$

(assuming that B is also an integer). Comparison of (7.29) with (4.62) reveals that the lowest intermodulation product which is in the pass-band of f_1,($B = 1$), is of a substantially higher order. At the same time leakage of the 'carrier', i.e. $N = 1$, is possible only in cases of extremely large B – the IF filter does not need to have a steep attenuation characteristic; in a lot of cases simple RC low-pass networks will be sufficient.

A further examination of (7.29) shows that for all integer values of $B \leqslant |q|$ the order of the lowest intermodulation product is an odd number. Because sidebands of this type are not present, in the ideal case, at the output of the ring modulator, see Table 4.5, its use reduces considerably the mixer output filter requirements.

The conclusion to be drawn is that the higher the original mixing ratio $|q|$ the simpler is the mixing problem. However, the variability of the input frequencies, $\pm \Delta f_1$ and $\pm \Delta f_2$, as well as the instability of the VCO itself, $\pm \Delta f_0$, put a limit to the upper bound of $|q|$. The criterion is to prevent locking onto the image frequency. The necessary conditions may be derived with the help of Fig. 7.11. From the peak deviations of the individual frequencies the necessary pull-in range is:

$$\Delta f_{P,\min} \geqslant \Delta f_0 + \Delta f_1 + \Delta f_2 \tag{7.30}$$

$$\Delta f_{P,\max} \leqslant 2f_1 - \Delta f_0 - \Delta f_1 - \Delta f_2 \tag{7.31}$$

Combination of the above inequalities leads to

$$\Delta f_0 + \Delta f_1 + \Delta f_2 \leqslant f_1 \tag{7.32}$$

Dividing the final relation by f_0 and using (7.28) we get

$$\frac{\Delta f_0 + \Delta f_1 + \Delta f_2}{f_0} \leqslant \frac{1}{|q+1|} \tag{7.33}$$

In the limiting case, if both f_1 and f_2 are invariable standard frequencies, $\Delta f_1 = \Delta f_2 = 0$, the maximum mixing ratio

$$|q+1| < \left|\frac{f_0}{\Delta f_0}\right| \tag{7.34}$$

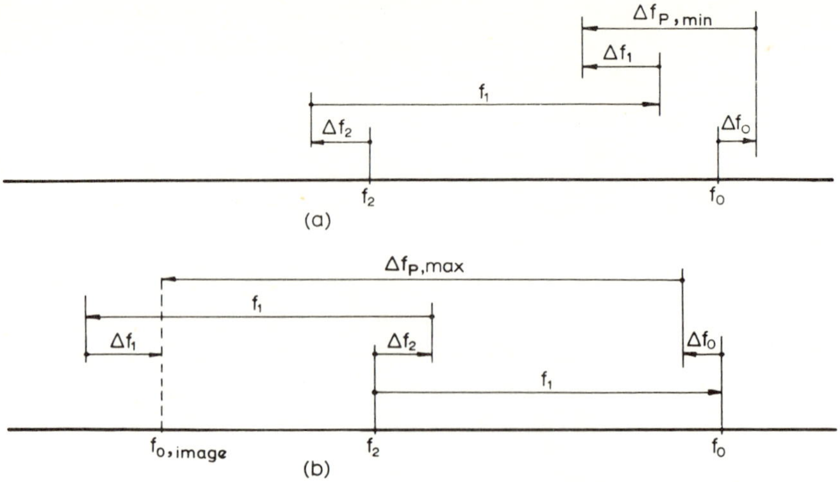

Fig. 7.11 Spectral diagrams illustrating pull-in problems in phaselocked mixers: (a) with respect to the tolerances encountered; (b) with respect to the image lock.

depends only upon the instability of the VCO used and may be very high, i.e. very small frequency increments can be added or subtracted in a single loop, particularly if crystal oscillators are employed.

However, in frequency synthesis applications either f_1 or f_2 is variable, and anti-image-lock precautions must eventually be employed, namely reductions of effective frequency variations, to achieve large mixing ratios (see Section 7.6, Fig. 7.31, and the description of the receiver in Section 8.2(B)(ii)).

(B) Addition and Subtraction of Several Frequencies

In some frequency synthesis applications, particularly of earlier design, two or more mixers may be encountered in the feedback path of the phaselock loop. Naturally, there are more possibilities for undesired locking modes to appear and spurious mixing signals to be generated. Considering the three-frequency system in Fig. 7.12, prevention of synchronization on the image frequency due to mixing in M_3 requires modification of (7.32) to

$$\Delta f_0 + \Delta f_1 + \Delta f_2 + \Delta f_3 < f_1 \tag{7.35}$$

If F_{m2} is a pass-band filter the correct input to M_3 is safeguarded and a less stringent condition

$$\Delta f_0 + \Delta f_1 + \Delta f_2 + \Delta f_3 < f_1 + f_3 \tag{7.36}$$

precludes locking to the image frequency due to mixing in M_2 and eventually permits 'switching' of the VCO in the case where f_2 is a harmonic spectrum (see Fig. 7.30). In a more general case f_3 may be another harmonic spectrum, consequently, F_{m3} must also be a pass-band filter. A practical application of this

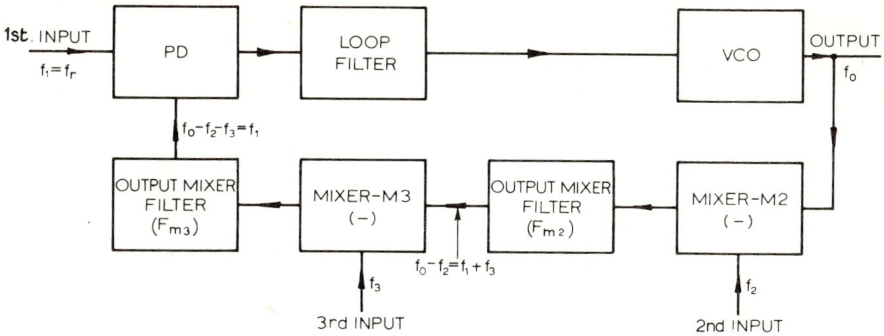

Fig. 7.12 Summing of three frequencies in a phaselock loop system.

technique, sometimes called frequency analysis, is the frequency synthesizer shown in Fig. 8.15, or an earlier one described in [7.5]. However, the cascade of mixers and filters in the feedback path cannot be long because of the limit imposed on the overall mixing ratio by (7.34) and the gradual degradation of the phaselock loop properties (stability, false lock, etc., see Appendix 7.1).

7.4 PHASELOCKED SINGLE-FREQUENCY SYNTHESIZERS

The design starts with eq. (5.1)

$$\frac{f_x}{f_s} = \frac{X_1}{Y_1} \tag{7.37}$$

and Fig. 5.1, but the frequency multiplier, X_1, is of the phaselocked type, and consequently much larger multiplying factors can be realized in a single stage.

(A) Frequency Synthesizer with Harmonic Generator

The simplest frequency synthesizer (Fig. 7.13) results when the selectivity and stability of the VCO itself guarantees the desired multiple X_1. A very large

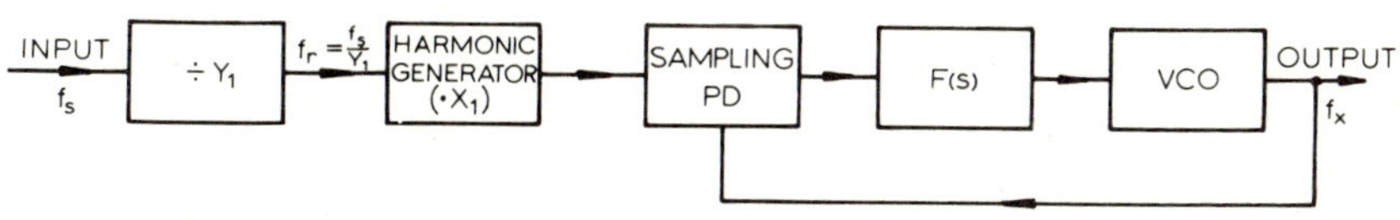

Fig. 7.13 Block diagram of simple frequency synthesizer with a divider (Y_1) and a harmonic generator (X_1).

number of practical cases can be solved in this way, particularly if voltage-controlled crystal oscillators are employed. When f_x is too high for any crystal resonator, simultaneous multiplication and division of (7.37) with a suitable factor, y_0, will help – see (5.42) and Fig. 7.14 of the following example.

Example 7.1

Redesign the television exciter from Example 5.1.

f_s = 100 kHz; f_x = 59·250 MHz; y_0 = 54 (power stage frequency multipliers)

$$\frac{f_x}{f_s} = \frac{54}{54} \cdot \frac{59250}{100} = \frac{54 \times 395}{36}$$

X_1 = 395 with a phaselocked crystal oscillator eq. (7.8) is easily fulfilled;
Y_1 = 36 with IC digital divider; there are no difficulties at 100 kHz input frequency (see section 3.5(B) and Example 3.1);
f_x' = 1·097$\bar{2}$ MHz.

The block diagram of this phaselocked exciter is in Fig. 7.14.

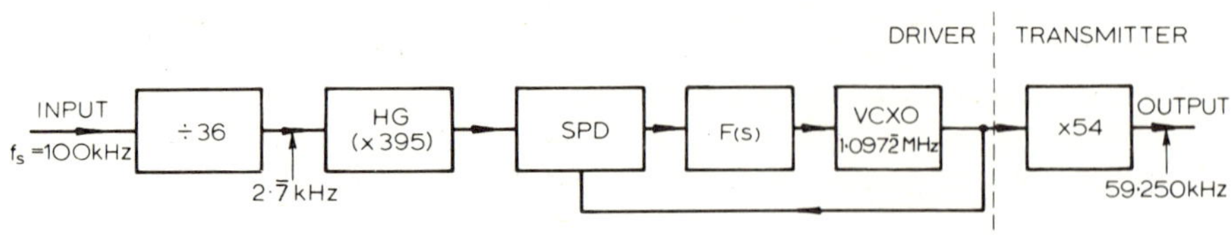

Fig. 7.14 Phaselocked synthesizer-exciter for a television transmitter with the carrier frequency f_x = 59·250 MHz.

(B) Frequency Synthesizer with Divider in the Feedback Path

If there are good reasons for using LC VC oscillators the simple scheme shown in Fig. 7.13 cannot generally be employed. However, the multiplying system with divider in the feedback path, described in Section 7.1(C), will help, and the block diagram changes as shown in Fig. 7.15, which is another type of the phaselocked version of Fig. 5.1. The main difference is that the multiplier X_1 has been replaced by a divider. Below a certain frequency there are no limitations on dividing factors (Section 3.5 – counting dividers), consequently, the block diagram in Fig. 7.15 can become the backbone of practical frequency synthesizers in a much larger number of cases than that of Fig. 5.1.

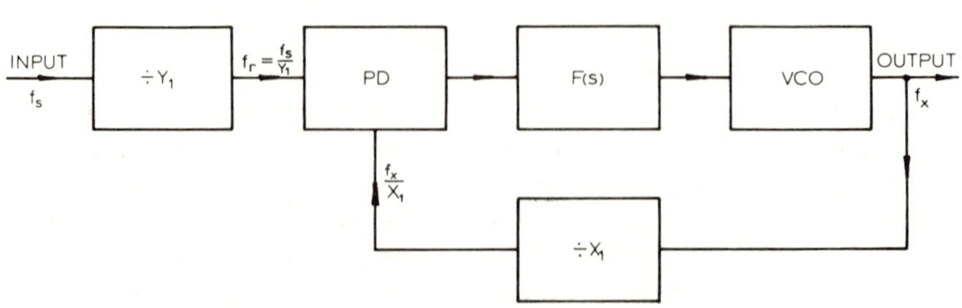

Fig. 7.15 Simple phaselocked frequency synthesizer for $f_x = f_s(X_1/Y_1)$ employing two dividers.

Closer examination of Fig. 7.15 reveals that the reference frequency is the largest common divisor

$$f_r = f_x / X_1 = f_s / Y_1 \tag{7.38}$$

(C) Frequency Synthesizer with Mixer in the Feedback Path

If the output frequency is beyond the range covered by commercial counting dividers the introduction of one synthesizing loop according to eq. (5.4) is a means of reducing the frequency level at the counter input, as illustrated in Fig. 7.16.

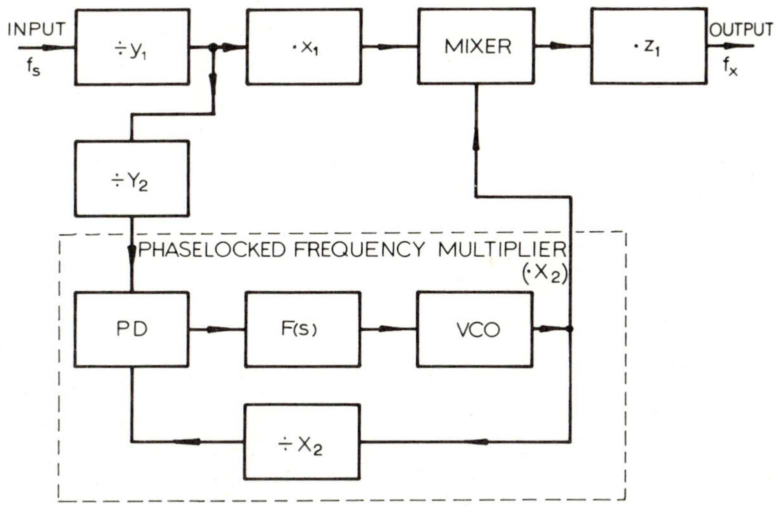

Fig. 7.16 Phaselocked frequency synthesizer combined with one ordinary synthesizing loop.

Transferring the term $z_1 x_1 / y_1$ in (5.4) to the left-hand side, thus

$$\frac{f_x}{f_s} - \frac{z_1 x_1}{y_1} = \frac{z_1}{y_1} \frac{X_2}{Y_2} \Rightarrow \frac{f_x / z_1 - f_s x_1 / y_1}{X_2} = \frac{f_s}{Y_1} \tag{7.39}$$

is equivalent to placing the mixer in the feedback path of the phaselock loop (Fig. 7.17). The advantages of this move have been discussed in Section 7.3.

Equation (7.39) is solved in the same manner as described in Section 5.2 with the exception that the upper bound of the mixing ratio, q_h, given by (7.34), may ultimately be quite high, and consequently nearly the whole area between the outer lines in Fig. 5.6 can be used. If the intersection lies in the upper half of the area, the auxiliary standard frequency, $f_s x_1 / y_1$, is subtracted from the VCO output frequency f_x / z_1. When the intersection lies in the lower half-area the sequence of subtracted terms is interchanged.

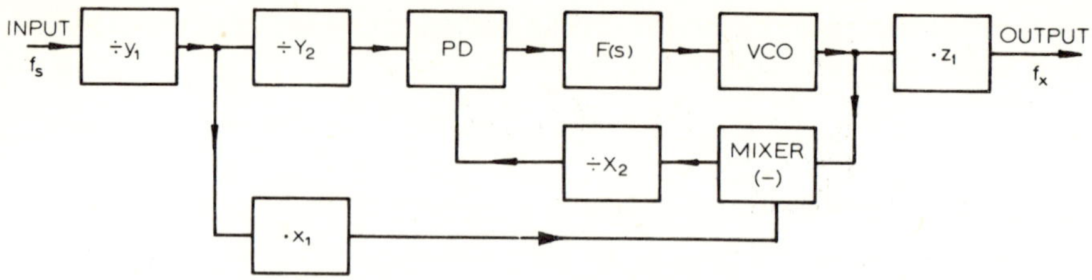

Fig. 7.17 Phaselocked frequency synthesizer having the mixer of the auxiliary synthesizing loop placed in the feedback path of the phaselock loop.

In communications the carrier frequencies in a given frequency band form, generally, an arithmetic progression. In this case a universal synthesizer-exciter suitable for all the channels can be designed by starting from the block diagram in Fig. 7.17. A substantial simplification can be achieved if mere replacement of the quartz crystal resonators in the VCO (VCXO) assists or provides for the 'coarse tuning'. Television exciters described in [7.6] and [8.6] are practical examples of this technique.

(D) Approximating Frequency Synthesizers

When performing phase comparison at the frequency level given by (7.38) it might happen that for very large Y_1 the respective reference frequency, $f_r = f_s/Y_1$, would be too low for the realization of a phaselock loop with reasonable parameters. This difficulty can be overcome by separating a partial divisor y_i from Y_1 and transferring it as multiplier into the feedback path.

The other solution of this problem, i.e. raising the reference frequency, depends upon the fact that the frequency synthesis process merely simulates the process of successive approximations of the real number X_1/Y_1 by a ratio of smaller integers A_k/B_k [5.2]. As soon as the residual error falls below an acceptable tolerance level the approximating procedure may be stopped:

$$|X_1/Y_1 - A_n/B_n| = R < \epsilon \tag{7.40}$$

Choice of the continued fraction algorithm

$$\frac{X_1}{Y_1} \approx \frac{A_n}{B_n} = b_0 + \cfrac{1}{b_1 + \cfrac{1}{b_2 + \cfrac{1}{\ddots \, b_{n-2} + \cfrac{1}{b_{n-1} + \cfrac{1}{b_n}}}}} \tag{7.41}$$

has the advantage that the residue, R, converges very rapidly to an arbitrarily small value, because

$$R \leqslant \frac{1}{B_n^2} \tag{7.42}$$

Successive approximations are given by the recurrence formulae

$$\begin{aligned} A_0 &= b_0 \quad A_1 = b_1 b_0 + 1 \quad A_k = b_k A_{k-1} + A_{k-2} \\ B_0 &= 1 \quad\;\, B_1 = b_1 \qquad\qquad B_k = b_k B_{k-1} + B_{k-2} \end{aligned} \tag{7.43}$$

where $b_0, b_1, \ldots, b_n$ are integers found by Euclid's algorithm as shown in the following example.

Another advantage of replacing X_1/Y_1 by the nth continued fraction approximation A_n/B_n is that A_n and B_n are eventually much smaller than X_1 and Y_1 and consequently the approximating synthesizer is much simpler.

Example 7.2

Design an approximating frequency synthesizer for the caesium atomic frequency standard.

$$f_x = 9\,192\,631\,770 \quad f_s = 5 \times 10^6 \quad \text{Hz}$$

Ordinarily one synthesizing loop with direct multiplication of f_s to f_{x1} = 9180 MHz, and addition of f_{x2} = 12 631 770 Hz is used. If the latter component is generated in an approximating synthesizer the continued fraction expansion of f_{x2}/f_s must be performed:

$$\frac{f_{x2}}{f_s} = \frac{X_2}{Y_2} = \frac{12\,631\,770}{5\,000\,000} = 2 + \frac{263\,177}{500\,000}; \quad b_0 = 2$$

the next step of Euclid's theorem leads to

$$\frac{500\,000}{263\,177} = 1 + \frac{236\,823}{263\,177}; \quad b_1 - 1$$

Proceeding in the same way $b_2, b_3, \ldots$ will be found and the successive approximations A_k/B_k calculated. Results are summarized in Table 7.1.

When stopping at the fifth approximation, the actual error is –1·596 Hz. This is well within the range of values encountered when Zeeman splitting by the magnetic field is imposed.

A block diagram of the synthesizer is shown in Fig. 7.18. The proper choice of the pull-in range of the voltage-controlled crystal oscillator (VCXO) prevents locking on undesired harmonics but at the same time enables small-step offsetting if the divider B_n is made variable (see also Section 8.3(A)).

The theory of approximating frequency synthesizers is not yet fully established. The problem is the same as that with which mechanical engineers were faced long ago when they had to approximate an ideal gear ratio with one realizable

Table 7.1

Successive continued fraction approximations of the rational number 1 263 177/500 000

k	0	1	2	3	4	5	6	7	8	9	10
b_k	2	1	1	8	1	71	1	1	1	1	72
A_k	2	3	5	43	48	3451	3499	6950	10 449	17 399	1 263 177
B_k	1	1	2	17	19	1366	1385	2751	4 136	6 887	500 000
f_s/B_k^2 (Hz)			$1{\cdot}25 \times 10^6$	17 301·0	13 850·4	2·679	2·607	0·661	0·292	0·105	–
$f_s(X_2/Y_2 - A_k/B_k)$ (Hz)			$+1{\cdot}32 \times 10^5$	−15 288·8	+191·1	−1·596	+1·047	−0·265	+0·174	$-1{\cdot}453 \times 10^{-3}$	0·0

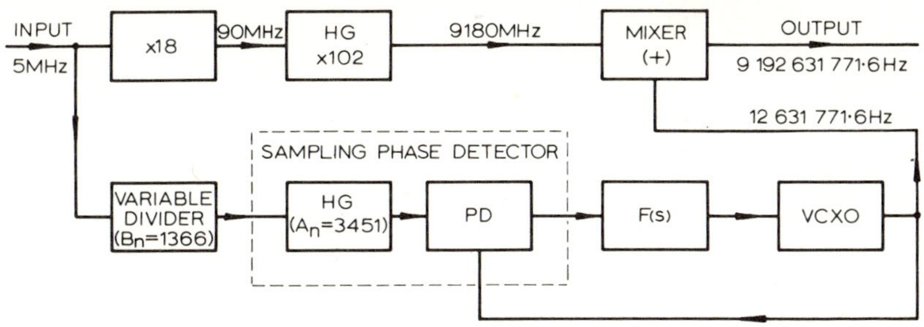

Fig. 7.18 Approximating frequency synthesizer for the caesium atomic frequency standard.

in practice i.e. with gears having a limited number of teeth. Frequency synthesis differs from the related mechanical problem solely in the boundary conditions which are applied; the permissible multiplication and division factors (number of 'teeth') can be about two orders larger and the acceptable tolerances are reduced in an even higher proportion.

Similarly, as in mechanical devices, continued fraction expansion may provide solution of approximating frequency synthesis problems in many cases (see the example above), but by no means generally. The main disadvantage is the uniqueness of the continued fraction expansion. The mechanical engineers overcame this difficulty by the compilation of tables in which the approximating numerators and denominators, A_n and B_n in eq. (7.40), are products of smaller integers belonging to a suitable interval.

The existing gear-ratio tables are too crude for frequency-synthesis applications. It would no doubt be possible to compute appropriate tables but size would be prohibitive. The other possibility is to work out a suitable programme and solve each individual case with the help of the computer in such a way that multiplication factors $a_0, a_1 \ldots, a_n = A_n$ and division factors $b_0, b_1 \ldots, b_n = B_n$ are subject to the limitation that neither can exceed some upper bound. In cases where phaselocked frequency multipliers are used this upper bound is quite high and depends on the smallest reference frequency allowed at the input of the phase detector (see Fig. 7.15) and on the maximum acquisition times which can be tolerated, etc. One can expect that multiplication factors will be of the order 10^4. Practical realization of this approximating frequency synthesizer is a chain of simple phaselocked frequency synthesizers shown in Fig. 7.15. Finally the hardware of the approximating frequency synthesizer would be simplified considerably if all partial ratios a_i/b_i $(i = 1, 2, \ldots, n)$ could be made to be nearly equal and very close to 1.†

† Czechoslovak patent application 5146-72 (V. Kroupa).

7.5 DIGITAL SYNTHESIZERS

When making the feedback path divider, X_1, in Fig. 7.15 variable, say from one N_{min} to one N_{max}, the output frequency may be changed in steps according to the encoded digital information:

$$f_x/X_1 = f_s/Y_1 = f_r \Rightarrow f_x = X_1 \frac{f_s}{Y_1} \equiv Nf_r \tag{7.44}$$

Considering the above equation, the main limitation of this system can be readily seen, namely that the smallest increment of the output frequency Δf_x is equal to the reference frequency f_r, and consequently the number of discrete output frequencies is restricted to $(N_{max} - N_{min})$. The physical practicability imposes further limitations. A serious one is that the frequency range of variable dividers does not exceed tens of MHz (see Section 3.5(C)).

(A) Digital Synthesizer with Prescaler

Sometimes a simple prescaler with a fixed division ratio, y_0, may help (Fig. 7.19). However the reference frequency f_r must be reduced in the same proportion; consequently there are more severe requirements on the loop filter, $F(s)$, leading, eventually, to unacceptably long acquisition times.

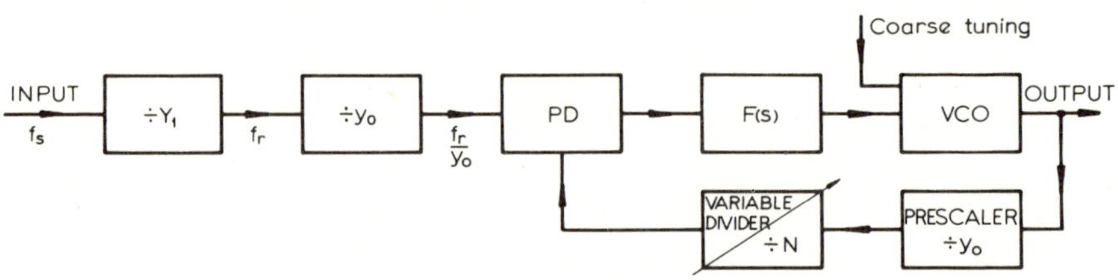

Fig. 7.19 Digital synthesizer with a fixed division ratio prescaler and a variable ratio divider in the feedback path of the phaselock loop.

(B) Digital Synthesizer with Mixer in the Feedback Path

Another solution of this problem is the use of an auxiliary synthesizing loop as explained earlier (Fig. 7.17). By this technique very high frequencies, even in GHz bands, are accessible to synthesis. A further advantage is that the multiplier x_1 (Fig. 7.20) can be made variable, which extends the output range and the number of discrete frequencies, f_x, as well. This system of digital synthesizers is common in practice [6.17], [7.1], [7.7], *et al.*

(C) Digital Synthesizer with Cascade of Phaselock Loops

When the unit output step, Δf_x, must be reduced without compromising the acquisition time, an auxiliary interpolating carrier generated in another phase-lock loop, with the same reference frequency, must be injected into the main

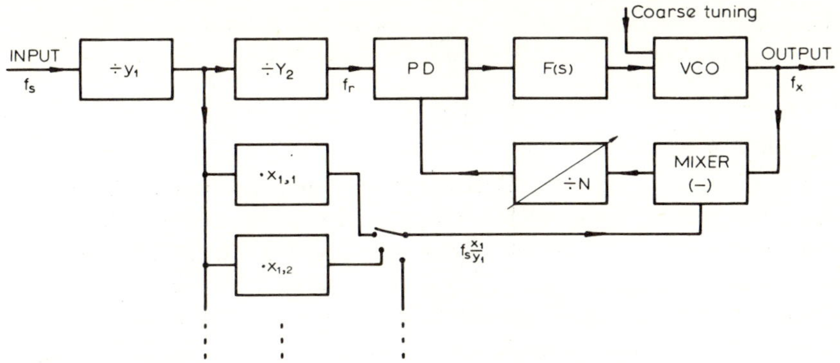

Fig. 7.20 Digital synthesizer with an auxiliary synthesizing loop (mixer in the feedback path).

loop after passing a fixed divider ($\div L$ in Fig. 7.21). In this way the smallest output frequency increment achieved is $\Delta f_x / L$ [6.12].

The mixing difficulties, which would be serious with a large L, may be removed by introducing another phaselock loop [7.8], [7.15]. A schematic arrangement is shown in Fig. 7.21.

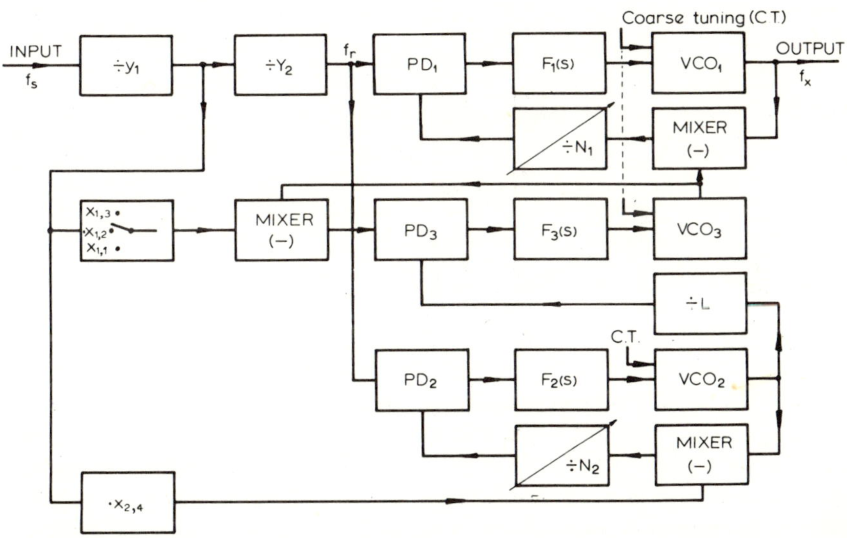

Fig. 7.21 Digital synthesizer with reduced output steps without compromising the acquisition time.

The importance of the cascaded phaselock loop systems lies in the fact that it forms the backbone of general decimal-variable synthesizers when choosing $L = 10$ (see the next section).

There might also be other reasons for introducing two-loop systems: e.g. when striving for better short-term stability or more effective suppression of undesired components, etc.

(D) Design Features

One of the most important problems in phaselock applications in frequency synthesis is effective suppression of leakage of the reference signal at the d.c. control-input of the VCO and maintaining at the same time a safe lock and reasonably small acquisition time. The sample-and-hold phase detector is appropriate to this type of frequency synthesizer; but even so there is a filtering problem, as illustrated by the following example.

Example 7.3

Design a loop filter for a digital synthesizer in which $N = 800$ to 1600, i.e. $N_{mean} = 800\sqrt{2}$; $K_{0,mean} = 0{\cdot}4 \times 10^6$ Hz/volt; $K_d = 6/\pi$ volts/radian; $f_r = 12\,500$ Hz and the reference signal causes at the output of the sawtooth sampler spikes with amplitude 0·2 volt and duration $t = 0{\cdot}6 \times 10^{-6}$ second. The spurious signal level at the VCO output should be below: (S/SpSB) > 80 dB.

Using (1.34) we may calculate the amplitude of the undesired reference signal to the sampler output $V_{r,\,sp} \leqslant 3$ mV. The spurious output phase modulation is given by (6.31) which simplifies, for ω_r large compared with ω_n, to

$$\Phi_{0,\,sp} \approx \frac{1}{\omega_r} K_0 |F(j\omega_r)| V_{r,\,sp}$$

and

$$\text{S/SpSB} \approx 2\omega_r / [K_0 V_{r,\,sp} |F(j\omega_r)|]$$

Substituting the above given values we get

$$80 \geqslant 20[\log 2{\cdot}5 \times 10^4 - \log 4 \times 10^5 - \log 3 \times 10^{-3} - \log |F(j\omega_r)|]$$
$$= 26{\cdot}4 - 20 \log |F(j\omega_r)|$$

Evidently the open loop gain at ω_r must be about -60 dB. Using a lag-lead filter with an additional independent RC section, a safe phase margin may be guaranteed because the phase asymptote for $\omega \to \infty$ is the $-180°$ line.

The graphic solution performed in Fig. 7.22 will supply in a straightforward manner poles and zero of the RC filter, and also the loop parameters (ω_n and ζ). Referring to Appendix 6.2 the open-loop gain asymptote must have at ω_r a slope of 40 dB/dec. Drawing the appropriate straight line we reach the corner frequency point ($1/T_3$) somewhere in the neighbourhood of zero gain. If we start the asymptote 20 dB/dec. at the -6 dB level ($1/T_3 = 4\omega_n$), the intersection with the axis is at $2\omega_n$ and the choice of the second corner frequency ($1/T_2 = 0{\cdot}5\ \omega_n$) ensures the damping factor $\zeta = 1$ for a high-gain loop (6.18). Below this frequency the slope of the asymptotic open-loop gain is again

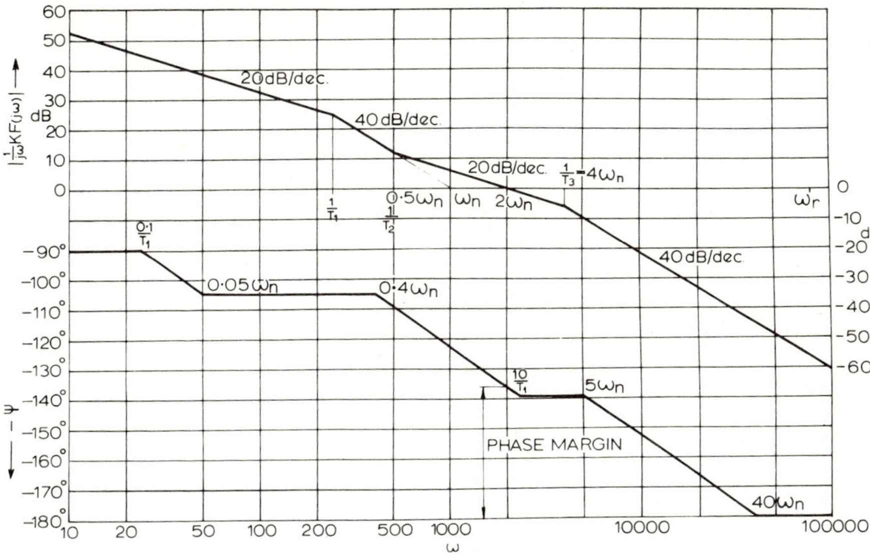

Fig. 7.22 Bode plot solution of the phaselock-loop filter parameters; numerical values are those from Example 7.3.

40 dB/dec. to the intersection, at $1/T_1$, with the low frequency part of this Bode plot drawn through the point $(1; K_{mean})$ with slope 20 dB/dec. The mean value of the gain K is

$$K_{mean} = K_0 K_d / N_{mean} = 4{\cdot}25 \times 10^3 \quad \text{or} \quad 72{\cdot}6 \text{ dB}$$

Finishing the Bode plot, the time constants of the loop filter can be read from Fig. 7.22:

$$T_1 = 4{\cdot}16 \times 10^{-3}$$

$$T_2 = 2{\cdot}0 \times 10^{-3} \qquad F(s) = \frac{1 + sT_2}{1 + sT_1} \cdot \frac{1}{1 + sT_3}$$

$$T_3 = 2{\cdot}5 \times 10^{-4}$$

The phase margin is about 45° ± 15°, which is adequate. The gain margin is, evidently, more than adequate. Note that in the construction of the Bode plot the time-delay term due to the sampling phase detector has been neglected because its contribution at $2\omega_n$ is very small ($2\pi 2\omega_n / 2\omega_r \approx 4{\cdot}5°$).

If another RC section is added to the loop filter without compromising the phase margin, its corner frequency ($1/T_5$) must be in the range of $40\omega_n$ The result would be only an increase of attenuation of about 6 dB at ω_r.

The loop parameters are:

$$\omega_n = 1000 \text{ rad/s} \quad (2\pi \text{ Hz})$$

$$\zeta = 1$$

Considering the primed system (see Section 7.1(C)) and the basic lag-lead filter, the hold-in, pull-in, and lock-in ranges (according to (6.64), (6.72), and to (6.78) multiplied by π, because of the sawtooth comparator) are:

$$\Delta\omega'_H = \pi K' = 1{\cdot}33 \times 10^4$$

$$\Delta\omega'_P = \Delta\omega'_H \sqrt{4T_2/3T_1} = 0{\cdot}8\Delta\omega'_H = 1{\cdot}06 \times 10^4$$

$$\Delta\omega'_L = \pi K' T_2/T_1 = 6{\cdot}4 \times 10^3 \qquad (2\pi \text{ Hz})$$

The lock-in time, using (6.82) is: $T_L \approx 2{\cdot}3/\zeta\omega_n = 2{\cdot}3 \times 10^{-3}$ second. Because $\Delta\omega'_P$ and $\Delta\omega'_L$ are larger than $1/T_3$ it must be expected that they will be changed in the presence of the additional RC section. Assuming a first-order approximation (d.c. component proportional to the fundamental of the beat-note amplitude [7.9]) the reduced pull-in range is:

$$\left(\frac{\Delta\omega'_{P,\,\text{red}}}{2}\right)^2 \approx \left(\frac{\Delta\omega'_P}{2}\right)^2 \frac{1}{1 + [(\Delta\omega'_{P,\text{red}}/2)\,T_3]^2} \tag{7.45}$$

After evaluation:

$$\Delta\omega'_{P,\text{red}} \approx 7{\cdot}65 \times 10^3 \qquad (2\pi \text{ Hz})$$

There might be good reasons for keeping ω_r/ω_n smaller than found in the above example (to reduce the lock-in time, or to exercise better control over the VCO (see eq. (6.98) etc.) in addition to striving for a larger ω_H (large tuning range of the VCO, dropping of the coarse tuning, etc.). In these circumstances the attenuation of a simple lag-lead filter with an additional RC section might be inadequate for some frequency synthesis applications. A rejection filter (a notch filter consisting of parallel-T RC networks etc.) suppressing selectively ω_r and its lowest harmonics presents a possible solution of the problem [6.1, chapter 7 and App. IX].

A twin T can supply about 60 dB attenuation at its rejection frequency,† causing no more than 20° phase-shift at 0·1 of that frequency (see Fig. 7.23) if matched so that the input/output impedance is small/large compared with the d.c. filter impedance (for the configuration of Fig. 7.23 $\omega_{\text{reject}} = 1/RC$).

Design of the VCO presents another crucial problem. It might seem desirable that the whole frequency band ($f_{\min}$ to $f_{\max}$) could be covered in one range without any switching of coils, oscillators, etc. Evidently for large ratios $f_{\max}/f_{\min}$ large $C_{\max}/C_{\min}$ are necessary and because of the limited tuning voltage interval ($-\Phi$; V_B), which cannot generally be fully exploited owing to saturation in the phase detector, operational amplifier, or VCO, varactors with

† The attenuation at the rejection frequency is theoretically infinite.

large γ (hyperabrupt) are preferable. However, in this case the simple eq. (6.107) no longer holds since the doping profile characteristic is voltage dependent [7.10]. Assuming piecewise linearization, (6.107) changes to

$$\left(\frac{f_H}{f_L}\right)_{\max} \leqslant \sqrt{\frac{C_{v,\max}}{C_{v,\min}}} \approx \prod_1^N \left(\frac{v_i}{v_{i-1}}\right)^{\gamma_i/2} \qquad v_0 = v_{\min};\ v_N = V_B \quad (7.46)$$

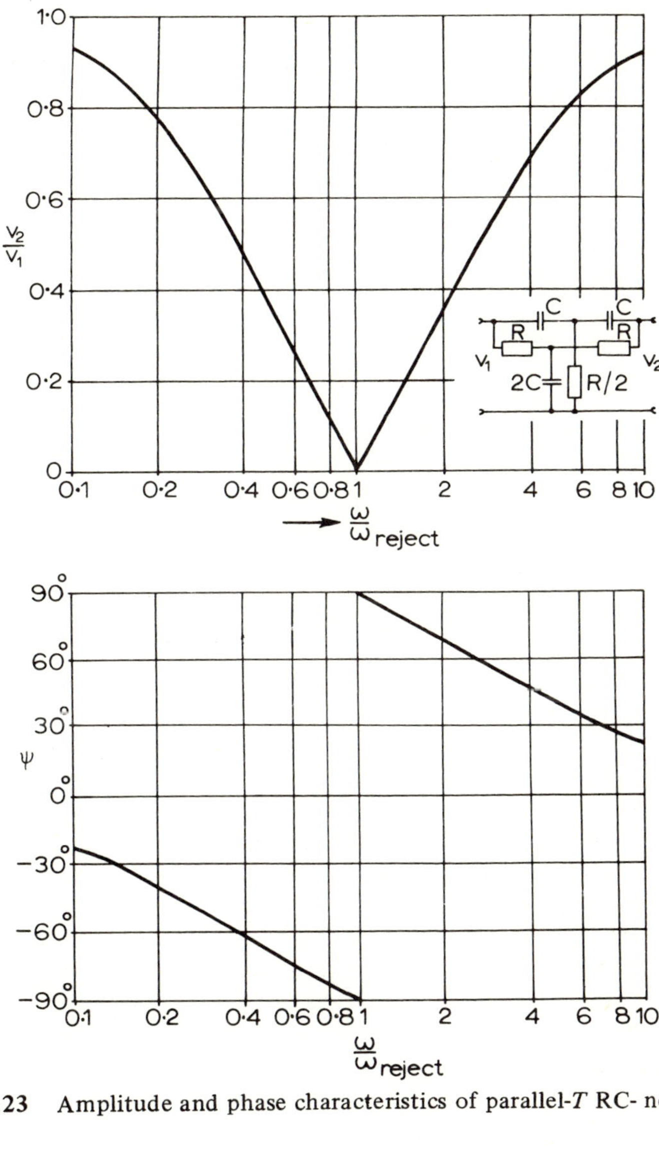

Fig. 7.23 Amplitude and phase characteristics of parallel-T RC- networks.

There are other features which must be taken into account, namely the Q factor of the voltage variable capacity, the large signal restrictions [2.5, App. E], and the nonlinearity of the frequency versus voltage characteristics of the VCO, i.e. of the respective K_0.

Considering the equivalent diode circuit in Fig. 2.3 it is clear that the reverse (saturation) current is responsible for lowering the Q at the low-frequency end, whereas at the high-frequency end the same effect is due to the series resistance R_s. Typical variation of Q with frequency is shown in Fig. 7.24. Further it should be emphasized that Q is a function of the junction capacitance and therefore of bias (see eq. (7.47) below).

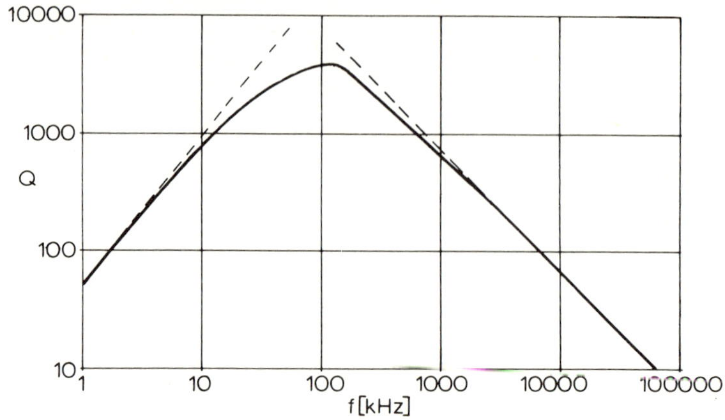

Fig. 7.24 Q factor of a varactor.

The cut-off frequency of the varactor depends upon the minimum Q of the circuit that can be tolerated.

$$Q = \frac{\omega L}{R} = \frac{1}{R}\sqrt{\frac{L}{C(v)}} \tag{7.47}$$

At high frequencies $R \approx R_s$, thus the application of (2.30) gives, for the case in which the circuit capacity is formed predominantly through $C(v)$

$$Q = \omega_c \sqrt{L \frac{C^2_{v,\,\min}}{C(v)}} \leqslant \frac{\omega_c}{\omega_H}\sqrt{\frac{C_{v,\,\min}}{C(v)}} \tag{7.48}$$

and the cut-off frequency of the varactor should be

$$\omega_c \geqslant Q_{\min}\,\omega_H \sqrt{\frac{C_{v,\max}}{C_{v,\min}}} > Q_{\min}\,\omega_H \left(\frac{f_H}{f_L}\right)_{\max} \tag{7.49}$$

If the RF voltage across the varactor is large then the maximum excursions of the tuning voltage must be limited in order that the sum of the d.c. and the instantaneous RF voltage lies between V_B and $-\Phi$. Further the average capacity C_{av} which determines the resonant frequency depends upon the RF excitation as well as on the d.c. tuning voltage. Thirdly, the power dissipation in the varactor caused by the RF voltage might be too large.

The condition for the resonance frequency f_x to be proportional to the tuning voltage v_x, in which case the gain K_0 would be constant, follows immediately from (6.107) or (7.46): the doping profile constant γ should be $\gamma = 2$ (assuming that the tank circuit capacity is formed predominantly by the voltage-dependent capacity, i.e. $C_1 \to \infty$ and $C_2 \to 0$ (Fig. 6.27). Unfortunately γ values larger than 0·5 (values just above 3 have been obtained) are voltage dependent, with initial and final values in the neighbourhood of 1/3 [7.10]. Evidently the gain K_0 is voltage dependent; typical variations for $\gamma = 1/3$ (voltage independent) and a $\gamma_{max} \approx 3{\cdot}2$ are shown in Fig. 7.25.

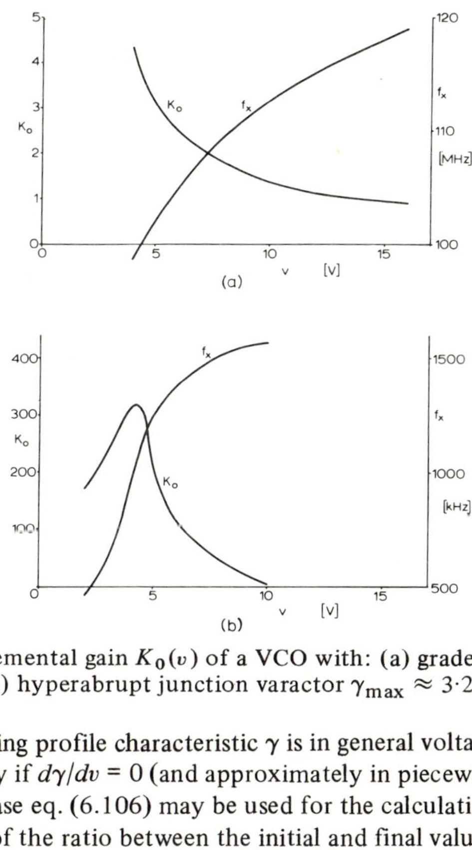

Fig. 7.25 Incremental gain $K_0(v)$ of a VCO with: (a) graded junction varactor $\gamma = 1/3$; (b) hyperabrupt junction varactor $\gamma_{max} \approx 3{\cdot}2$.

Because the doping profile characteristic γ is in general voltage dependent, eq. (6.106) is valid only if $d\gamma/dv = 0$ (and approximately in piecewise linearized intervals). In this case eq. (6.106) may be used for the calculation of the upper and lower bounds of the ratio between the initial and final values of K_0:

$$\frac{K_{01}}{K_{02}} = \left(\frac{f_1}{f_2}\right)^3 \left(\frac{C_1 + C_2 + C_{v2}}{C_1 + C_2 + C_{v1}}\right)^2 \left(\frac{v_2}{v_1}\right)^{\gamma+1} \leqslant \left(\frac{f_1}{f_2}\right)^3 \left(\frac{v_2}{v_1}\right)^{\gamma+1} \tag{7.50}$$

and

$$\frac{K_{01}}{K_{02}} = \frac{f_2}{f_1}\left(\frac{C_2 + C_{v2}}{C_2 + C_{v1}}\right)^2 \left(\frac{v_2}{v_1}\right)^{\gamma+1} \geqslant \left(\frac{f_2}{f_1}\right)^{(2-\gamma)/\gamma} \tag{7.51}$$

The inequalities are valid if $f_1 < f_2$.

In cases when an effectively voltage-independent K_0 is required (digital coarse pre-tuning, search oscillators in frequency synthesizers, etc.) steps for linearization must be undertaken. This may be achieved with hyperabrupt junction varactors if the operation is limited in the neighbourhood of γ_{max} (Fig. 7.25(b)), whereas with abrupt or graded junction varactors large biasing voltage must be used (see eq. (7.50) and Fig. 7.25(a)). But generally the designer must use nearly the whole range $(-\Phi; V_B)$ for tuning the VCO. In these cases the piecewise linearization of frequency-voltage characteristic helps. Starting with Fig. 7.25(a) two straight lines are drawn in such a way that all maximum departures $|\Delta f_x|_{max}$ from the actual f_x-curve are nearly the same. Now, noting the intersection voltage, $v_{intersec}$, and the slopes, K_{01} and K_{02}, of the approximating straight lines, the linearization circuit can be designed. A diode biased to $v_{intersec}$ causes the necessary abrupt change between the input voltage and the effective varactor voltage. If the required accuracy is not too high, say ± 5%, one linearizing diode is sufficient [6.17], but about fifteen diodes would be required to achieve a 0·5% accuracy. Another possibility is the introduction into the phaselock loop of an operational amplifier with a nonlinear gain that compensates changes in K_0.

Finally the temperature dependence of the voltage variable capacity is of interest; two parameters must be considered, the contact potential Φ (about 0·6 volts for silicon) decreasing at a rate of 1·5 to 2·0 mV/°C, and the dielectric constant ϵ with a normalized increase of about 2×10^{-4}/°C (both for silicon and germanium). The resulting fractional capacitance change is approximately

$$\frac{\Delta C}{C} = \left(\frac{1-\gamma}{\epsilon}\frac{\partial \epsilon}{\partial T} - \frac{\gamma}{v+\Phi}\frac{\partial \Phi}{\partial T}\right)\Delta T$$

$$\approx [0{\cdot}2(1-\gamma) + 1{\cdot}7\gamma/(v+\Phi)] \times 10^{-3} \quad /°\mathrm{C} \tag{7.52}$$

A simple temperature compensation to the VCO may provide a diode circuit which forces v to increase when Φ decreases [6.17].

To help with the choice of a suitable type of voltage variable capacity, the main properties of different varactors are summarized in Table 7.2.

7.6 PHASELOCKED DECIMAL-VARIABLE FREQUENCY SYNTHESIZERS

Simple digital synthesizers, described in the previous section, cannot be successfully used when large frequency bands extending from zero to hundreds

Table 7.2

Main properties of different varactors

Varactor type	Alloyed	Diffused	Planar diffused epitaxial	Hyperabrupt
Doping profile characteristic (Capacitance slope factor) γ	0·5	0·33	0·33 to approx. 0·46	Maximum above 3 (voltage dependent)
Maximum capacitance ratio	High	Medium–high	Mediocre–high	Very high
Varactor Q factor	Low	Mediocre	Very high	Medium–high
Reverse breakdown	Mediocre	High	Low	Mediocre
Typical application areas	VCO, FM, APC	VCO, FM, APC	APC and tuning in FM, VHF, UHF frequency band; VCO	AM tuning, VCO
Typical capacitance range available	6 to 100 pF	6 to 100 pF	5 to 50 pF	40 to 1000 pF

(From Sokoloff, Proc. Nat. Electronic Conf. 1967)

of MHz, with arbitrarily small frequency steps, must be covered. Cascading by means of appropriate dividers may be a partial remedy. The proper solution is the arrangement relying upon eq. (5.49), as explained in Section 5.3, and employing phaselock loops to simplify the selectivity problems. This particularly applies to the generation or separation of harmonic multiples (5.61), i.e. x_0, $x_0 + 1$, etc., and leads to a variety of designs.

(A) Decimal Modules with Phaselock Loops

The common feature is that each decimal module contains the phaselock loop which filters out the desired multiple, $x_0 + x_{0i}$, according to the tuning of the VCO or the feedback circuits respectively.

(i) Switching by Correct Setting of the VCO Frequency

When the decimal module must be switched merely by setting of the VCO frequency then the order of harmonic multiples is limited by (7.8). In a conservative design $x_0 < 50$ for an LC-oscillator. Fulfilling of condition (5.60) can be achieved when x_0 is split into two parts, $x_0 = x_0' + x_0''$, as in Example 5.3. The principle and arrangement of this decimal module is shown in Fig. 7.26. Note the very close resemblance to Fig. 5.10. The indicated numerical values are taken from an actual synthesizer [7.11]; because the

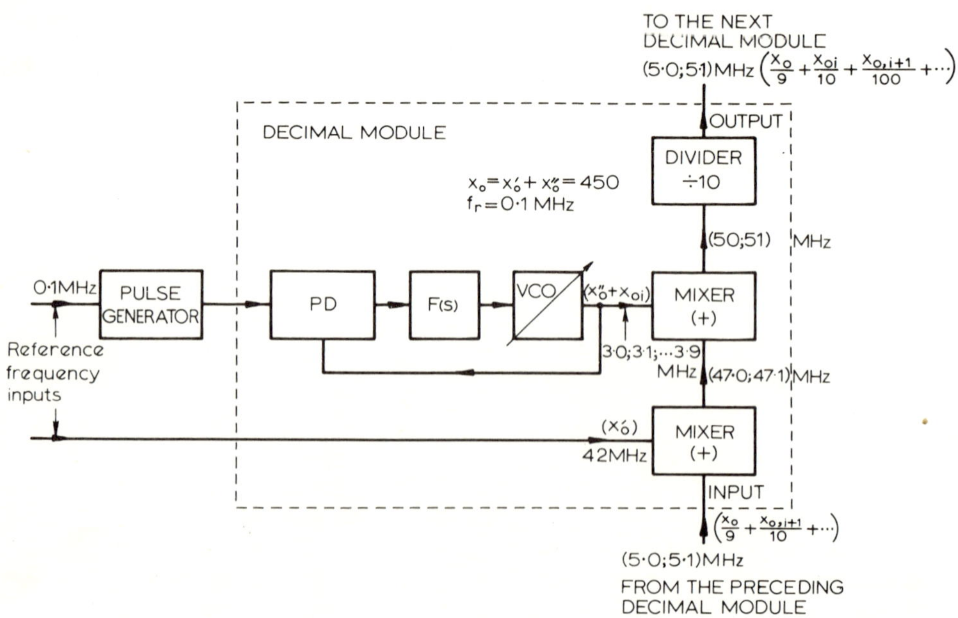

Fig. 7.26 Phaselocked decimal module; the desired harmonic is selected by tuning the VCO.

output frequency (50 to 51 MHz) is beyond the range of commercial counting dividers, phaselocked, and later injection-locked, oscillator dividers have been employed.

(ii) Counter with Variable Division Ratio in the Feedback Path

A phaselock loop having in the feedback path a counter with division ratio variable from $x_0, x_0 + 1$, to $x_0 + 9$ provides another solution of the problem (see Fig. 7.27). In this case no limitation is imposed on x_0. Numerical values in Fig. 7.27 are taken from an actual commercial synthesizer [7.12]. It may be noted that $x_0 = 180$ is just above the theoretical value given by (5.60). No additional tuning of the VCO is used and the pull-in procedure is facilitated by shunting diodes across the resistor in the loop filter (see Fig. 6.13). The advantage of this arrangement is simple remote tuning and programming of the decimal module and of the synthesizer as a whole.

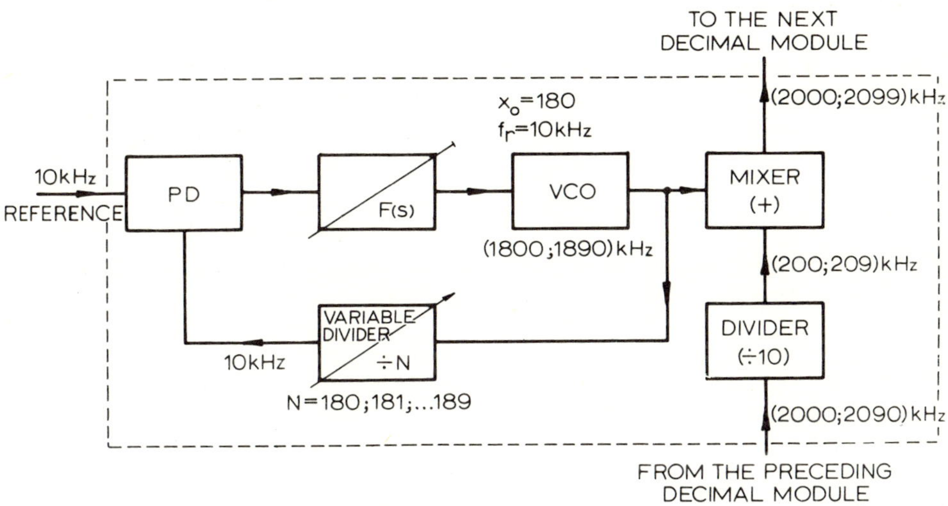

Fig. 7.27 Phaselocked decimal module; selection of the desired harmonic is secured by setting the division ratio of the feedback divider.

(iii) The Output Frequency Delivered by the VCO

The feature common to the following group of phaselocked decimal modules is that the output frequency is delivered by the VCO itself. This can be readily achieved by placing the mixer in Fig. 7.27 into the feedback path of the loop as shown in Fig. 7.28. An important property of this system is that the output signal is delivered by an oscillator which guarantees freedom from distant spurious components. Another advantage is that the output filter of the mixer

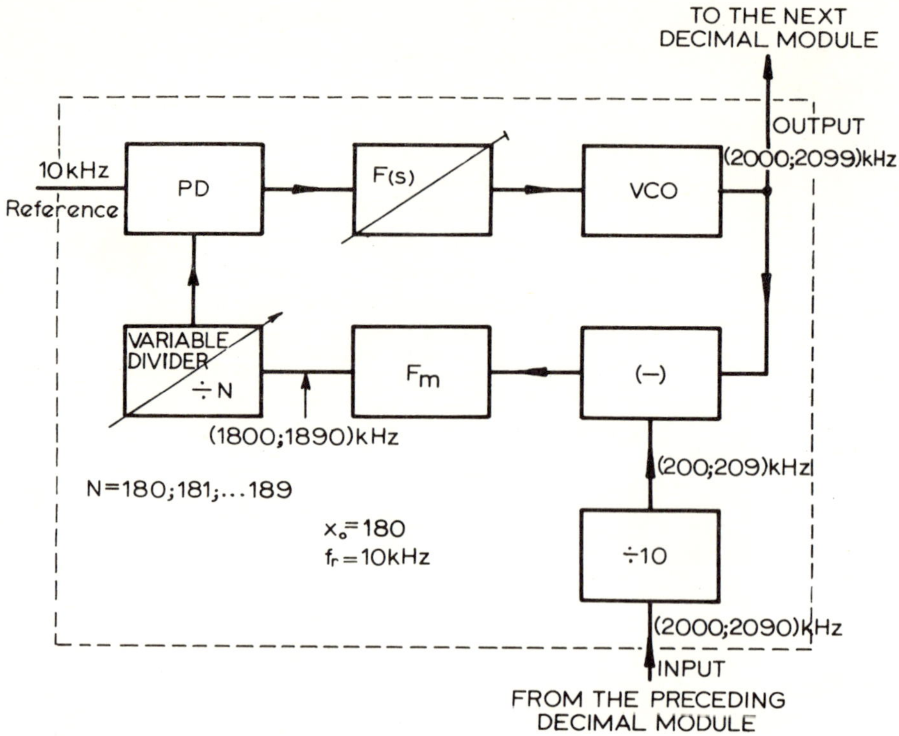

Fig. 7.28 Phaselocked decimal module; the feedback divider version with the output frequency delivered by the VCO itself.

F_m can be simpler because of the filtering properties of the loop. The desired harmonic, $x_0 + x_{0i}$, is chosen by setting the variable divider as explained above.

Many commercial synthesizers use decimal modules of this type; the individual concepts are slightly different, but the main properties are similar and will be illustrated with the aid of Fig. 7.29. The output frequency of the VCO is mixed with a suitable spectral line of f_r and in this way transposed in the pass-band of F_m; because its width is f_r another signal at this distance may eventually pass. This could lead to ambiguity in setting of the VCO (the ith) in the case where $x_{0,i+1}$ has a higher value (5, 6, etc.). Therefore additional information dependent upon the tuning of the preceding decimal unit, i.e. the $(i + 1)$th, must be supplied by mechanical or electrical coupling between the two oscillators; another complication is that the applied spectrum of f_r must be limited to prevent either direct transmission through the pass-band of the filter F_m or generation of image frequencies. Some producers solve the problem by simply tuning the harmonic generator output filter, BPF1, in accordance with the respective VCO (shown dashed in Fig. 7.29). Use of double-mixing is another possibility, as in an early design of the 'frequency decade' ND5 of the

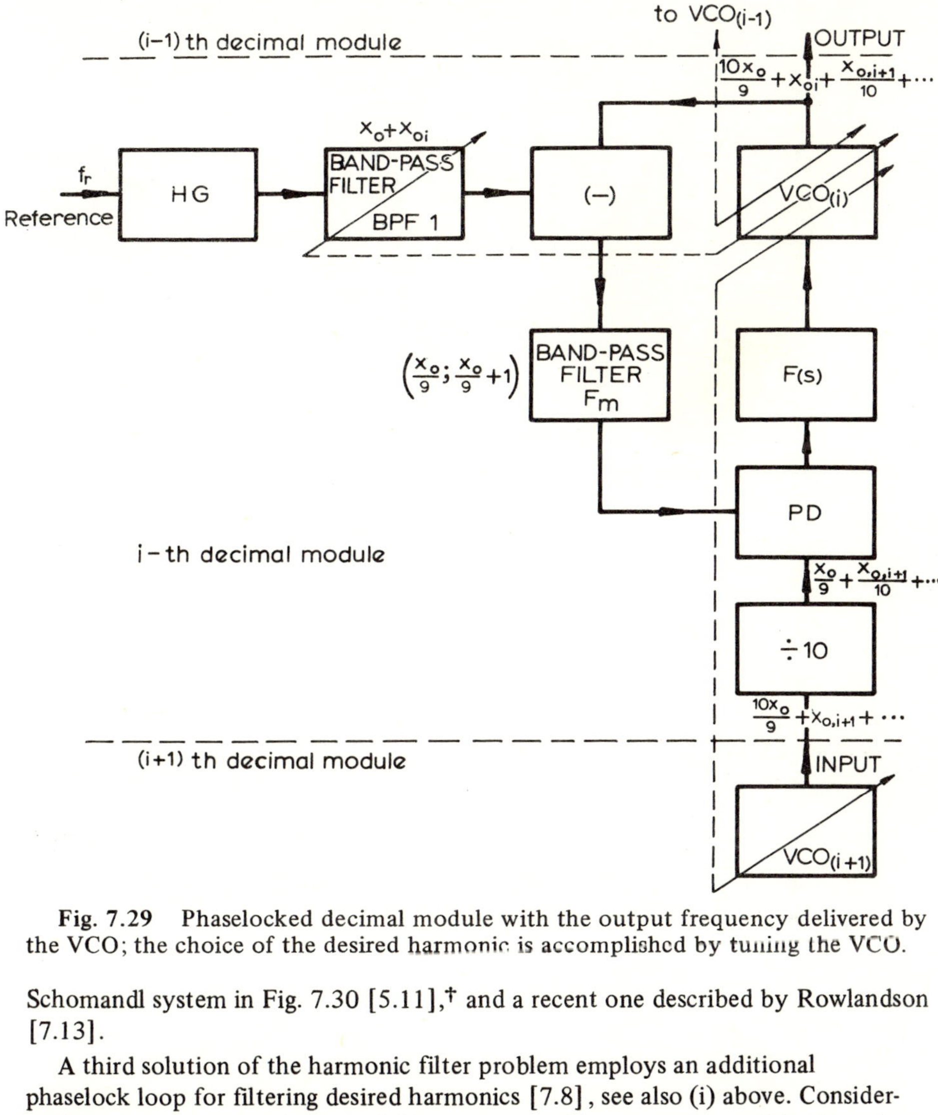

Fig. 7.29 Phaselocked decimal module with the output frequency delivered by the VCO; the choice of the desired harmonic is accomplished by tuning the VCO.

Schomandl system in Fig. 7.30 [5.11],† and a recent one described by Rowlandson [7.13].

A third solution of the harmonic filter problem employs an additional phaselock loop for filtering desired harmonics [7.8], see also (i) above. Considering Fig. 7.31 we conclude that all filters may be of low-pass type and further, that in agreement with (7.32), $x_0 > 41$ is necessary to exclude the synchronization of the VCO-1 on an image frequency. If this condition cannot be fulfilled the simultaneous tuning of both VCOs is the solution (shown dashed). The advantage

† Note that the permitted mixing ratio at M1 put the lower bound on x_0 (for $q_d \geqslant 7; x_0 > 27$). Another criterion is the tuning range of the VCO (assuming 1 to 1·2, we get $x_0 = 45$).

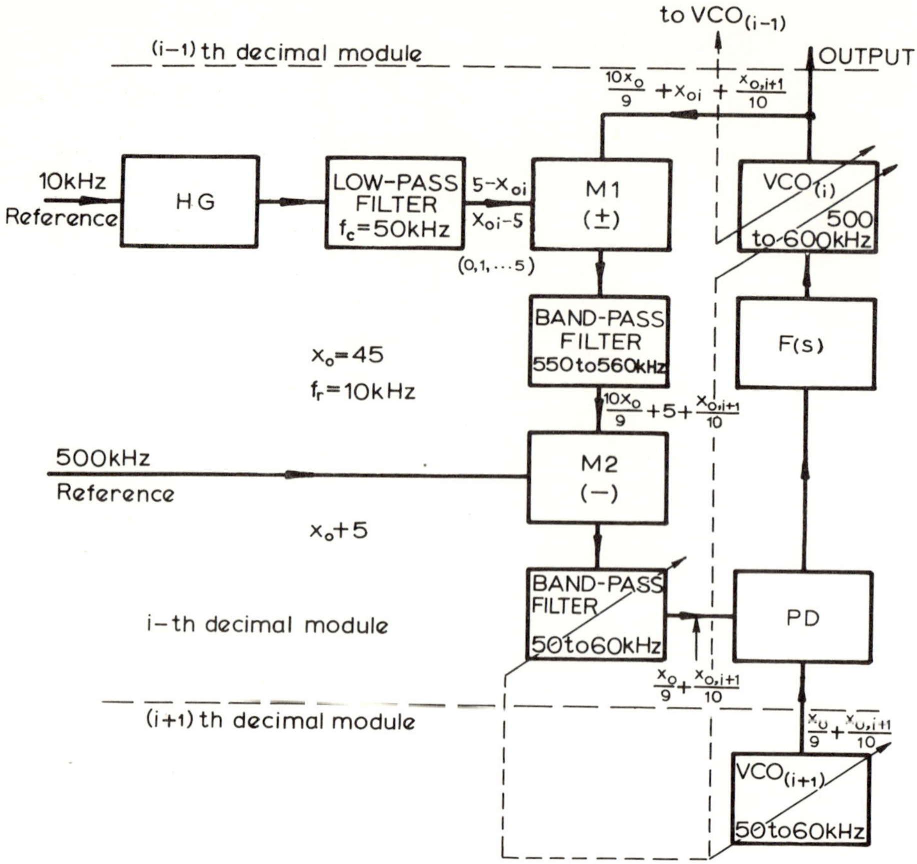

Fig. 7.30 Phaselocked decimal module (10 kHz) from the Frequency Decade ND5 of the early Schomandl system.

of this arrangement of the decimal module is that it is compatible with IC technology.

(B) Output Circuits

Concepts of decimal-variable phaselocked frequency synthesizers are based on ideas explained in Section 5.3 and illustrated in Example 5.3.

Smaller frequency steps, usually up to 100 kHz, are generated in a cascade of equal decimal modules; then in the simplest case the output mixer, in which the auxiliary frequency $f_r X_0$ is subtracted, follows (Fig. 7.32). But generally a special decimal module, having a similar concept to those in the cascade, but with ten times higher frequency levels, is inserted. Further, in systems where x_0 is much larger than required by (5.58) additional steps may be gained by

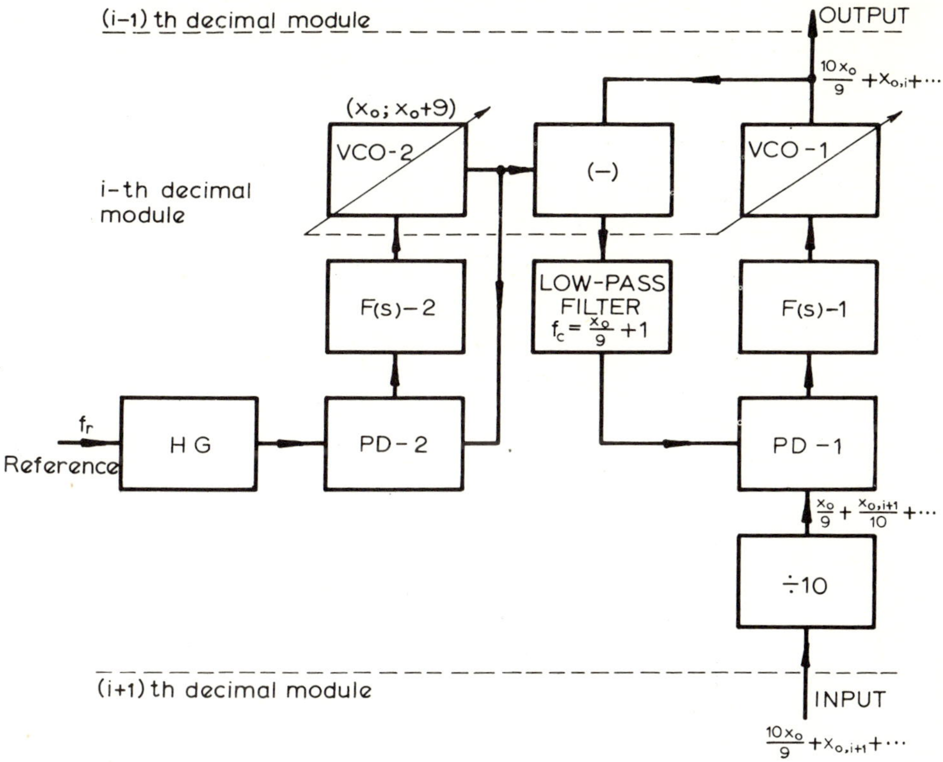

Fig. 7.31 Decimal module formed by a cascade of phaselock loops.

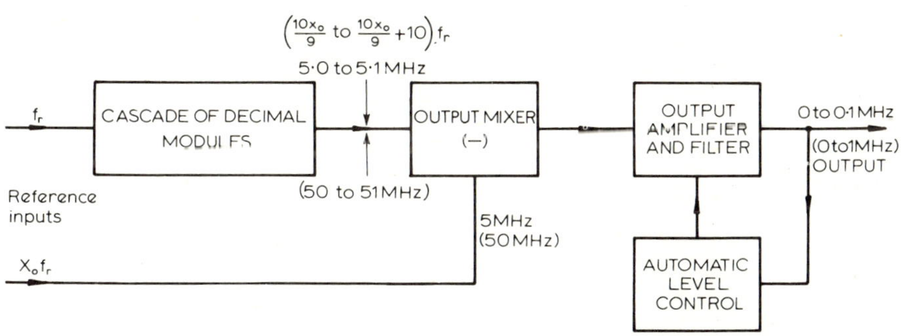

Fig. 7.32 Block diagram of simple decimal-variable frequency synthesizer composed of equal decimal modules only (numerical example is according to Fig. 7.26 – values in brackets are for the case where the last divider 10 : 1 is omitted).

subtracting $f_r(X_0 - x_{01})$ instead of f_rX_0 (see Fig. 5.11). There are systems in which the variable part of x_{01} extends from 0 to 10 or even to a much larger number; for example, to 20 in an actual synthesizer shown in Fig. 7.33.

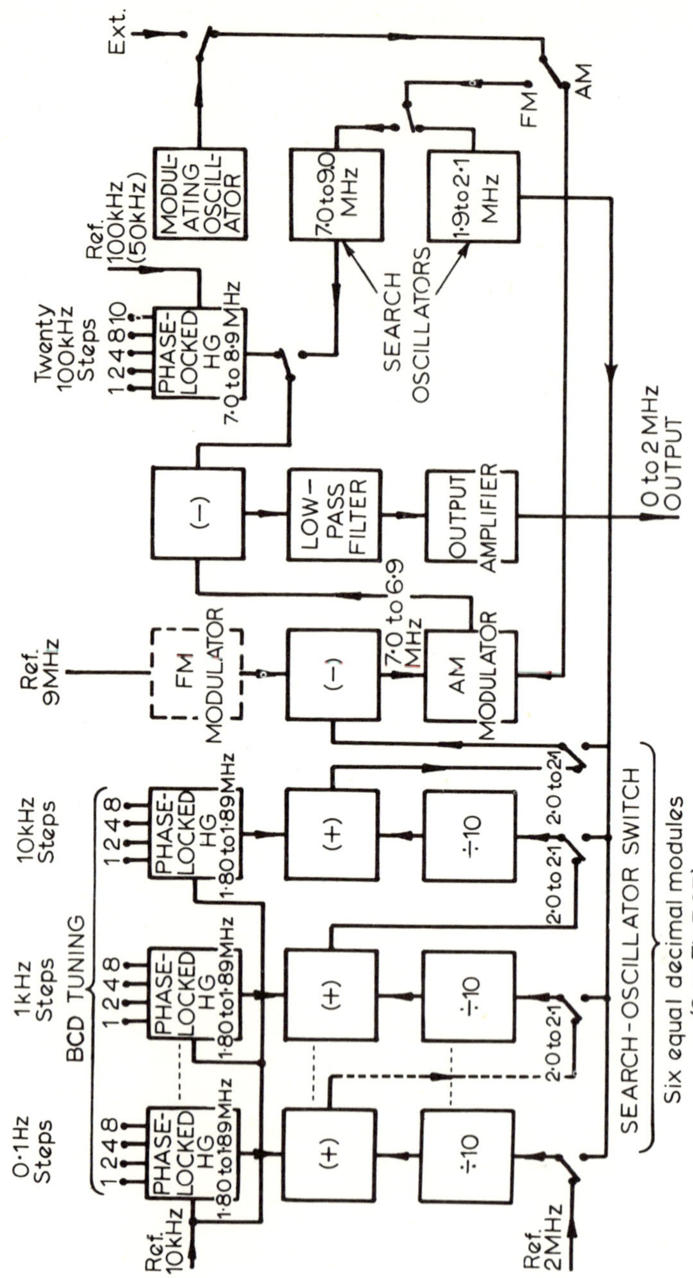

Fig. 7.33 Decimal-variable phaselocked frequency synthesizer with the extended variability of $f_r(X_0 - x_{01})$. (Courtesy of Adret-Electronique [7.12].)

Some applications require a modulated signal from the frequency synthesizer (when used as signal generator, transmitter exciter, etc.). This can be provided by suitably modified output circuits. One solution is indicated in Fig. 7.33. Unfortunately in this case the FM modulation index depends on the position of the 'search-oscillator switch', and in addition the output mean-frequency is not coherent with the reference frequency. These difficulties can be overcome by placing a separate FM modulator in the 9 MHz reference branch (shown dashed).

Generally, addition of a modulated auxiliary carrier solves the modulation problem [7.5, 7.13]. Because no nonlinear operation (frequency multiplication or division, amplitude limitation, etc.) can follow thereafter, the compensating branch of the frequency synthesizer (see $X_0 f_r$ in Fig. 7.32 or $(10x_0/9)f_r$ in Fig. 5.9) is preferred for the addition of the modulated auxiliary carrier. AM modulation can be introduced although at the output of the frequency synthesizer (the AM modulator can eventually be combined with the output-level control [7.13]).

7.7 CONCLUSIONS

When comparing a purely additive frequency synthesizer, as described in chapter 5, with versions employing phase-controlled oscillators, the most striking feature is the reduced size, weight and power consumption of the latter [6.12]. The reasons are both theoretical (i) and technological (ii):

(i) The phaselock loop technique makes possible frequency multiplication by very large factors in one stage only. The same is true for the upper bound of mixing ratios. Consequently single-frequency or narrow-band variable-frequency synthesizers can often be reduced, in practice, to a basic chain of one frequency divider and multiplier only; exceptionally additional synthesizing loops may be needed.

(ii) Generally, the VCO is the only circuit that contains an LC network; evidently, the widespread use of IC and hybrid technology with all the advantages of small size, cheap mass production, etc. is possible.

However, large-band decimal-variable frequency synthesizers cannot be solved with one phaselock loop only, even though successful units have been made [7.14], because with small frequency steps the switching times become intolerably long. This difficulty is overcome by employing a system of phaselock loops (Fig. 7.21, [7.15], [8.12] etc.) or by a mixed approach: the basic building blocks – decimal modules – are designed as phaselocked narrow-band frequency synthesizers, whereas their cascading proceeds as described in chapter 5.

Finally, it must be emphasized that owing to the phaselock technique and digital IC technology, the frequency synthesizer is no longer an expensive piece of laboratory equipment; it is an extensively used electronic device.

APPENDIX 7.1 IF FILTER IN THE PHASELOCK LOOP

When studying effects of the IF filter in the feedback path of the phaselock loop we shall assume that a small input phase disturbance $\varphi_i(t)$ causes an output phase disturbance $\varphi_0(t)$. The actual voltages in different points of the phaselock loop (with reference to Fig. 7.34) are:

$$v_i(t) = V_i \sin[\omega_i t + \varphi_i(t)] \quad |\varphi_i(t)| \ll 1 \tag{7.53}$$

$$v_0(t) = V_0 \cos[\omega_0 t + \varphi_0(t)] \quad |\varphi_0(t)| \ll 1 \tag{7.54}$$

$$v_a(t) = V_a \cos(\omega_0 - \omega_{0i})t \tag{7.55}$$

The lower sideband at the mixer output is

$$v_m(t) = V_m \cos[\omega_{0i} + \varphi_0(t)] \tag{7.56}$$

Limiting further discussions to a simple sinusoidal disturbance i.e.

$$\varphi_0(t) = -\ m_p \cos \nu t \tag{7.57}$$

the $v_m(t)$ will be

$$v_m(t) = V_m \left\{ \cos \omega_{0i} t + \frac{m_p}{2} [\sin(\omega_{0i} + \nu)t + \sin(\omega_{0i} - \nu)t] \right\} \tag{7.58}$$

If the following IF filter is symmetrical around ω_{0i} both sidebands undergo the same phase shift ψ_ν, but of opposite signs, and suffer an attenuation

$$|F_M(\omega_{0i} + \nu)| = |F_M(\omega_{0i} - \nu)| = |F_M(\nu)| \tag{7.59}$$

Supposing $F_M(\omega_{0i}) = 1$ the resulting output voltage is

$$v_{if}(t) = V_m \left(\cos \omega_{0i} t + \frac{m_p |F_M(\nu)|}{2} \{ \sin[(\omega_{0i} + \nu)t - \psi_\nu] + \sin[(\omega_{0i} - \nu)t + \psi_\nu] \} \right)$$

$$= V_m \cos[\omega_{0i} t - m_p |F_M(\nu)| \cos(\nu t - \psi_\nu)] \tag{7.60}$$

or when returning to the previous symbolism

$$v_{if}(t) = V_m \cos[\omega_{0i} t - |F_M(\nu)| \varphi_{0\nu}(t - T_M)] \tag{7.61}$$

In the locked state $\omega_{0i} = \omega_i$ and

$$v_d(t) = K_d[\varphi_{i\nu}(t) - |F_M(\nu)| \varphi_{0\nu}(t - T_M)] \tag{7.62}$$

The loop filter causes, at the frequency ν, an attenuation $|F(\nu)|$ and introduces a time delay T, i.e.

$$v_2(t) = K_d |F(\nu)| [\varphi_{i\nu}(t - T) - |F_M(\nu)| \varphi_{0\nu}(t - T - T_M)] \tag{7.63}$$

When substituting

$$v_2(t)\,K_0 = \frac{d\varphi_0(t)}{dt} \tag{7.64}$$

and generalizing to an arbitrary number of sinusoidal components in $\varphi_0(t)$, the application of the Fourier transform with the following transition to the Laplace transform reveals the loop equation

$$\frac{\Phi_0(s)}{\Phi_i(s)} = \frac{K_d\,K_0\,F(s)}{s + K_d\,K_0\,F_M(s)\,F(s)} \tag{7.65}$$

The result is that with respect to the small disturbing signal (small phase disturbance) the mixer and the IF filter behave as a black box with the transfer function $F_M(s)$, often called the 'modulation transfer function', which is approximately obtained by translating the actual filter transfer function to the zero frequency and retaining the part of positive frequencies only. The simple resonant circuit may serve as an example. Its transfer function is [1.12, Section 3.1]

$$\frac{\text{Actual current}}{\text{Current at resonance}} = \frac{R_0}{R + j(\omega L - 1/\omega C)} \approx \frac{1}{1 + jQ\,2\nu/\omega_0}$$
$$= \frac{1}{1 + 2j\nu/\omega_B} \tag{7.66}$$

Q is the circuit Q, ω_0 is the resonant frequency, $\nu = \omega - \omega_0$ is the off-resonance detuning, and ω_B is the 3-dB bandwidth. The respective modulation transfer function is evidently

$$F_M(s) \approx \frac{1}{1 + s/(\omega_B/2)} \tag{7.67}$$

which is equivalent to a simple RC filter with the time constant $2/\omega_B$ (the respective corner frequency is $\omega_B/2$).

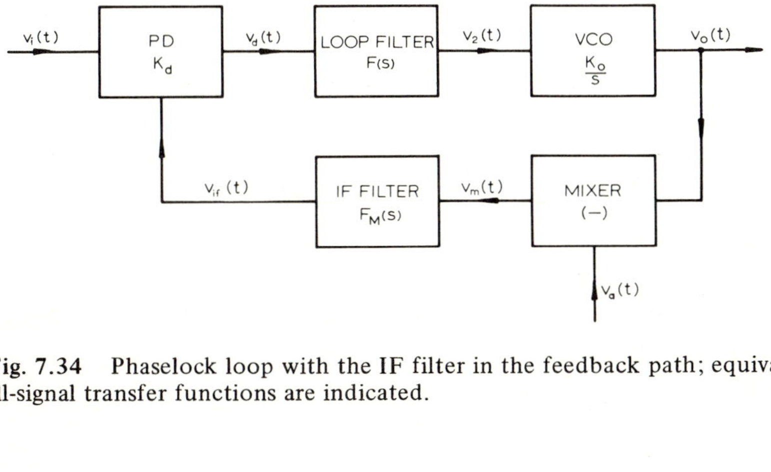

Fig. 7.34 Phaselock loop with the IF filter in the feedback path; equivalent small-signal transfer functions are indicated.

However, there are other problems. If the modulation transfer function $F_M(s)$ has four or more poles, a false lock may be encountered. Considering an unlocked loop (with $\varphi_i(t) = 0$) and the detuning exceeding the lock-in range but not the pull-in range

$$\Delta\omega_L < (\omega_i - \omega_{0i}) = \Delta\omega_i < \Delta\omega_P \tag{7.68}$$

then

$$v_d(t) = K_d\,(\sin\Delta\omega_i t + \text{high frequency comp.}) \tag{7.69}$$

Assuming the validity of (6.71)

$$v_2(t) \approx K_d \frac{T_2}{T_1} \sin\Delta\omega_i t \tag{7.70}$$

After substituting in (7.64), and integrating, we get the output phase disturbance

$$\varphi_0(t) \approx -\frac{K_d K_0}{\Delta\omega_i}\cdot\frac{T_2}{T_1}\cos\Delta\omega_i t = -\frac{\Delta\omega_L}{\Delta\omega_i}\cos\Delta\omega_i t \tag{7.71}$$

The voltages $v_m(t)$ and $v_{if}(t)$ are effectively the same as above, eq. (7.58) and (7.60). One of the sidebands

$$\frac{1}{2}\frac{\Delta\omega_L}{\Delta\omega_i}|F_M(\Delta\omega_i)| \sin[(\omega_{0i} \pm \Delta\omega_i)t \mp \psi_{\Delta\omega_i}]$$

has a frequency equal to the input frequency ω_i and generates at the PD output a d.c. component, v_P, which is responsible for the pull-in effect (see Section 6.3(C)):

$$v_P = \frac{1}{2} K_d \frac{\Delta\omega_L}{\Delta\omega_i}|F_M(\Delta\omega_i)| \cos\psi_{\Delta\omega_i} \tag{7.72}$$

Two phenomena must be noted, namely, reduction of the pull-in voltage in proportion to the IF-filter attenuation, and a further reduction in proportion to the cosine of the additional phaseshift angle. It is evident that for $\psi_{\Delta\omega_i} = \pi/2, 3\pi/2$, etc. v_P disappears completely. This problem is discussed in more detail by Gardner [6.2, chapter 7]; of practical importance is the statement that if the difference frequency exceeds the frequency of the first null of $\cos\psi_{\Delta\omega_i}$ (unstable), the loop will never pull-in to the correct lock.

The last phenomenon to which attention must be drawn is the reduction of the original pull-in range $\Delta\omega_P$. In the case of the simple resonant circuit, eq. (7.45) may be employed (for the first approximation).

REFERENCES

[7.1] BREIDING, R. J. and VAMMEN, C. RADA frequency synthesizer. *Proc. of the 21st Annual Frequency Control Symposium*, pp. 308-30. Fort Monmouth, N.J., 1967.

[7.2] CLAPP, J. K. and LEWIS, F. D. A unique standard-frequency multiplier. *I.R.E. Nat. Conv. Record*, 1957, Part 5, pp. 131-6.

[7.3] STITCH, M. L., ROBINSON, N. O. and SILVEY, W. Parametric diodes in a maser phaselocked frequency divider. *I.R.E. Trans. on Microwave Theory and Technique* **MTT-8**, 218-21, March 1960.

[7.4] VESSOT, R., *et al.*, An intercomparison of hydrogen and caesium frequency standards. *I.E.E.E. Trans. on Instrumentation and Measurement*, **IM-15**, 165-76, Dec. 1966.

[7.5] VALDORF, H. and KLINGER, R. Die Entwicklung einer hochkonstanten dekadischen Kurzwellensteurstufe für den Bereich 1,5 . . . 30 MHz. *Frequenz*, **14**, 335-43, Oct. 1960.

[7.6] MOOSER, L. Precision offset exciter equipment XZO for suppression of TV common channel interference. *News from Rohde-Schwarz,* **7**, 40-3, May 1967.

[7.7] HUGHES, R. J. and SACHA, R. J. The LOHAP frequency synthesizer. *Frequency,* **6**, 12-21, Aug. 1968.

[7.8] NOORDANUS, J. Frequency synthesizers – a survey of techniques. *I.E.E.E. Trans. on Communication Technology*, **COM-17**, 257-71, April 1969.

[7.9] WOODBURY, J. R. Phase-locked loop pull-in range. *I.E.E.E. Trans. on Communication Technology*, **COM-16**, 184-6, Feb. 1968.

[7.10] SOKOLOFF, L. IC voltage variable capacitors (VVC). *I.E.E.E. Trans. on Broadcast and Television Receivers,* **BTR-15**, 33-40, Feb. 1969.

[7.11] NOYES, A., Jr., BYERS, W. F. and LOHRER, G. H. 'Coherent decade frequency synthesizers', a set of articles in *General Radio Experimenter*, Sept. 1964, May 1965, Nov.–Dec. 1965, Sept. 1966 (Summarized in the G.R. Company reprint E119), May-June 1969, Jan.–Feb. 1970.

[7.12] Adret-Electonique. 2 MHz signal generator-synthesizer – Codasyn 201. *Instruction Manual,* Trappes, July 1969.

[7.13] ROWLANDSON, G. A. G. Frequency synthesis techniques. *Industrial Electronics,* **6**, pp. 320-23, Aug. 1968, and 355-9, Sept. 1968.

[7.14] GILLETTE, G. C. The digiphase synthesizer. *Frequency Technology,* 7, 25-9, Aug. 1969.

[7.15] SHANAHAN, J. C. Uniting signal generation and signal synthesis. *Hewlett-Packard Journal*, **23**, 2-13, Dec. 1971.

8 Applications of Frequency Synthesizers

This survey primarily concerns decade (or decimal) frequency synthesizers. The versatility of these devices has been long recognized but their practical employment was limited mainly to frequency measurements because of the relatively high level of spurious signals at the output. However, growing demands for frequency stability in science, research, and communications in the last decade, together with technological progress, has accelerated development of frequency synthesizers to such a degree that today widespread use is possible not only for precise laboratory measurements but also for production-line testing of various kinds.

Another important field of application is in communications. Nowadays, generation of carrier and heterodyning frequencies in decade or special-purpose synthesizers is not only common practice in transmitting and receiving centres but is also applicable in portable, aircraft-borne, and even in spacecraft-borne equipment. Consequently spectrum-saving modulation methods may be widely used and a high degree of traffic automation can be achieved.

Finally, applications of frequency synthesizers in frequency and time laboratories will be discussed. It is no exaggeration to state that the easy access and the widespread utilization of the primary standard (atomic frequency), an unusual occurrence in metrology, would not be possible without the help of frequency synthesis.

8.1 APPLICATIONS IN MEASUREMENT TECHNIQUES

A decade frequency synthesizer is an active source which generates discrete frequencies in an extremely wide frequency band with long-term stability equal to that of the reference or local standard. The 'tuning accuracy' is not dependent on the output frequency and incremental steps can be made as small as desired and can be selected by either manual or electronic command. The frequency synthesizer is a superb measuring device, the introduction of which in the mid

50s has revolutionized many measuring procedures: simplification of operation has been combined with improved precision and resolution. It is not surprising, therefore, that frequency synthesizers are now used not only as laboratory instruments but also as production-line tools. And in production their compatability with remote (digital) programming and control is of considerable importance.

(A) Measurement of Frequency

Direct digital measurement of frequency with electronic counters, introduced in the early 50s, can hardly be simplified. Hence the use of frequency synthesizers will be justified only:

(i) in exceptional circumstances when no suitable counter is available,
(ii) when the input signal is weak, noisy, keyed, or otherwise disturbed,
(iii) when the input frequency is beyond the counting capabilities of direct digital measurement,
(iv) when an analogue record is desired,
(v) when very precise frequency measurements are required (multiplication of frequency differences, etc.).

(i) The Frequency Synthesizer as a Simple Frequency Meter

The digital measurement (comparison) of an unknown frequency with the use of a decade synthesizer is very simple (see Fig. 8.1). First the search oscillator is

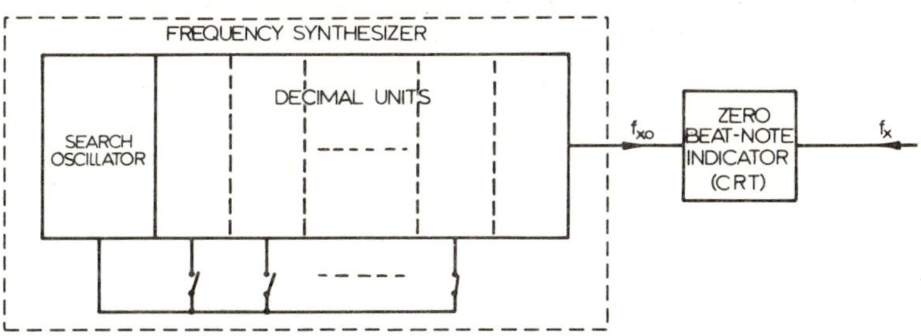

Fig. 8.1 Digital measurement (comparison) of frequencies with the frequency synthesizer.

connected to the most significant decimal module (see also Fig. 5.11 or 7.33) and the beat-note with f_x is found; audio indication at this stage of measurement is helpful. According to the first digits displayed on the search-oscillator scale the setting of the preceding decimal modules is made, and the narrowing procedure is repeated until no further increase in precision is possible. In the last stages of measurement the oscilloscope is a convenient zero-beat-note indicator if the short-term stability of f_x is such that a nearly-standing Lissajous figure can be established.

When estimating the final accuracy one must consider:

(a) the fractional frequency deviation of the standard frequency $(\Delta f_s/f_s)_{rms}$
(b) the frequency deviation introduced by the frequency synthesizing process $(\Delta f_0)_{rms}$ (Fig. 9.5(a)) in terms of the output frequency,
(c) the uncertainty in reading of the interpolating (search) oscillator $(\Delta f_i)_{rms}$ in terms of the output frequency,
(d) the uncertainty in establishing the beat-note frequency $(\Delta f_b)_{rms}$ in terms of the output frequency.

$$\overline{\left(\frac{\Delta f_x}{f_x}\right)^2} = \overline{\left(\frac{\Delta f_s}{f_s}\right)^2} + \frac{\overline{(\Delta f_0)^2} + \overline{(\Delta f_i)^2} + \overline{(\Delta f_b)^2}}{f_x^2} \tag{8.1}$$

At low frequencies the two last terms predominate as is illustrated in Fig. 8.2.

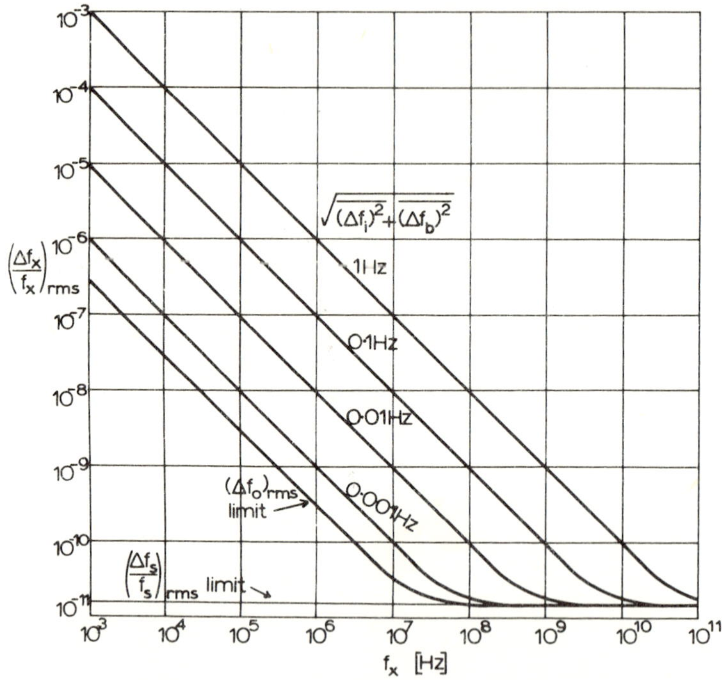

Fig. 8.2 The accuracy of digital frequency comparison using a typical frequency synthesizer (see Fig. 9.5); the assumed averaging time is about 1 sec., $(\Delta f_s/f_s)_{rms} = 1{\cdot}5 \times 10^{-11}$, $(\Delta f_0)_{rms} = 3 \times 10^{-4}$, and $\sqrt{\overline{\Delta f_i^2} + \overline{\Delta f_b^2}} = 1, 0{\cdot}1, 0{\cdot}01, 0{\cdot}001$ Hz.

(ii) Frequency Comparisons of Disturbed Signals

Measurement of carrier frequencies of remote transmitters is a typical example. Into the antenna input of the accurately tuned communications receiver the output signal from the frequency synthesizer is injected and the beat-note established (Fig. 8.3). The coarse audio indication is simple and helps in making the final optical display of the zero beat-note on the tuning indicator or on an auxiliary CRT. The local signal should be attenuated to be comparable with the

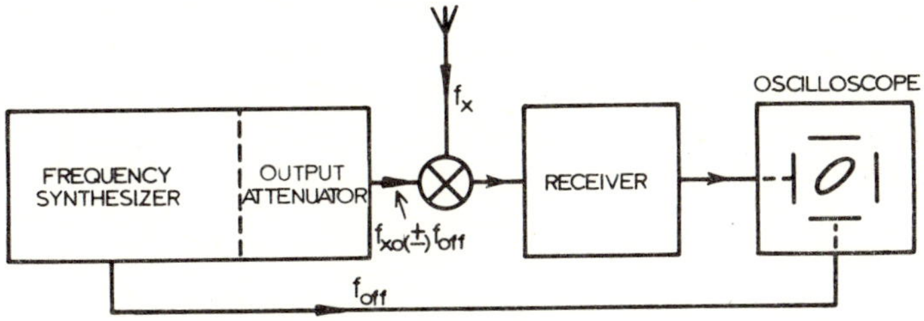

Fig. 8.3 Arrangement for carrier frequency measurements of remote transmitters.

received one. If rapid fading interferes with measurement the frequency synthesizer is offset, say by 1 kHz, and the audio output of the receiver compared on an oscilloscope with the standard frequency of 1 kHz. This arrangement provides additional narrow-band filtering and thus reliable establishment of the zero-beat-note even in the presence of interference, modulation, keying, etc. (Space and mark frequencies in phase-shift keying can be measured in this way.)

(iii) Extension of Counting Ranges by Frequency Synthesis

In the simplest case the ordinary frequency synthesizer solves the problem, as illustrated in Fig. 8.4. If the input frequency, f_x, is not known approximately, the harmonic multiple (Nf_s) employed must be established. To do this the synthesizer output frequency f_s is offset by Δf_s in such a way that the counted frequency increases; the difference in the counter readings Δf_x divided by Δf_s gives N

$$N = \Delta f_x / \Delta f_s \tag{8.2}$$

Δf_s must be sufficiently large to exclude any ambiguity in N. This difficulty is overcome in commercial versions of Fig. 8.4 called 'frequency converters'. The frequency synthesizer is reduced to a mere harmonic generator and mixer in

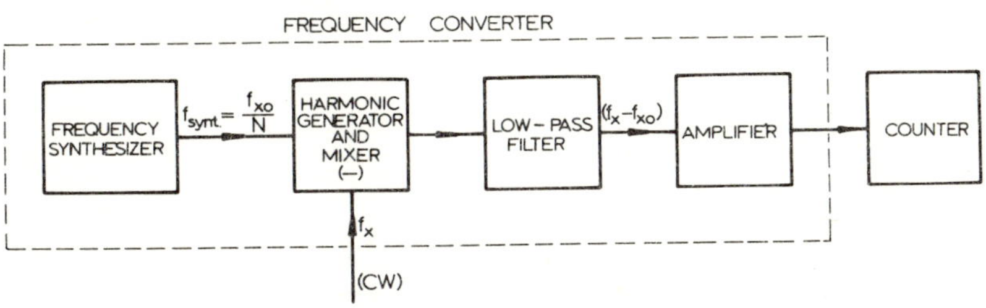

Fig. 8.4 Frequency synthesizer extending the range of the electronic counter.

which known multiples of 10, 50, (etc.) MHz are subtracted (depending upon the counter cut-off frequency). In this way the available 'plug-in' units extend the frequency range of counters (with resolution below 100 MHz) into the microwave region (18 GHz).

Another approach is prescaling of the input frequency until the counter range is reached. A weakness of this method is that the precision of measurement is reduced in proportion to the dividing factor. On the other hand the simple and automatic operation, peculiar to counters, is retained. Application of phaselocked dividers makes possible extension into the microwave region [8.1].

(iv) Analogue Recording of Frequency Deviations

When monitoring variations (stability) of a frequency f_x the digital record supplied by a counter-printer assembly does not provide an easy survey of the behaviour. This difficulty can be overcome by analogue measurement of the difference between f_x and the reference, f_{x0}, generated in a decade synthesizer (Fig. 8.5). It is advisable to use the output low-pass filter (LPF) with the smallest possible cut-off frequency according to the f_x or the expected variations of f_x.

Frequency measuring and deviation recording assemblies of this type are commercially available covering ranges up to 18 GHz and even above.

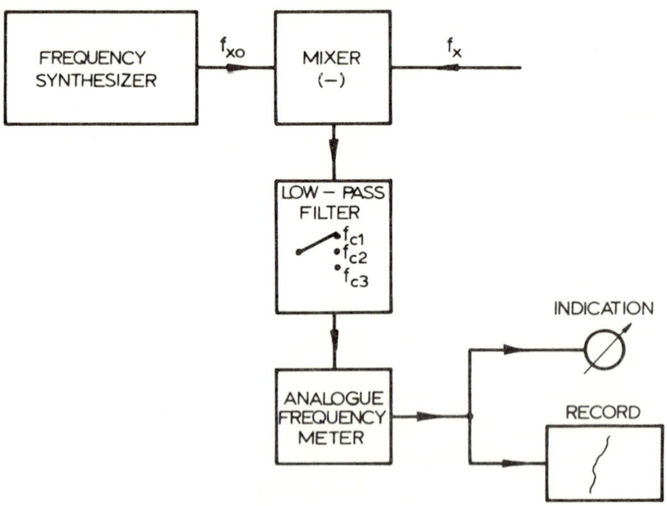

Fig. 8.5 Frequency monitoring assembly providing an analogue record.

(v) Very Precise Frequency Comparisons

Frequency synthesizers, both special purpose and commercial, enable very precise frequency comparisons to be made in reasonably short time without having to use beat-notes between extremely high harmonics in the microwave range. The method is known as the 'error multiplication technique' and will be

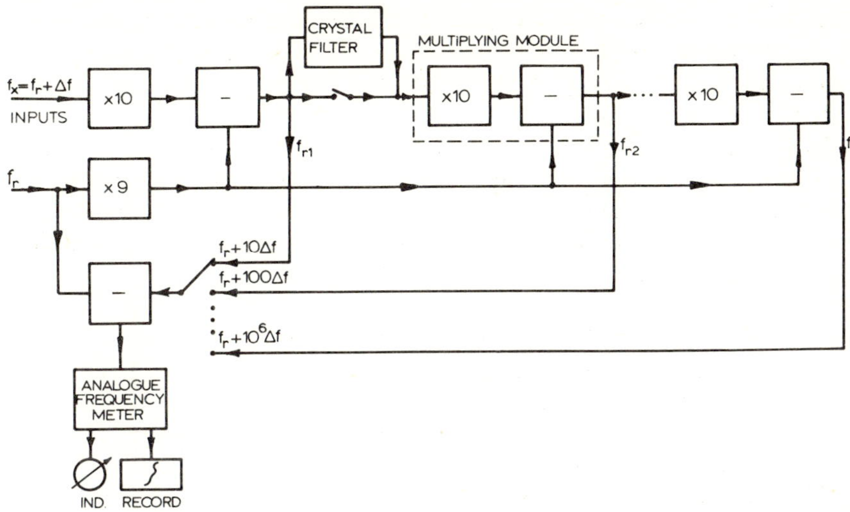

Fig. 8.6 Frequency difference multiplication with a special-purpose synthesizer. (Courtesy of Parzen Research, Inc.)

explained with the help of the block diagram in Fig. 8.6. One of the input frequencies is multiplied by 10 and the other by 9, and subtracted.

$$f_{r1} = 10(f_r + \Delta f) - 9f_r = f_r + 10\Delta f \tag{8.3}$$

It is clear that the frequency difference has risen ten times. Repeating the procedure once more the 'error' will be increased a hundred times, and so on. Extreme care must be exercised to prevent spurious couplings between individual 'multiplying modules' because all are tuned to the same frequencies and are operated at high signal levels. Nevertheless commercial 'Frequency comparators' of this type with multiplication factors as high as 10^6 are marketed.†

A phaselocked version of the 'frequency error multiplier' may readily be realized with a conventional decade frequency synthesizer if it is composed of equal decimal modules containing dividers (as described in Section 5.3 and 7.6). One of the frequencies to be compared is used as the local standard for the synthesizer, either directly or through a phaselock, and the other is phaselocked with its output frequency via the search oscillator as shown in Fig. 8.7 [8.2]. An important advantage is that the input frequencies, f_{x1} and f_{x2}, need not be in a simple quasiharmonic relation. The necessary conversion is performed by the setting of the frequency synthesizer. Further, it may be noted that with respect to the small difference between the input frequencies f_{x1} and f_{x2} the equipment behaves as a phaselocked frequency multiplier. The describing equation is that of (7.12) with $N = 10^n$, n being the number of (÷10) dividers in the synthesizer involved.

† Parzen model 106 A.

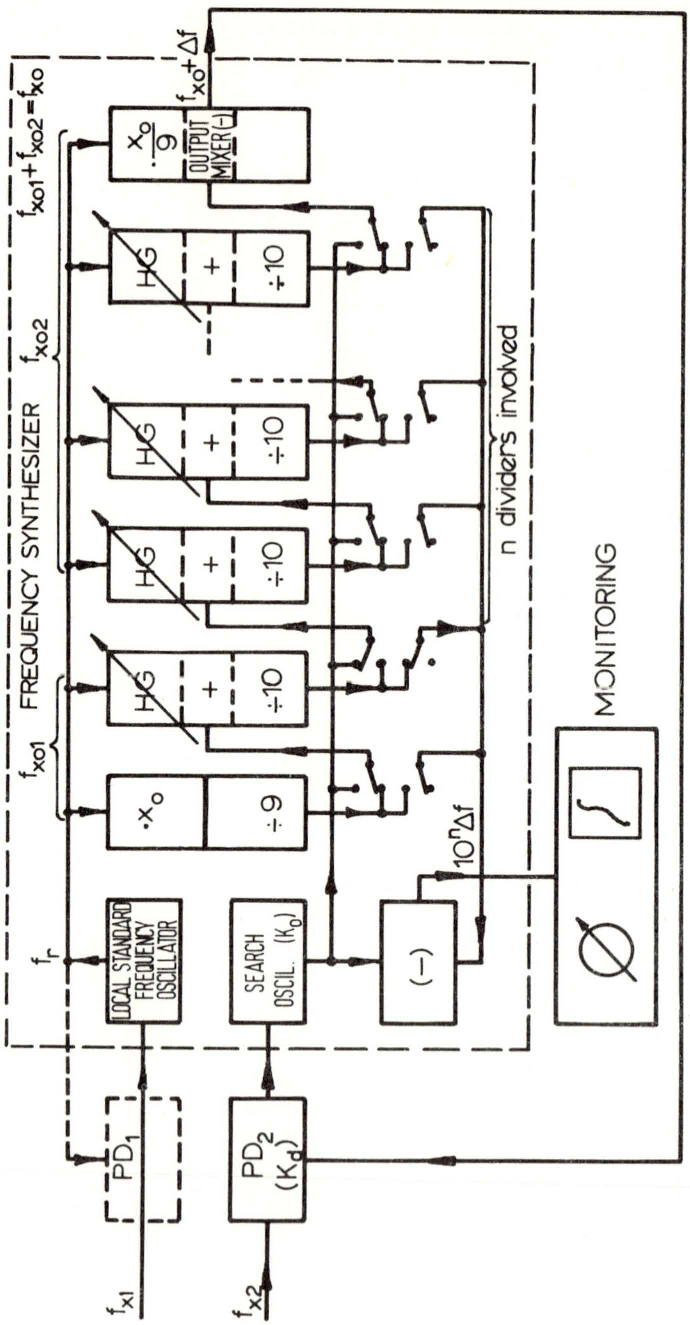

Fig. 8.7 Phaselocked frequency difference multiplication employing a commercial decade synthesizer.

Example 8.1

Estimation of the natural frequency ω_n and the noise bandwidth B_L of the phaselocked difference frequency multiplier in Fig. 8.7. Assuming:

$K_0 = 10^5/3$ (Hz/volt) (found in frequency synthesizer specifications)

$K_d = 6/\pi$ (volt/radian)

$N = 10^5$

If no loop filter is used

$$\omega_n = K' = \frac{K_d K_0}{N} = 4 \quad (2\pi \text{ Hz})$$

$$B_L = K'/4 = 1 \quad (\text{Hz})$$

(B) Measurement of Phase

If two identical a.c. signals, differing in phase by the amount ψ, are mixed with a common 'local oscillator' the phase information is transferred, unchanged, to the resulting IF signals. Making the IF frequency very low the phase difference can easily be measured as the respective time delay. A direct digital indication is achieved when the IF period T_{if} and the time measuring unit T_M are in a suitable coherent or quasicoherent relation. For a hundredth degree resolution

$$T_{if}/T_M = 36\,000 \tag{8.4}$$

Using a 10 MHz counter the IF frequency should be

$$f_{if} = 1/T_{if} = 10^7/36\,000 = 277{\cdot}\bar{7} \quad (\text{Hz}) \tag{8.5}$$

Evidently the 'local oscillator' must be offset from the input frequency, f_x, by a relatively small value f_{if}, whereas the additional phase error should be kept to the order of 10^{-3} degrees. Application of the frequency synthesis technique enables these severe requirements to be met, as shown in the block diagram in Fig. 8.8. The local oscillator frequency is generated in a frequency synthesizer which is roughly tuned to $f_{x(\pm)}f_{if}$, and the coherence condition (8.5) is achieved by phaselocking via its search oscillator. Note that, in many cases, only a few decimal units are needed in the 'local-oscillator-synthesizer'.

When phase measurements of the highest precision have to be made, the input frequency f_x must be correspondingly stable. The advantage of generation of the input frequency in another frequency synthesizer is evident. This suggests a modified solution in which the phaselock loop can be omitted. However, the 'local-oscillator-synthesizer' must have enhanced resolution to provide a sufficiently close approximation of f_x + 277·$\bar{7}$ Hz. According to Maxwell [8.3], using $f_{LO} = f_x + 277{\cdot}78$ Hz, a 10 MHz counter, and averaging over 1000 successive phase intervals, results in reading-to-reading variations of less than 0·002°.

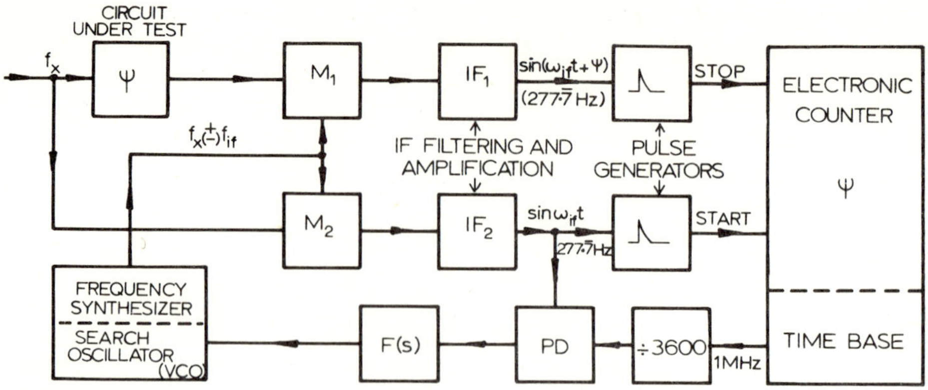

Fig. 8.8 Application of the frequency synthesis technique for digital phase measurements.

(C) Other Measurement Applications

If a variable frequency synthesizer is equipped with AVC together with coarse and fine output attenuators and AM, FM, PM, etc. modulators (see Fig. 7.32 and 7.33) it becomes a true signal generator, with which all types of sensitivity and selectivity measurements can be performed.

In the paragraphs below some particular applications of frequency synthesizers are mentioned. But the list is by no means complete.

(i) Measurement of Sharply Tuned Devices

The ease of finding transmission or insertion loss characteristics of filters with extremely steep flanks must be emphasized. Here, the stability and resolution provided by a synthesizer are essential. The same is true for examination of resonance curves and for finding resonant frequencies of circuits with very high Q. Testing of crystal filters and resonators may be mentioned as a typical example.

(ii) Spectrum Analysis

High resolution of audio analysers can easily be exploited for the narrow-band spectral analysis of RF signals when heterodyned with suitably offset frequencies generated in a frequency synthesizer with sufficiently low output noise. Measurement of modulation distortion, spurious discrete signals, including power frequency sidebands, suppression of the carrier in SSB systems, etc., may be made in this way. The actual measuring arrangement can be that of Fig. 8.7 if the audio analyser examines the 'd.c. output' of the phase detector PD_2; see also Fig. 9.8.

(iii) Frequency Synthesizer as a Sweep-Frequency Generator

When the frequency of the search oscillator varies periodically, the frequency synthesizer becomes a sweep-frequency generator capable of being swept with precision over ranges as small as a fraction of a Hz and as large as many kHz (according to the position of the search oscillator switch, see Fig. 8.9). In each case a marker, indicating the centre frequency etc., will help with orientation in the displayed frequency characteristic. A possible solution of this problem is

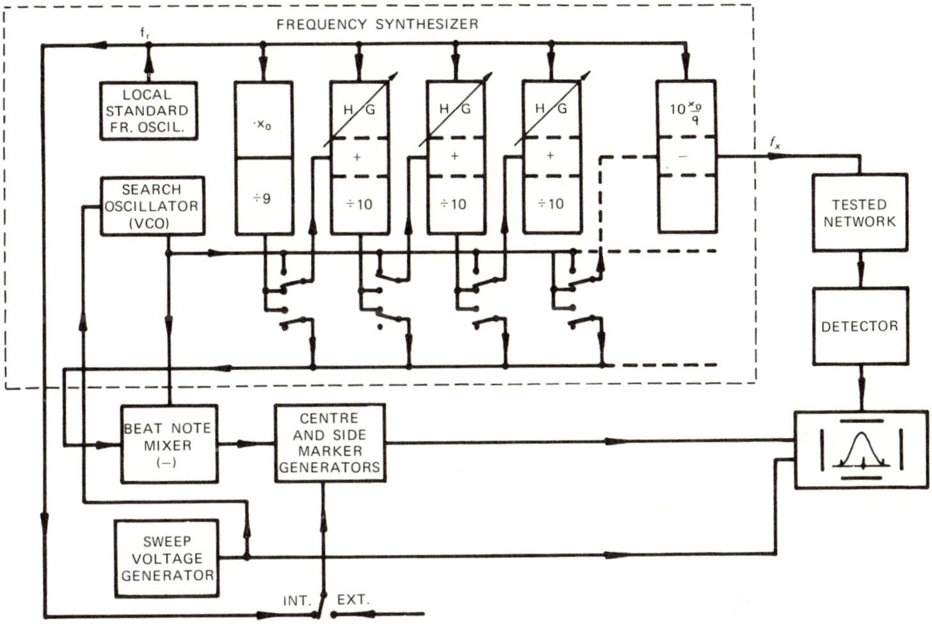

Fig. 8.9 Decade frequency synthesizer as a precision sweep generator.

shown in Fig. 8.9. The centre-frequency marker is generated as the beat-note between the search oscillator and the output of the disconnected part of the frequency synthesizer. The advantage of this arrangement is that the position of the marker depends upon the state of all decimal modules. This is not the case with many commercial decade synthesizers. Often the best which can be done is to heterodyne the search oscillator frequency with the auxiliary standard frequency $(x_0/9)f_r$, see equation (5.50). In this case the generated marker indicates the frequency set only on decimal modules to the right from the actuated search oscillator switch. When modulating one of the heterodyned frequencies suitable side markers can be added [8.4, 7.11].

Particularly easy orientation is achieved by digital sweeping when the output frequency of the synthesizer is set in discrete steps according to a programme. In this case short switching times are of advantage.

(iv) Check-Out of Very-Low-Frequency Equipments

Very-low-frequency synthesizers (below the audio range) are valuable in many applications, including the check-out of medical and geophysical equipment, servo-mechanism analysis, etc. The usefulness of these synthesizers is enhanced if they provide a second output which can be phase-shifted in known amounts with respect to the primary output [8.5]. The principle is illustrated in Fig. 8.10. The synthesizer has two output mixers which are fed, in parallel, by a common cascade of decimal modules (determining the desired output frequency). The desired phase difference is introduced by the subtracted auxiliary carriers $f_r X_0$. In the simplest case a 360-times higher frequency is divided by a counter which has two outputs, one fixed, and the other responsive to each of the 360 logic states in accordance with the information stored in the 'phase programmer'.

Another approach to the phase variability of the synthesizer output is the 'Digiphase' system described in [7.14].

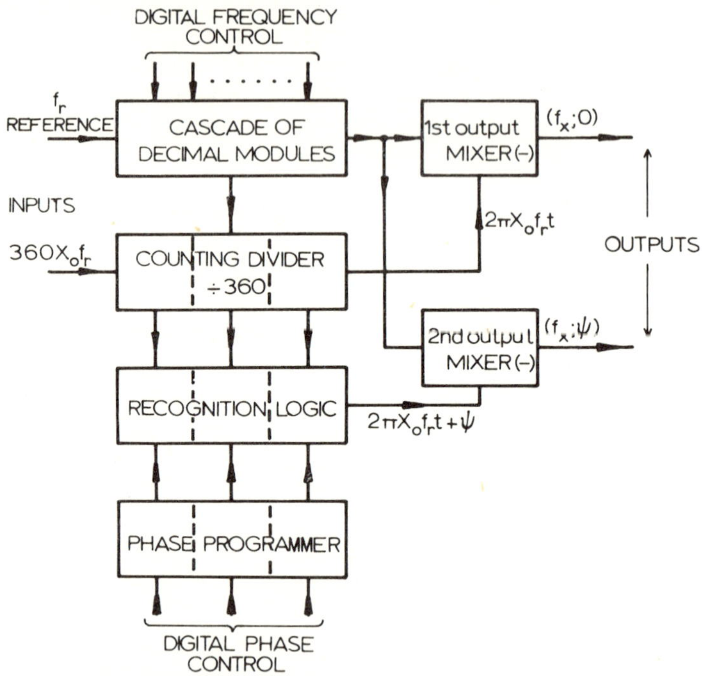

Fig. 8.10 Frequency synthesizer with an additional phase-programmable output.

(v) Velocity Measurements

Velocity measurements of moving objects (space vehicles etc.) involve determination of small frequency shifts (Doppler shifts) of signals in microwave bands, often deeply embedded in noise. Therefore receivers having bandwidths of but a few Hz are used [6.2]. Evidently to keep the incoming signal in the centre of the IF bandwidth, even when the velocity of the vehicle changes, the local oscillator must track the frequency shift. Here the synthesizer is ideal because its frequency can be changed in known selectable increments.

(vi) Resonance Spectroscopy

The growing areas in physical research that are coming to depend upon high-precision frequency determination for studying physical events (nuclear magnetic resonance (NMR), electron spin resonance (ESR), etc.) make good use of the wide frequency range, narrow-band sweeping, high resolution, etc. of frequency synthesizers.

To illustrate the problems further we shall have a look at nuclear magnetic resonance spectroscopy, which achieves not only identification of elements but also the qualitative and quantitative structure of molecules. In this method the strength of an applied d.c. magnetic field and the frequency of a simultaneously applied RF field needed to produce nuclear resonance in the investigated material must be known. Clearly the use of a frequency synthesizer for generation of RF excitation frequency simplifies the procedure in many ways and at the same time lends itself to automation of the testing process.

8.2 APPLICATIONS IN COMMUNICATIONS

Parts of the electromagnetic spectrum employed for point-to-point communication of information form a sort of common property which must be carefully utilized. The prevailing congestion in HF bands makes it out of the question to seek additional frequency allocations. The only effective way to increase the number of the communications channels is to improve the spectrum utilization by reducing frequency tolerances, adopting single-sideband (SSB) modulation and telegraph multiplexing techniques.

Nevertheless, in the most congested communications bands even several services, transmitting the same or different programmes, are placed in a common channel (synchronized broadcasting on medium waves, co-channel television systems, etc.).

All efforts for better utilization of the electromagnetic spectrum, and for reduction of mutual interference, end with claims for increased stability of the respective carrier frequencies. The general solution to this problem is the application of frequency synthesizers, fed from local, or if necessary off-air, atomic standards, on the transmitting, and often also the receiving ends of the information channels.

(A) Frequency Synthesizer as Transmitter Exciter

Choice of an appropriate frequency synthesizer depends on many factors: above all on the number of 'discrete' frequencies to be generated and in which frequency bands, and further, on the type of the communications services to be provided, on the tolerances to be secured, and on the specified 'tuning' or switching times, etc. Often initial cost will play an important role.

(i) One Discrete Frequency

There are services the assigned frequencies of which are not changed for years or even ever. In these instances the use of a commercial decade synthesizer is unnecessarily costly. The proper solution is to design a suitable single-frequency synthesizer as explained in Sections 5.2 and 7.4 (see also Examples 5.1 and 7.1).

A new idea is the 'all-band single-frequency synthesizer' [8.6], with the 'hardware' (see Fig. 8.11) practically identical for all communications bands. The actual single-frequency design is in the 'software', based on the concept of approximating frequency synthesizers as explained in Section 7.4(D). The following numerical example illustrates the procedure.

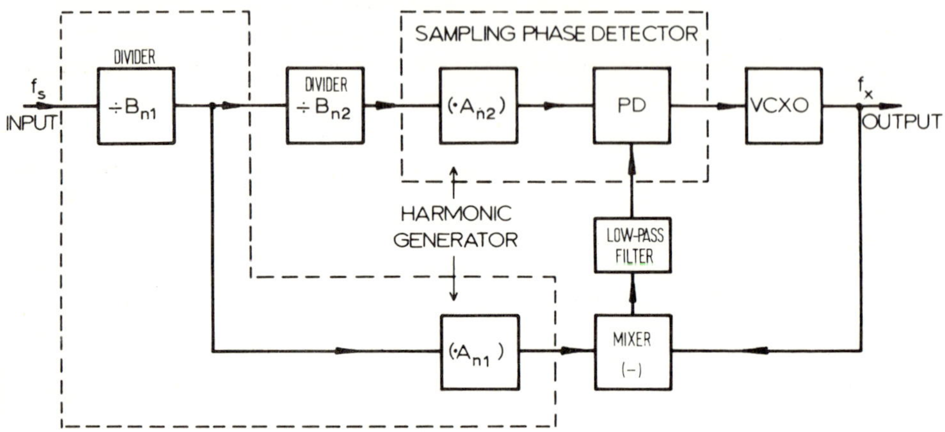

Fig. 8.11 An all-communications-bands 'single-frequency' synthesizer. (Kroupa [8.6].)

Example 8.2

Design a single-frequency synthesizer for a system of three co-channel television transmitters working in the precision offset mode, i.e. mutual instability must be below ±2·5 Hz.

Choosing f_{x0} = 495·250 MHz and assuming the multiplication factor in the power stages of the transmitter to be y_0 = 72, then with f_s = 5 MHz eq. (5.1) gives

$$f_{x0}/y_0\, f_s = X_1/Y_1 = 1981/1440$$

Progressing to eq. (5.4) through the continued fraction algorithm the fourth approximation gives (z_1 = 1)

$$x_1/y_1 = A_{4,1}/B_{4,1} = 11/8 \qquad \text{and} \qquad X_2/Y_2 = R_1\, B_{4,1} = 1/180$$

In the case of the offset carriers (±10·4 kHz in the television systems with 625 lines and 25 frames) the coarse approximation, $A_{4,1}/B_{4,1}$, is the same but the residual errors are different, viz.

$$R_1^{\pm} B_{4,1} = 1/180 \pm 13/56\,250$$

Their expansion according to the continued fraction algorithm leads to the fine approximations

$$A_{5,2}^{+}/B_{5,2}^{+} = 37/6394 \quad \text{and} \quad A_{5,2}^{-}/B_{5,2}^{-} = 16/3005$$

When computing the residual errors one will find that

$$\Delta f^{+} = 0{\cdot}38 \quad \text{Hz} \quad \text{and} \quad \Delta f^{-} = 0{\cdot}67\,\text{Hz}$$

which is well within the indicated tolerance range ±2.5 Hz.†

(ii) Restricted Number of Discrete Frequencies

Typical of this category is the mercantile-marine radiocommunications [8.7], to which a small number of channels is allocated in the HF and VHF frequency bands. Pressure for more channels leads to the gradual adoption of SSB systems and consequently of frequency synthesizers for transmitter driving because frequency tolerances cannot be otherwise met. The effort to keep purchase cost as low as possible results in systems based on partial synthesis only (see Fig. 8.12).

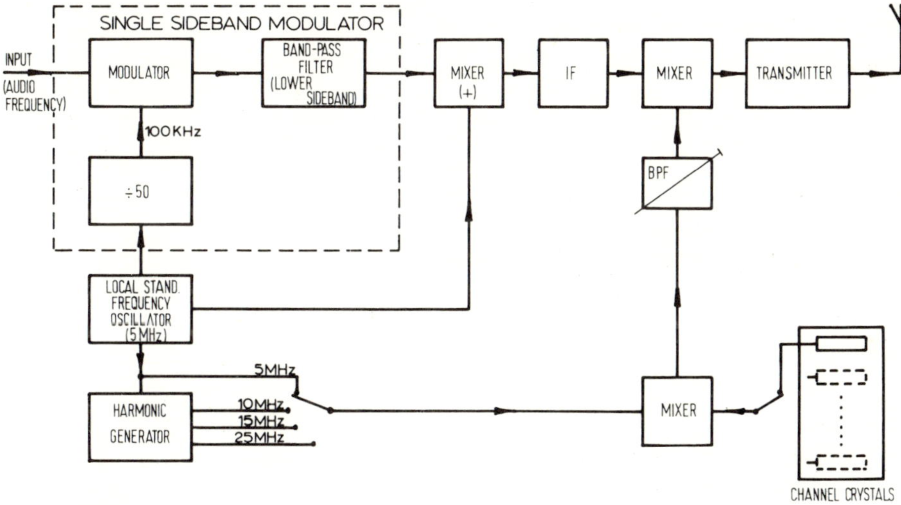

Fig. 8.12 A partial-frequency-synthesis system. (From McDonald & Burnham [8.7].)

The obvious disadvantages of this arrangement are that the output frequency is composed of two independent components and that a large number of channel crystals is used. The aging problems experienced in the past are also not encouraging

† Another solution of the above problem is described in [7.6].

for this simple system. Digital and approximating frequency synthesizers described in Sections 7.5 and 7.4(D) respectively, relying on IC technology, provide another and competitive solution.

(iii) Large Number of Discrete Frequencies

This is the case encountered with short wave (3 to 30 MHz) communications services. Long-distance point-to-point links can be established with relatively small HF power, but the changing conditions of propagation require the use of the carrier frequency best suited to the time of day and to the season. The need to make the transmitter as flexible as possible led to the adoption of variable LC oscillators (synchronized, on chosen discrete frequencies with individual crystals) as driving generators. A serious drawback of this solution is limited frequency stability, which is insufficient for some communications systems (SSB etc.). It is not surprising, therefore, that as soon as the problem of spurious signals in the synthesizer output was mastered to a degree compatible with CCIR requirements the widespread application of frequency synthesizers for driving short wave transmitters started – in the mid 50s.

About ten years later the advent of the frequency-following type of transmitter, which in association with a frequency synthesizer will automatically tune itself on remote instructions, made possible a high degree of traffic automation [8.8], [8.9]. In principle the incoming modulation signals may be switched to any transmitter of the selected group and the chosen transmitter associated with the appropriate aerial system (see Fig. 8.13). The last point deserving attention is the

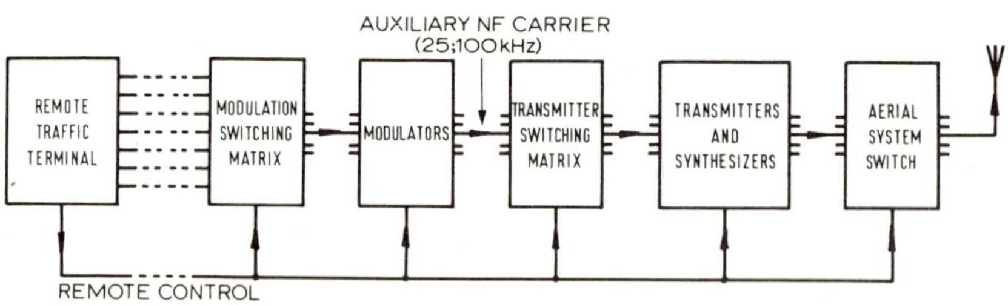

Fig. 8.13 Simplified arrangement of the remote control of a short-wave radio station.

actual modulation process in the transmitter. If an ordinary decade synthesizer is used an additional synthesizing loop must be added at the transmitter input [8.9]. A difficulty is that a desirable mixing ratio ($q > 7$) cannot always be maintained. Another solution provides a special frequency synthesizer with modulation input as discussed in Section 7.6(B), and in [7.5], [8.8], and others. In this case the synthesizer output frequency is the modulated carrier frequency itself, and the output power level and the impedance must be appropriate.

(B) Frequency Synthesizer as Local Oscillator in Receivers

This particular application requires the extreme of spectral purity, some −100 dB for discrete spurious components and correspondingly low phase noise. This extreme requirement may best be met if one retains the conventional super-hetrodyne-receiver structure but takes care to cancel the local oscillator drift, either by means of direct frequency synthesis or by the application of phaselock. Both approaches, however, result in digital setting, which precludes the normal ease of receiver tuning, and this some users find undesirable.

(i). Drift-Cancelling Loops

Drift cancelling loops, described in Section 5.4, may be adopted for receiver frequency control [8.7], [8.10]. The idea will be explained with reference to Fig. 8.14. Two modifications of the conventional HF receiver should be noted: the local oscillator is tuned in steps (f_s = 100 kHz, 1 MHz), and the pass-band of the first IF filter is that of one incremental step plus twice the largest drift of the local oscillator that might be encountered. Further there is an auxiliary branch, a simplified version of the main receiver, driven by the harmonics of the 'step-frequency' f_s supplied by the local standard frequency oscillator. Its IF frequency is offset by the amount of the second IF of the main receiver and that of the fixed part of the 'interpolation' or the 2nd local oscillator which in its turn tunes between the drift-compensated frequency steps and thus becomes the focus of the stability problem. Clearly the use of a frequency synthesizer here also is desirable (note the second drift-cancelling loop in Fig. 8.14). When choosing the frequency of the 'first local oscillator' well above the received HF band (see the numerical values in brackets in Fig. 8.14) a mere low-pass filter at the receiver (and the auxiliary branch) input will be sufficient to suppress image reception and direct IF leakage.

(ii) Phaselocked Local Oscillators

Digital synthesizers described in Section 7.5 present an excellent solution when the received frequency bands are relatively narrow, with a restricted number of communications channels, and if the VCO frequency after all divisions (i.e. the reference frequency) can be kept reasonably high to secure both spectral purity and acceptable switching times. Receivers for special services in the VHF and UHF frequency bands (see also Section 8.2(C)) are particularly suitable for this type of design because the phaselocked local oscillator can be the simple digital synthesizer shown in Fig. 7.19. Moreover it has been demonstrated that compact, electronically tuned, AM (550 to 1620 kHz) and FM (88 to 108 MHz) home receivers are feasible using this exact system and IC circuits [3.12].

On the other hand frequency synthesizers in communications receivers covering the relatively large HF band (3 to 30 MHz) cannot be so simple. Generally some sort of frequency analysis is employed. The idea has been explained in Section 7.3(B). Here we shall refer to an actual equipment [8.11],

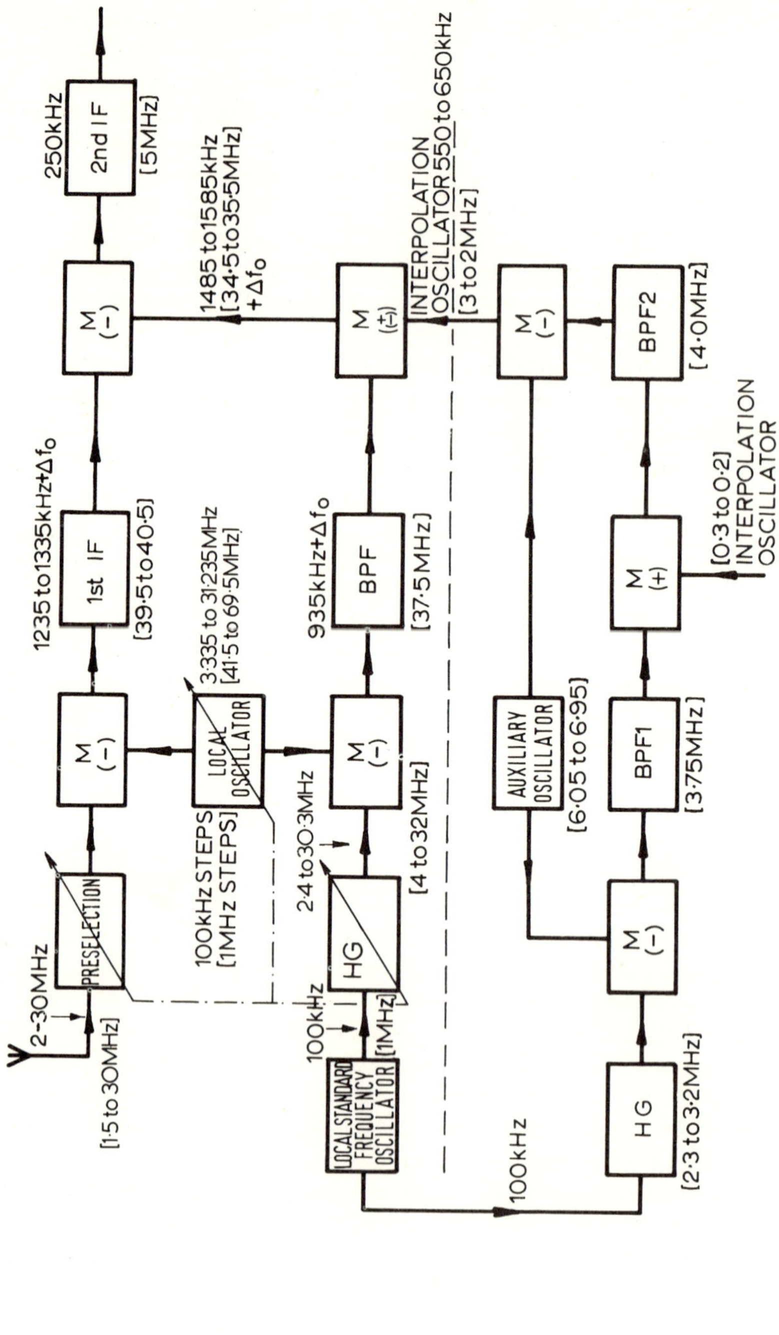

Fig. 8.14 Frequency synthesizing loop cancelling drift of the local oscillator in an HF receiver. (Numerical details from [8.7] and [5.12].)

using the systematic decade arrangement (Fig. 8.15), and point out the practical precautions that have been taken to prevent undesired VCO image-lock. The whole range (3 to 13 MHz) has been subdivided into four with widths of 1 or 2 MHz. At the output of the mixer M_5 is a band-pass filter of 1 to 2 MHz which is equivalent to a reference frequency of 1·5 ± 0·5 MHz; according to (7.35) the VCO must be pretuned to better than ±1 MHz. This tolerance is narrowed down to ±200 kHz because of the following stage, which has the output pass-band between 200 to 300 kHz. The respective filter, BPF_4, secures the preselection to the next 'analysing decade' and prevents image-locking if the effective reference frequency is $f_{r,\,\mathrm{eff}} > 50$ kHz; actual values are from 70 to 80 kHz. The resulting pass-band of BPF_5 through BPF_2 is very narrow and phaselock acquisition must be assisted by tracking and sweeping (not shown) of the VCO.

Frequency analysis can also be performed with a cascade of phaselock loops (see Fig. 7.21). The advantage of this approach is the application of digital technique and consequently of IC circuits. Further, owing to the fixed divider ($\div L$), smaller frequency increments, without compromising the acquisition time and anti-image-lock properties, can be realized. In Fig. 8.16 is shown a simplified block diagram of an actual VLF, LF, and HF receiver (10 kHz to 30 MHz) utilizing the idea of cascaded phaselock loops for frequency analysis [8.12]. A peculiarity of this system is that the first IF (73·03 MHz) is large compared with the highest received frequency (30 MHz). The phaselocked local oscillator tunes from 73·04 to 103·03 MHz. Clearly the necessary relative frequency change was drastically reduced, to 1·41, which is compatible with digital IC technology. To keep the output noise low the VCO_3 is formed by three individual oscillators each covering a 10-MHz band.

(C) Shared Frequency Synthesizers in Transceivers

In communications stations on board craft and vehicles the transmitter and receiver are generally so close together that combination in one equipment – the transceiver – can have little disadvantage and is certainly profitable in respect of weight, size, power consumption, etc., because some parts are common to both. One of the circuit elements that can be shared is the synthesizer.

Usually a digital synthesizer is employed because it complies well with conditions of an on-board equipment and provides easily for the automatic frequency offset between the transmitter and receiver operations. However, care must be exercised when designing the phaselock loop system to keep the settling time sufficiently short (below 500 ms) to ensure that the communications system is immediately available after every switching in or over.

Any digital synthesizer described in Section 7.5 may be adapted for transceiver operation. And it was this particular application, at VHF and UHF, that contributed so much to the perfection of these particular circuits [3.11], [6.12], [7.1], [7.7], etc.

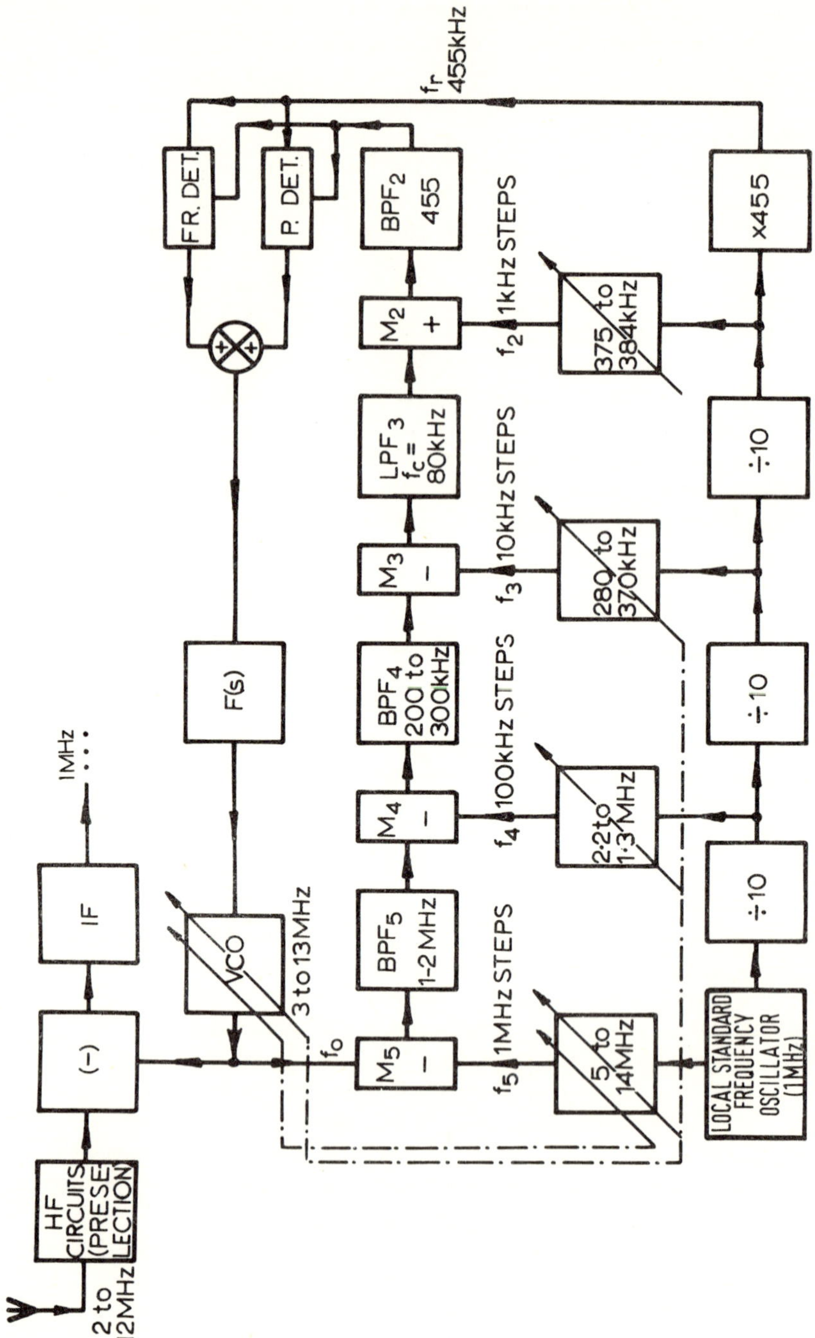

Fig. 8.15 Phaselocked local oscillator in an HF receiver employing the systematic decade-frequency-analysis arrangement. (From Muller & Lisimaque [8.11].)

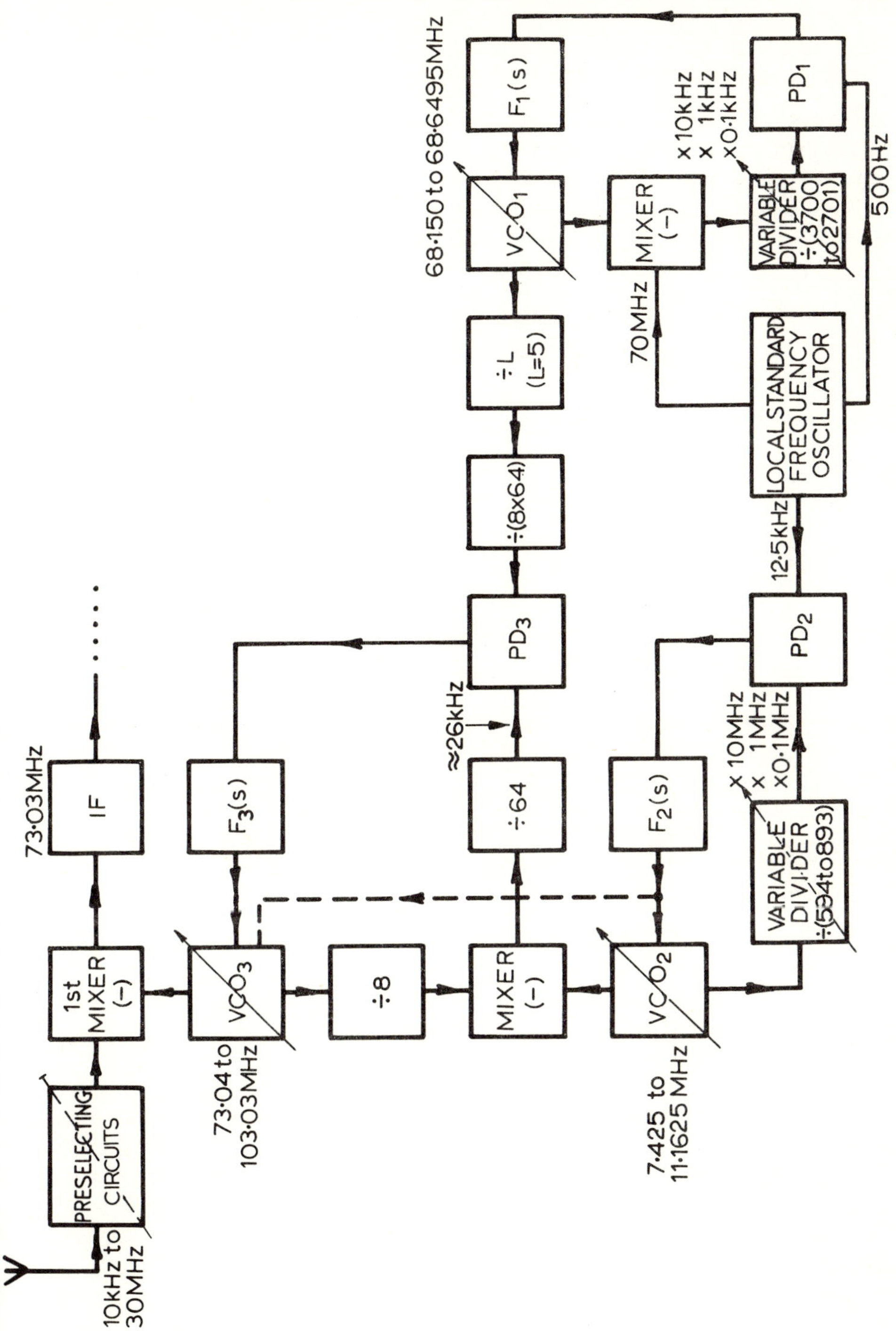

Fig. 8.16 Phaselocked local oscillator of a communications receiver utilizing a cascade of phaselock loops for frequency analysis. (From Gerhold & Pilz [8.12].)

8.3 APPLICATIONS IN FREQUENCY AND TIME STANDARDS

Today's primary standard frequencies have excellent stability and reproducibility, but only for very few values. They are based on a small number of atomic or molecular transitions placed in the frequency spectrum of electromagnetic radiation between 1 and 30 GHz. It is evident that in order to compare frequencies of other sources with those of atomic standards, or to generate atomic and other time scales, frequency-changing circuits or frequency synthesizers are needed.

(A) Reduction of Atomic Standard Frequencies

The general concept of atomic frequency standards is shown schematically in Fig. 8.17(a) and (b). The main difference is that atomic beam devices utilize a passive atomic resonator to steer a conventional crystal oscillator via a frequency

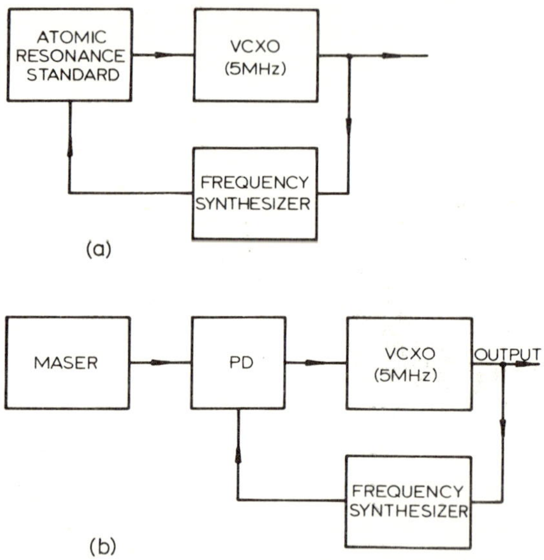

Fig. 8.17 (a) Frequency-locked atomic frequency standards (caesium atomic beam, rubidium gas cell, etc.); (b) Phaselocked atomic frequency standards (hydrogen maser, rubidium maser, etc.).

locking loop, whereas in the case of masers, which are true generators of microwave energy, an actual phaselock can be established (see Section 7.2 and Fig. 7.9) The majority of frequency synthesizing circuits, which transpose the primary microwave frequency to the more convenient RF frequency, are generally placed in the feedback path. It has been shown in Examples 5.2 and 7.2 that despite the definition of the caesium 133 atomic transition frequency between the hyperfine levels (4, 0) and (3, 0) to ten decimal places (9 192 631 770 Hz) the associated single-frequency synthesizers are relatively simple. One reason for this unexpected outcome is the *C*-field tuning.

Atomic frequency standards require a uniform (d.c.) magnetic field (*C*-field) in the transition region. Consequently the spectral lines are split into their Zeeman components and the transition frequencies depend on the strength of this field, but for central levels $m_F = 0$, the dependence is only a small second-order effect. The sensitivity, $\Delta f/f_0 H^2$ (H in A/m), is $3{\cdot}06 \times 10^{-10}$ for the hydrogen maser; $1{\cdot}33 \times 10^{-11}$ and $7{\cdot}26 \times 10^{-12}$ for the rubidium 87 and caesium 133 standards respectively. It is this property of atomic frequency standards which makes possible precise frequency and time-scale adjustments also in cases where simple approximating synthesizers are used (see Example 7.2 and Ref. [8.13]).

(B) Offsetting of Standard Frequencies

Commercial atomic clocks may, in general, be offset if desired, as mentioned in the previous section. However many users do not wish to alter the original setting of their primary standards, whilst others may require different time scales simultaneously. In these instances external offsetting devices based on the frequency synthesis technique must be employed.

(i) Electromechanical Translator

A motor-driven phase-shifter (synchronous resolver etc.) retarding or advancing continuously the phase of the output signal with respect to the input signal, i.e.

$$\varphi_0(t) = \omega_s t \pm \Omega t \tag{8.6}$$

where Ω is proportional to the angular velocity of the shaft, raises or lowers in this way the input frequency ω_s to

$$\omega_0 = \omega_s \pm \Omega \tag{8.7}$$

The amount of the offset Ω may be changed according to the gear ratio between the synchronous driving motor and the rotating phase-shifter (see Fig. 8.18). In the most general case a true 'revolution synthesizer' is used. However, because of the output phase jitter, due to inevitable imperfections of mechanical components, gear-wheel play, etc., and of the impact of reliable solid-state IC circuits, electromechanical devices are becoming of less importance.

(ii) Error-Division Technique

Offsetting generators based on the 'error-division' technique (the inverse process to error multiplication described in Section 8.1) exhibit excellent short-term stability. This was verified experimentally by Yasuda *et al.* [8.14]. The block diagram in Fig. 8.19 reveals that the arrangement is in principle that of a decade synthesizer (Fig. 5.9) if all its switches are set to the zero position; consequently the respective decimal modules (Sections 5.3 and 7.6) are reduced to simple 'dividing modules'. The desired offset is introduced by the appropriate

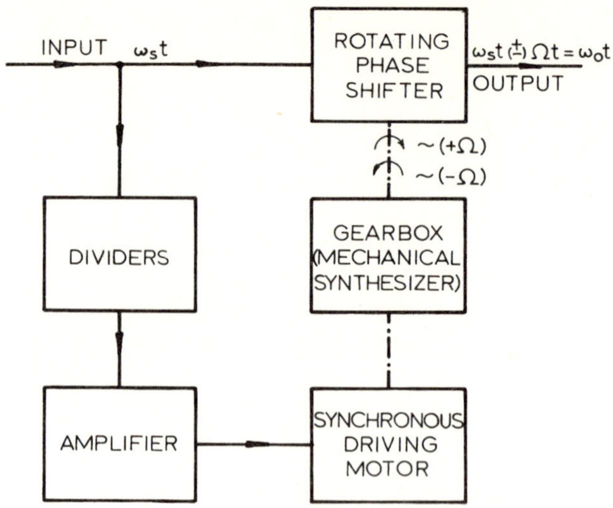

Fig. 8.18 Electromechanical frequency offsetting generator.

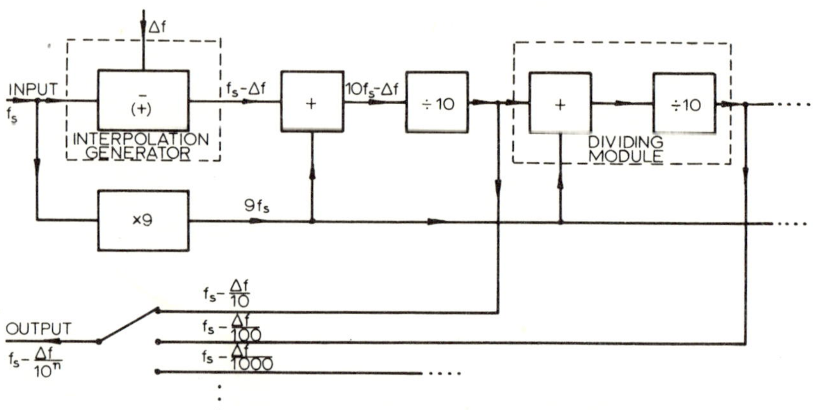

Fig. 8.19 Offsetting generator based on the 'error division technique'.

setting of the 'Interpolation generator'. When exactly defined frequency deviations are required, schemes incorporating common decimal variable synthesizers must be used [5.2].

(iii) Decade Synthesizer as Offsetting Generator

All commercial frequency synthesizers with decimal modules containing dividers may be adapted for offsetting purposes by phaselocking the search oscillator to the desired Δf (Fig. 8.20). The advantage, apart from using an off-the-shelf equipment, is that the nominal output frequency, f_{x0}, may be quite different from the reference frequency f_r, and can be effectively set to any

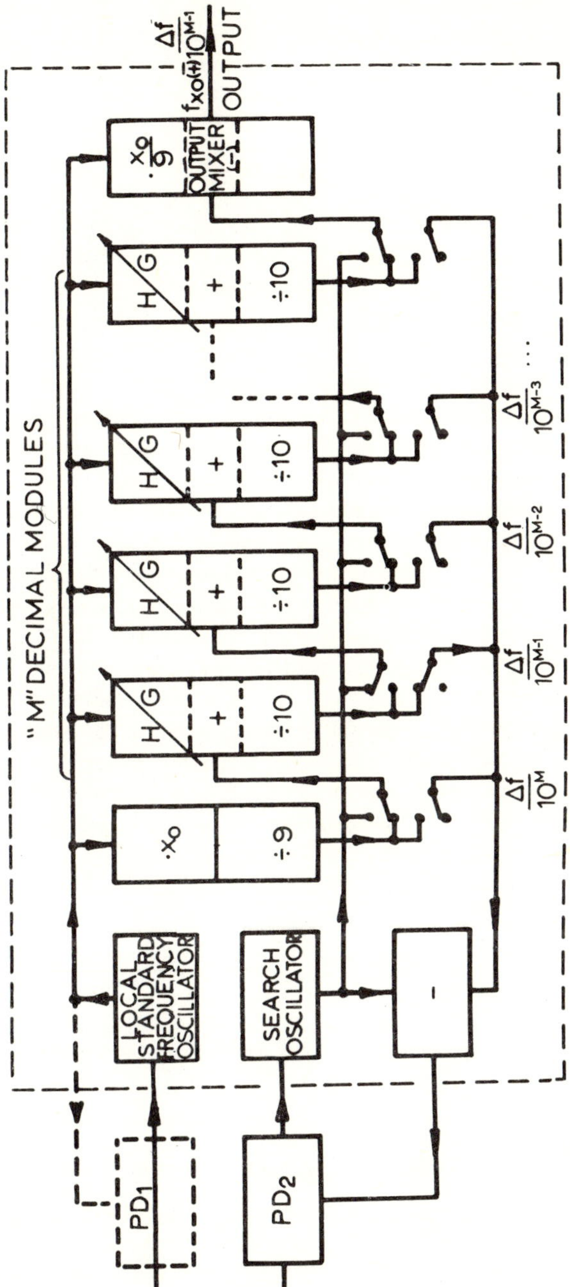

Fig. 8.20 Commercial decade frequency synthesizer employed as offsetting generator.

desired value. Note that in commercial synthesizers of the type shown in Fig. 7.33 the beat-note is made between the search oscillator and the reference frequency (5.50); in this case the f_{x0} can be set on decimal modules to the right of the actuated switch only.

CONCLUSIONS

When considering applications of frequency synthesizers one must bear in mind that primarily the frequency synthesizer is a generator, a generator with exceptional qualities: with long-term stability equal to that of the reference, with perfect resettability, with extremely large frequency range (decade synthesizers), and with increment tuning independent of the output frequency, etc.

Frequency synthesizers can do all the jobs that signal generators do, and often do them better. The outstanding advantage, which is appreciated in these times of automation, is its adaptability to remote and digital tuning and programming compatible with computer operation.

The precision and high resolution of frequency synthesizers are appreciated in many measuring applications (frequency comparisons, phase measurements, etc.) as well as in standard frequency and time laboratories, where decade synthesizers together with some special purpose synthesizers form an indispensable part of the instrumentation.

In communications the driving of transmitters with frequency synthesizers is a long-established technique. Reduction of undesired signals and phase noise together with the introduction of digital synthesizers and IC technology paved the way, in the mid-60s, for widespread use in receivers and transceivers. At the present state of the art, transition to communications systems in which all carrier and heterodyning frequencies are derived from the primary (atomic) standard frequency are conceivable [8.15].

REFERENCES

[8.1] ALLEN, R. L. Frequency divider extends automatic digital frequency measurements to 12·4 GHz. *Hewlett-Packard J.*, **18**, 2-7, April 1967.

[8.2] NOYES, A., Jr. The use of frequency synthesizer for precision measurements of frequency stability and phase noise. *General Radio Experimenter,* **41**, 15-21, Jan. 1967.

[8.3] MAXWELL, D. E. A 5 to 50 MHz direct-reading phase meter with hundredth-degree precision. *I.E.E.E. Trans. on Instrumentation and Measurement,* **IM-15**, 304-10, Dec. 1966.

[8.4] MOYNIHAN, R. L. A sweeper for GR synthesizers. *General Radio Experimenter,* **41**, 15-21, Jan. 1967.

[8.5] Adret-Electronique. Nouveaux générateurs de signaux électriques programmables. *Note d'information* 03.

[8.6] KROUPA, V. An all-band 'single-frequency' synthesizer. *International Broadcasting Convention, London, I.E.E. Conference Publication* No. 69, pp. 117-19 Sept. 1970.

[8.7] McDONALD, G. J. and BURNHAM, C. S. Review of progress in mercantile-marine radiocommunication. *Proc. I.E.E.*, **116**, 1807-20, 1969.

[8.8] RUHRMANN, A. The remotely controlled transmitter center at Elmshorn. *Telefunken-Zeitung,* **35**, 284-298, Dec. 1962.

[8.9] WATT-CARTER, D. E., *et al.* 'The new Leafield radio station' set of articles in The Post Office Electrical Engineers' J., **59**, pp. 130-4, 178-81, 196-8, 267-70, 283-7; 1966.

[8.10] Unconventional communications receiver. *Wireless World,* **63**, 388-9, Aug. 1957.

[8.11] MULLER, J. J. and LISIMAQUE, J. Portable single-sideband high-frequency transceiver with military applications. *Electrical Communication,* **43**, 360-8, Dec. 1968.

[8.12] GERHOLD, J. and PILZ, G. The EK 47 – A communications receiver with digital tuning facilities for the range 10 kHz to 30 MHz. News from Rohde-Schwarz, **9**, 8-12, Oct./Nov. 1969.

[8.13] THRONE, D. H. A report of the performance characteristics of a new rubidium vapor frequency standard. *Frequency Technology,* **8**, 16-9, Jan. 1970.

[8.14] YASUDA, Y., YOSHIMURA, K. and SAITO, Y. One of the methods of frequency offsetting. *J. of the Radio Research Laboratories (Tokyo),* **13**, 211-25, July/Sept. 1966.

[8.15] KROUPA, V. Single-frequency synthesis and frequency-coherent communication systems. *Conference on Frequency Generation and Control for Radio Systems, London I.E.E. Conference Publication* No. 31, pp. 96-9, May 1967.

9 Phase and Frequency Fluctuations in Frequency Synthesizers

Physical phenomena encountered in everyday life are seldom ideal, and often unwanted perturbations, sometimes called noise, are present. This is naturally also true for the output signals of frequency synthesizers. These perturbations may be of different kinds and origin: there are discrete spurious components, due to intermodulation or to imperfect filtering in frequency multipliers or mixers on the one hand, and random unpredictable perturbations, noise, on the other hand.

For many applications the noise properties of the output signal of a frequency synthesizer are important and must be known. Direct measurement of the phase noise spectrum, or of the fractional frequency stability, supplies nearly all the necessary information. Unfortunately, current practice often tends towards single-value indications of noise levels, r.m.s. phase or frequency deviations, etc. A common fault of these specifications is that a substantial part of the pertinent information is missing.

9.1 SMALL AND BAND-LIMITED PERTURBATIONS OF SINUSOIDAL SIGNALS

It is in the nature of frequency synthesis that the amplitude and phase perturbations of the output-signal are small and band-limited. For the first statement the logical reasoning, that if the perturbations were not small the frequency synthesizer would lose its attractiveness for many applications, is sufficient. The second statement has a solid physical background because the synthesized-frequency signal must propagate through a large number of the narrow-band selective circuits (with the exception of the final output network).

(A) Superposition of Band-Limited Sinusoidal Signals

The composite signal may be written as

$$f(t) = \sum_{n=1}^{N} A_n \cos(\omega_n t + \varphi_n) \tag{9.1}$$

Introducing

$$\omega_n t + \varphi_n = \omega_1 t + (\omega_n - \omega_1) t + \varphi_n = \omega_1 t + \Phi_n(t) \tag{9.2a}$$

and

$$\frac{A_n}{A_1} = \alpha_n \tag{9.2b}$$

$$f(t) = A_1 \sum_{n=1}^{N} \alpha_n \cos[\omega_1 t + \Phi_n(t)]$$

$$= A_1 \cos\omega_1 t \sum_{n=1}^{N} \alpha_n \cos\Phi_n(t) - A_1 \sin\omega_1 t \sum_{n=1}^{N} \alpha_n \sin\Phi_n(t) \tag{9.3}$$

or

$$f(t) = A_1 \alpha(t) \cos[\omega_1 t + \Phi(t)] \tag{9.4}$$

Where $\alpha(t)$ is the normalized instantaneous amplitude and $\Phi(t)$ is the instantaneous phase departure

$$\alpha^2(t) = \left[\sum_{n=1}^{N} \alpha_n \cos\Phi_n(t)\right]^2 + \left[\sum_{n=1}^{N} \alpha_n \sin\Phi_n(t)\right]^2 \tag{9.5}$$

$$\Phi(t) = \arctan \frac{\sum_{n=1}^{N} \alpha_n \sin\Phi_n(t)}{\sum_{n=1}^{N} \alpha_n \cos\Phi_n(t)} \tag{9.6}$$

Without any loss of generality it is possible to choose the time scale in such a way that

$$\varphi_1 = \Phi_1(t) = 0 \tag{9.7}$$

If only small perturbation is assumed one may put, further,

$$\alpha_1 \gg \alpha_n \quad (n = 2, \ldots N); \quad (\text{note } \alpha_1 = 1) \tag{9.8}$$

In these circumstances

$$\alpha(t) \approx 1 + \sum_{n=2}^{N} \alpha_n \cos\Phi_n(t) \tag{9.9}$$

and

$$\Phi(t) \approx \arctan \sum_{n=2}^{N} \alpha_n \sin\Phi_n(t) \approx \sum_{n=2}^{N} \alpha_n \sin\Phi_n(t) \tag{9.10}$$

Example 9.1

For the superposition of one strong signal $A_1 \cos \omega_1 t$ and one weak signal $A_2 \cos(\omega_2 t + \varphi_2)$, i.e. $A_1 \gg A_2$ we get

$$f(t) \approx A_1 \left[1 + \frac{A_2}{A_1} \cos(\Omega t + \varphi_2)\right] \cos\left[\omega_1 t + \frac{A_2}{A_1} \sin(\Omega t + \varphi_2)\right] \quad (9.11)$$

Evidently in the first approximation we face a simultaneous amplitude and phase modulation of the stronger signal at the rate of the difference frequency $|\Omega| = |\omega_2 - \omega_1|$. For larger values of A_2/A_1, terms including higher harmonics of Ω must be added to the expansion (9.11) [9.1].

(B) Narrow-Band Noise

If the noise power is concentrated in a relatively narrow band around the frequency ω_1 the noise voltage may be expressed as

$$n(t) = n_c(t) \cos \omega_1 t - n_s(t) \sin \omega_1 t \quad (9.12)$$

which resembles (9.3); however, $n_c(t)$ and $n_s(t)$ are independent, i.e. the mean value

$$\overline{n_c(t)\, n_s(t)} = \overline{n_c(t)} \,.\, \overline{n_s(t)} = 0 \quad \text{if} \quad \overline{n(t)} = 0 \quad (9.13)$$

and also

$$\overline{n_c(t)} = \overline{n_s(t)} = 0; \quad \overline{n^2(t)} = \overline{n_c^2(t)} = \overline{n_s^2(t)} \quad (9.14)$$

Further properties are:

Both $n_c(t)$ and $n_s(t)$ are slowly varying functions and their spectra contain only low-frequency components.

If $n(t)$ is gaussian, $n_c(t)$ and $n_s(t)$ are also gaussian.

In the case where a strong signal is associated with narrow-band noise, as at the synthesizer output, it is helpful to write

$$v(t) = A(t) \cos[\omega_0 t + \varphi(t)] \,. \quad (9.15)$$

where $v(t)$ represents a voltage or current, $A(t)$ and $\varphi(t)$ are slowly varying real functions of time, and ω_0 is a constant. $A(t)$ is the variable amplitude of the signal and does not contribute directly to the frequency fluctuations. Of course, in frequency multipliers and other nonlinear devices, conversion of amplitude modulation to phase modulation and vice versa can occur. Nonetheless, at any given point in the system, (9.15) is valid.

9.2 PRINCIPLES OF THE SHORT-TERM FREQUENCY-STABILITY THEORY

It is beyond the scope of this book to go into great detail. The interested reader is referred to the special issue of the Proceedings of the I.E.E.E., February 1966, and particularly to the paper by L. S. Cutler and C. L. Searle [9.2] and to an earlier paper by E. J. Baghdady *et al.* [9.3], on the results of which our explanations are built.

(A) Time Domain and Frequency Domain Specifications

The signal from the frequency synthesizer can be described by (9.15). The pertinent frequency stability can be specified either in the time domain in terms of autocorrelation functions (also called autocovariance) or in the frequency domain using spectral densities. The transition from one system to the other is through the Fourier transform [9.4].

The autocorrelation of the random phase departure $\varphi(t)$ is defined as

$$R_\varphi(\tau) = \overline{\varphi(t+\tau/2)\,\varphi(t-\tau/2)} = \lim_{T\to\infty} \frac{1}{T} \int_{-T/2}^{T/2} \varphi(t+\tau/2)\,\varphi(t-\tau/2)\,dt \qquad (9.16)$$

where the bar signifies statistical average, and the ergodicity, that ensemble and time averages are equal, is assumed.

The spectral density of the phase (using two-sided spectra) is

$$S_\varphi(\omega) = \int_{-\infty}^{\infty} R_\varphi(\tau)\exp(-j\omega\tau)\,d\tau = 2\int_0^{\infty} R_\varphi(\tau)\cos\omega\tau\,d\tau \qquad (9.17)$$

Because $R_\varphi(\tau)$ is the Fourier transform of $S_\varphi(\omega)$, it follows

$$R_\varphi(\tau) = \frac{1}{2\pi}\int_{-\infty}^{\infty} S_\varphi(\omega)\exp(j\omega\tau)\,d\omega = \frac{1}{\pi}\int_0^{\infty} S_\varphi(\omega)\cos\omega\tau\,d\omega \qquad (9.18)$$

Experimentally $S_\varphi(\omega)$ may be found by passing $v(t)$ through an ideal phase detector and measuring the spectrum.† In frequency synthesizers the AM power spectral density is generally lower (about 10 or more dB) than that due to the PM; further, for frequencies above some minimum value, the mean square value of the phase jitter is much less than one radian^2. If both conditions are fulfilled the RF power spectral density (two-sided) of (9.15) is

$$S_{RF}(\omega) \approx \frac{P_s}{2}[S_\varphi(\omega+\omega_0) + S_\varphi(\omega-\omega_0)] \qquad (9.19)$$

where P_s is the carrier power. The approximation is valid except for $\pm\,\omega_0$ and the surrounding narrow bands containing all the high modulation-index, low-frequency sidebands.

The frequency stability of the signal described by (9.15) may also be defined in terms of the instantaneous frequency, i.e.

$$\omega(t) = \frac{d}{dt}[\omega_0 t + \varphi(t)] = \omega_0 + \dot{\varphi}(t) \qquad (9.20)$$

† The units of $S_\varphi(\omega)$ are radian^2/Hz.

Similarly, as above, $R_{\dot{\varphi}}(\tau)$ and $S_{\dot{\varphi}}(\tau)$ may be introduced. Recalling that the differentiation in the time domain corresponds to multiplication by $j\omega$ in the frequency domain and hence multiplication by ω^2 in terms of spectral densities, it is evident that

$$S_{\dot{\varphi}}(\omega) = \omega^2 S_{\varphi}(\omega) \tag{9.21}$$

(B) Principal Specifying Formulae

For convenience we shall summarize all formulae used in specifying frequency stability. Variances, i.e. mean square averages, σ^2, are used because of the random variables.

(*i*) *Average m.s. frequency error in the time interval* τ

$$\sigma^2[\langle\dot{\varphi}\rangle_{t,\tau}] = \frac{2}{\tau^2}[R_{\varphi}(0) - R_{\varphi}(\tau)] \tag{9.22a}$$

$$= \frac{2}{\tau}\int_0^{\tau} R_{\dot{\varphi}}(\tau')\left(1 - \frac{\tau'}{\tau}\right) d\tau' \tag{9.22b}$$

and after substituting for the autocorrelation from (9.18)

$$= \frac{2}{\pi\tau^2}\int_{-\infty}^{\infty} S_{\varphi}(\omega)\sin^2 \omega\tau/2\, d\omega \tag{9.23a}$$

$$= \frac{1}{2\pi}\int_{-\infty}^{\infty} S_{\dot{\varphi}}(\omega)\frac{\sin^2 \omega\tau/2}{(\omega\tau/2)^2}\, d\omega \tag{9.23b}$$

(*ii*) *R.M.S. fractional frequency stability in the time interval* τ is the root square of the above value normalized in respect to ω_0, i.e.

$$\frac{\sigma[\langle\dot{\varphi}\rangle_{t,\tau}]}{\omega_0} = \frac{1}{\omega_0}\sqrt{\frac{1}{2\pi}\int_{-\infty}^{\infty} S_{\dot{\varphi}}(\omega)\left(\frac{\sin \omega\tau/2}{\omega\tau/2}\right)^2 d\omega} \tag{9.24}$$

The advantage of this characterization is that the same number holds after processing the signal through a perfect frequency multiplier or divider. Another advantage is that the digital counting measurements in many cases are sampled in a very similar manner. Therefore the quantity (9.24) is often indicated in specifications.

(*iii*) *M.S. frequency error or variance of frequency*

$$\sigma^2[\dot{\varphi}(t)] = R_{\dot{\varphi}}(0) = \frac{1}{\pi}\int_0^{\infty} S_{\dot{\varphi}}(\omega)\, d\omega \tag{9.25a}$$

$$= \frac{1}{\pi}\int_0^{\infty} \omega^2 S_{\varphi}(\omega)\, d\omega \tag{9.25b}$$

Because of the ω^2 in (9.25b) and the possibility of the presence of frequency-independent phase noise (white noise) the upper limit of integration (the bounded LF frequency range) must be specified.

(*iv*) *Variance of average phase*

$$\sigma^2[\langle\varphi\rangle_{t,\tau}] = \frac{2}{\tau}\int_0^\tau R_\varphi(\tau')(1-\tau'/\tau)\,d\tau' \tag{9.26}$$

(*v*) *Variance of accumulated phase*

$$\sigma^2[\Delta_\tau\varphi] = \tau^2\sigma^2[\langle\dot{\varphi}\rangle_{t,\tau}] = 2[R_\varphi(0) - R_\varphi(\tau)] \tag{9.27a}$$

$$= \frac{2}{\pi}\int_{-\infty}^{\infty} S_\varphi(\omega)\sin^2(\omega\tau/2)\,d\omega \tag{9.27b}$$

(*vi*) *Variance of phase, i.e. m.s. phase error*

$$\sigma^2[\varphi(t)] = R_\varphi(0) = \frac{1}{\pi}\int_0^\infty S_\varphi(\omega)\,d\omega \tag{9.28}$$

For a non-zero frequency spectral density $S_{\dot{\varphi}}(\omega)$ at $\omega = 0$, the phase noise spectral density $S_\varphi(\omega)$ is, according to (9.21), infinite (the property of the random walk process in oscillators), therefore in (9.28) the lower limit ($\omega_L > 0$) of integration must be indicated; this is often true also for the upper limit ω_H.

(*vii*) *The phase noise power in the LF frequency band* (ω_L; ω_H)

$$P_N = P_S\overline{\varphi^2(t)} = P_S\frac{1}{\pi}\int_{\omega_L}^{\omega_H} S_\varphi(\omega)\,d\omega = 2P_S\int_{f_L}^{f_H} S_\varphi(f)\,df \tag{9.29}$$

(*viii*) *Signal-to-Phase-Noise-Ratio*

It follows from (9.29) by dividing both sides by P_S

$$\frac{1}{SN_\varphi R} = 2\int_{f_L}^{f_H} S_\varphi(f)\,df = \int_{f_L}^{f_H} S'_\varphi(f)\,df \tag{9.30}$$

where $S'_\varphi(f) = 2S_\varphi(f)$ is the so-called 'one-sided spectral density'.

(C) Significant Types of Noise

Spectral densities $S_{\dot{\varphi}}(\omega)$ and $S_\varphi(\omega)$ may be described by power series in ω

$$S_{\dot{\varphi}}(\omega) = \frac{A}{|\omega|} + B + C\omega + D\omega^2 \tag{9.31}$$

and

$$S_\varphi(\omega) = \frac{A}{|\omega|^3} + \frac{B}{|\omega|^2} + \frac{C}{|\omega|} + D \tag{9.32}$$

where the individual terms may be related to noise processes that, to date, have some significance (see, for example, [9.5])

$$S_{\dot{\varphi}A}(\omega) = \frac{A}{|\omega|} \quad \text{'flicker' frequency noise}$$

$$S_{\dot{\varphi}B}(\omega) = B \quad \text{'white' frequency noise}$$

$$S_{\dot{\varphi}C}(\omega) = C\omega \quad \text{'flicker' phase noise}$$

$$S_{\dot{\varphi}D}(\omega) = D\omega^2 \quad \text{'white' phase noise}$$

Generally, however, spectral densities $S_{\dot{\varphi}}(\omega)$ and $S_{\varphi}(\omega)$ may contain terms with higher and lower powers of ω; e.g. after passing through a filter with the transfer function $H(j\omega)$ the original spectral density changes to

$$S_{\varphi H}(\omega) = S_{\varphi}(\omega)|H(j\omega)|^2 \tag{9.33}$$

9.3 NOISE IN FREQUENCY SYNTHESIZERS

All frequency synthesis systems may be reduced effectively to the block diagram in Fig. 5.2. By incorporating the divider y_1 and the multiplier z_1 into the synthesizing loop, the universal block diagram in Fig. 9.1 results. The multiplication factors N_1 and N_2 are rational numbers. When comparing

$$f_s|N_1 - N_2| = f_x \tag{9.34}$$

with (5.1) it is seen that

$$|N_1 - N_2| = X_1/Y_1 = N^{\dagger} \tag{9.35}$$

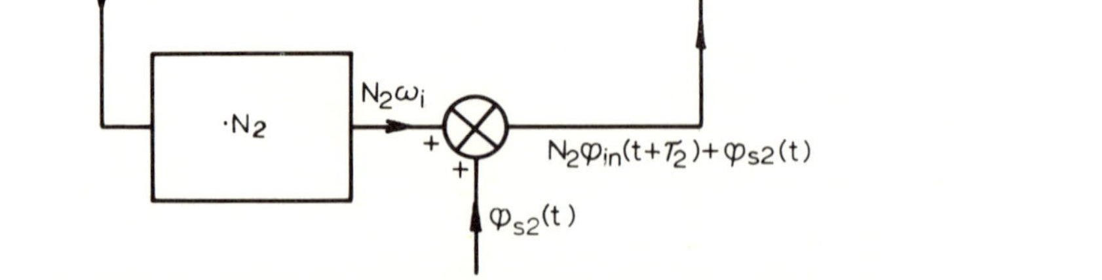

Fig. 9.1 Universal block diagram of frequency synthesis systems.

† For a decade synthesizer (5.49) reveals

$$N_1 = \frac{z_1}{y_1}\left(x_0 + x_{01} + \frac{x_0 + x_{02}}{10} + \cdots\right) \text{ and } N_2 = \frac{z_1}{y_1} X_0$$

Considering phase noise one readily deduces that the input noise $\varphi_{in}(t)$ is multiplied in the first branch by N_1 and increased by an amount $\varphi_{s1}(t)$ of the system noise, and the same happens in the second branch.† Taking into account that the propagation delays through the sub-synthesizers 'N_1' and 'N_2' are τ_1 and τ_2, the output noise composed of the low-frequency offset components passed by all filters may be written as

$$\varphi_{xn}(t) = N_1\,\varphi_{in}(t+\tau_1) + \varphi_{s1}(t) - N_2\,\varphi_{in}(t+\tau_2) - \varphi_{s2}(t) + \varphi_{s3}(t) \quad (9.36)$$

where $\varphi_{s3}(t)$ is the noise added in the mixer and the following output circuits. Summarizing all the additive system noise into $\varphi_s(t)$, the output phase noise spectral density is, according to App. 9.1,

$$S_{\varphi_{xn}}(\omega) = S_{\varphi_{in}}(\omega)\,[N_1^2 + N_2^2 - 2N_1\,N_2\cos\omega(\tau_1 - \tau_2)] + S_{\varphi_s}(\omega) \quad (9.37a)$$

$$\approx S_{\varphi_{in}}(\omega)\,[N^2 + N_1\,N_2\,\omega^2\,\tau^2] + S_{\varphi_s}(\omega) \quad (9.37b)$$

if the differential time delay $\tau = |\tau_1 - \tau_2|$ and the offset frequency ω are small.

Now the aim of the designer is to reduce the output phase noise $\varphi_{xn}(t)$ to a minimum. It may be taken for granted that the additive system-noise $\varphi_s(t)$ cannot be reduced below a certain limit. Evidently, the lowest possible output noise is $\varphi_s(t)$. This ideal case can be approached closely by reducing N_1 and N_2 by choosing a standard frequency as high as is compatible with stability requirements (a 5-MHz crystal oscillator is a better choice than 1 MHz) and further when subjecting the reference signal to narrow-band filtering before starting the synthesis process. For moderate and high N the above procedures may not be fully effective. At this point there remains only one possibility, reduction of N_1 and N_2, i.e. reduction of the value of the transporting carrier x_0.

When a decimal variable frequency synthesizer is based on eq. (5.49), N_1 and N_2 are generally more than one order higher than N_{max}‡. And even in the case where both N_1 and N_2 are made variable (see Example 5.3 and Fig. 5.11), N_{max} is still nearly one order lower. In consequence the contribution from $N_1 N_2 \omega^2 \tau^2$ cannot be neglected, particularly at higher frequencies. On the other hand in drift-cancelling systems the transporting carrier is $x_0 + x_{ai}$, see (5.70); the larger part x_{ai} being derived in the independent auxiliary oscillator. The universal block diagram of the frequency synthesizer changes, in this case, to that shown in Fig. 9.2(a). The multiplication factors N_1' and N_2' are of the same order as N_{max} (see the synthesizer output circuit in Fig. 9.2(b) which is compatible with the system shown in Fig. 5.11) and the second term, in brackets, on the right hand side of (9.37b) loses its importance. However, the output phase noise spectral density is modified by the auxiliary oscillator noise to

$$S_{\varphi_{xn}}(\omega) = S_{\varphi_{in}}(\omega)\,N^2 + 2S_{\varphi_{an}}(\omega)\,[1 - \cos\omega\tau_{cancel}] + S_{\varphi_s}(\omega) \quad (9.38a)$$

† As to spurious amplitude modulation it is generally not enhanced in multiplying stages (Section 2.4) but reduced by the limiting action in counting dividers etc.

‡ $N_{max} \leqslant 10\frac{z_1}{y_1}$; $N_2 = X_0\frac{z_1}{y_1} \geqslant 190\frac{x_1}{y_1}$; see (5.60).

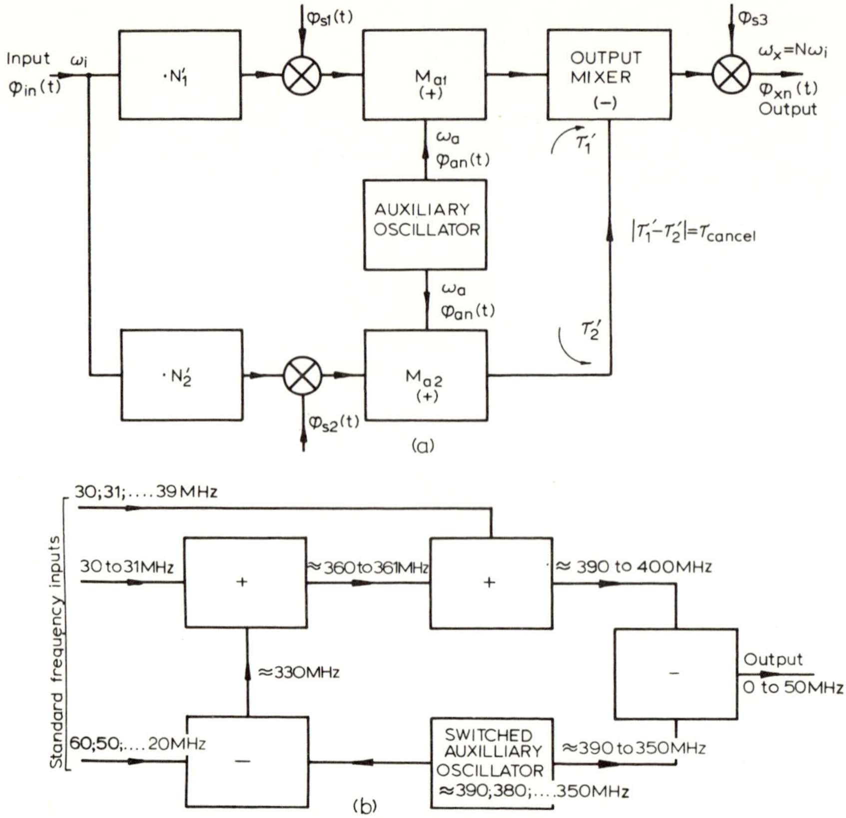

Fig. 9.2 Reduction of synthesizer output phase noise by a drift-cancelling loop: (a) universal block diagram; (b) practical arrangement with numerical example.

where $S_{\varphi an}(\omega)$ is the phase noise spectral density of the auxiliary oscillator (in the drift-cancelling loop $N_{1a} = N_{2a} = 1$). The above equation can be simplified for small values of $\omega\tau_{cancel}$ to

$$\approx S_{\varphi\,in}(\omega)\,N^2 + S_{\varphi\,an}(\omega)\,\omega^2\,\tau_{cancel}^2 + S_{\varphi s}(\omega) \qquad (9.38b)$$

In other words the phase noise of the auxiliary oscillator cancels out, but not the frequency noise ($\omega^2 S_{\varphi an}(\omega)$– see (9.21)) if the propagation delays in both branches of the drift-cancelling loop are not equal. The simplified equation (9.38b) was originally derived, in the time domain, by Tykulski [9.6].

Examination of the phase noise spectral densities of actual frequency synthesizers (see Fig. 9.3) reveals the predominance of the approximately $1/f$ phase noise characteristic in ranges up to 1 or 10 kHz, and the 'white' phase noise characteristic, may be modified by filtering actions, at higher frequencies.

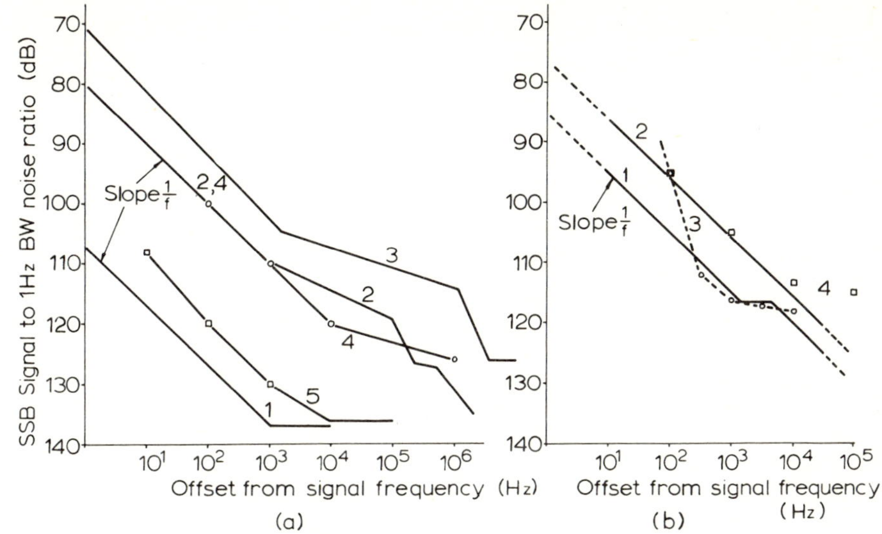

Fig. 9.3 Spectral densities $S_{\varphi}(f)$ ('two-sided') of the output phase-noise in different frequency synthesizers.

(a) additive decimal modules:
- (1) H.-P. Model 5102A (100 kHz range) 100 kHz out;
- (2) H.-P. model 5100B (0·01 Hz to 50 MHz) 49 MHz out;
- (3) H.-P. model 5105A (100 kHz to 500 MHz) 499 MHz out;
- (4) R.-Sch. model SMDH (1 MHz and 50 MHz range) guaranteed values;
- (5) Fluke model 645A (d.c. to 50 MHz) 49·9 MHz out;

(b) phaselocked decimal modules:
- (1) G.R. type 1163 (30 Hz to 12 MHz);
- (2) G.R. type 1164 (10 kHz to 70 MHz);
- (3) ○ R.-Sch. type ND 100M (300 Hz to 100 MHz);
- (4) □ R.-Sch. type ND 25M (20 Hz to 30 MHz).

(Adopted from technical data published by Hewlett-Packard, Rohde & Schwarz, General Radio Company, and Fluke.)

There remains one point which must be stressed. The output phase noise of a well-designed frequency synthesizer is composed mainly of the additive system-noise, and is nearly independent of the output frequency (see Fig. 9.7). Consequently when the best signal-to-noise ratio, $(SNR)_{out}$, is desired, frequency synthesizers with appropriate ranges should be used and multiplication of the output frequency avoided as much as possible.

9.4 MEASUREMENT TECHNIQUES

There are several well-known procedures for making measurements of short-term frequency stability, particularly of fraction frequency deviations and phase noise spectral densities. The reader looking for a more theoretical background is referred to [9.2] whereas in [9.7] and [9.8] more practical aspects are discussed.

(A) Multiple-Period Technique

The basic arrangement is illustrated in Fig. 9.4 and it works as follows. The output frequency of the synthesizer is offset by a small amount Δf_0 from a

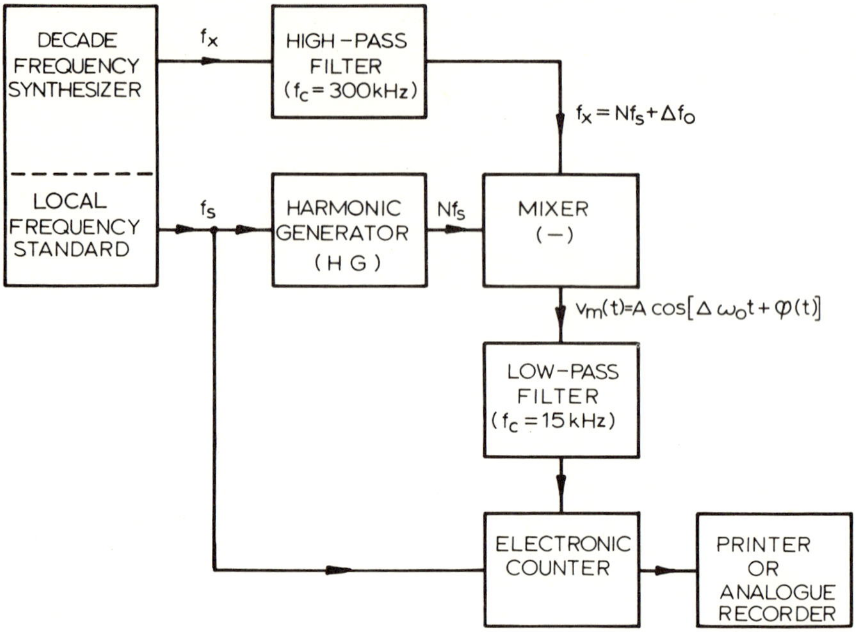

Fig. 9.4 Frequency deviation measurement – multiple period technique. (Courtesy of Hewlett-Packard [9.7].)

chosen multiple of the internal standard frequency Nf_s and mixed with it. Then duration τ, of N_0 periods of the mixer output waveform

$$v_m(t) = A\cos\left[\Delta\omega_0\, t + \varphi(t)\right] \tag{9.39}$$

is measured using a counter; evidently

$$2\pi N_0 = \Delta\omega_0\,\tau + \varphi(\tau) - \varphi(0) \tag{9.40}$$

Introducing

$$\tau_0 = 2\pi N_0/\Delta\omega_0 \quad \text{and} \quad \tau = \tau_0 - \Delta\tau \tag{9.41}$$

eq. (9.40) reduces to

$$\varphi(\tau) - \varphi(0) = \Delta\omega_0\,\Delta\tau \tag{9.42}$$

Making the approximation $\varphi(\tau) \approx \varphi(\tau_0)$ and writing $\varphi(t)$ instead of $\varphi(0)$ we get

$$\frac{\varphi(t+\tau_0) - \varphi(t)}{\tau_0} = \langle\dot{\varphi}\rangle_{t,\tau_0} \approx \frac{\Delta\omega_0\,\Delta\tau}{\tau_0} \tag{9.43}$$

The standard deviation of $\langle \dot{\varphi} \rangle_{t,\,\tau_0}$ can be found by making repeated measurements, say M:

$$\sigma[\langle\dot{\varphi}\rangle_{t,\tau_0}] = \frac{\Delta\omega_0}{\tau_0}\sqrt{\frac{1}{M}\sum_{m=1}^{M}(\Delta\tau_m)^2} \tag{9.44}$$

Note that the instability of the internal standard frequency is effectively correlated out and further that the low-pass filter at the output of the mixer is necessary, otherwise the term $D\omega^2$ in (9.31) would prevent convergence of the integral in (9.23b).

Typical results adopted from published specifications are shown in Fig. 9.5.

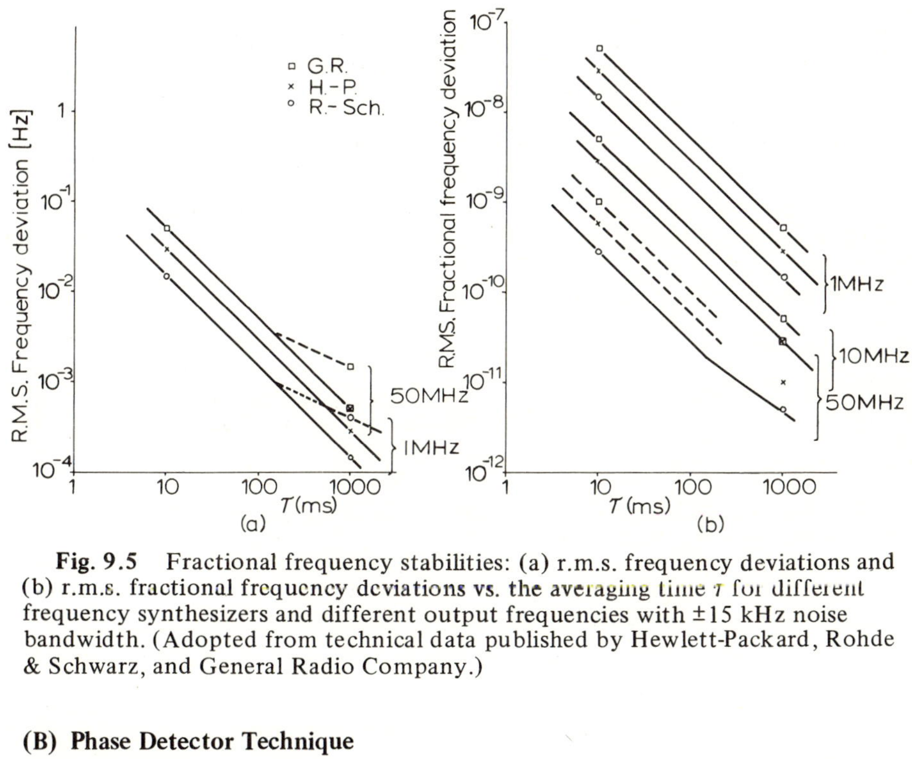

Fig. 9.5 Fractional frequency stabilities: (a) r.m.s. frequency deviations and (b) r.m.s. fractional frequency deviations vs. the averaging time τ for different frequency synthesizers and different output frequencies with ±15 kHz noise bandwidth. (Adopted from technical data published by Hewlett-Packard, Rohde & Schwarz, and General Radio Company.)

(B) Phase Detector Technique

A very useful instrumentation set-up is shown in Fig. 9.6 [8.2], [9.8]. Two frequency synthesizers of the same type are set to an equal frequency and phaselocked via one of the search oscillators. The gain of the loop is very small because frequency dividers (÷ 10) between individual decimal modules are effectively in the feedback path and eq. (7.13) applies (see Example 8.1). The natural frequency, ω_n, and the pass-band of this phaselock arrangement are correspondingly small and consequently only the lowest frequency portion of the phase jitter at the detector output degenerates, see eq. (6.13), as

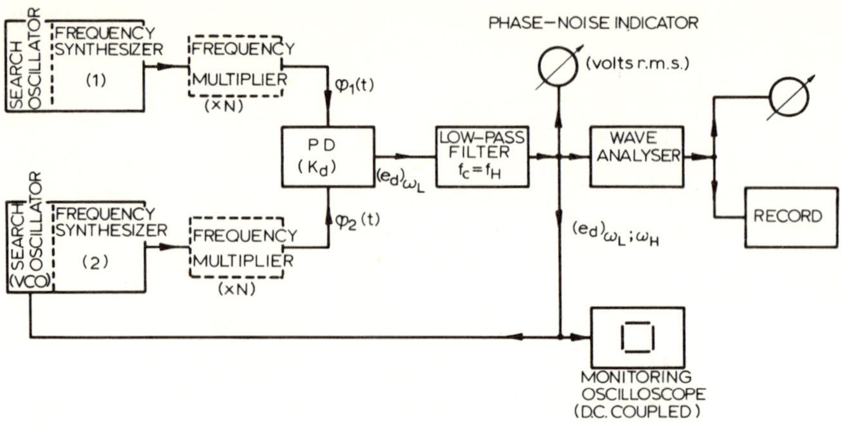

Fig. 9.6 Phase-noise measurement set-up using two phaselocked frequency synthesizers.

required above in Section 9.2(B) (vi) and (vii). The following low-pass filter suppresses phase noise components above the chosen upper-bound frequency, f_H. The relation between the synthesizer phase jitter and the detector output voltage is

$$e_d(t) \approx K_d\,[\varphi_1(t) - \varphi_2(t)] = K_d\,\varphi(t) \tag{9.45}$$

The phase detector gain K_d is easily determined by detuning one of the synthesizers, say by $\Delta\omega$; the beat-note voltage is

$$e_{d,b}(t) = K_d \sin \Delta\omega t \tag{9.46}$$

According to (9.29) the signal-to-phase-noise ratio is

$$SN\varphi R = 1/\overline{\varphi^2(t)} = K_d^2/\overline{e_d^2(t)} = 2e_{d,b}^2/\overline{e_d^2(t)} \tag{9.47}$$

Converting (9.47) to the single-side-band (SSB) signal-to-phase-noise ratio, as usually used in frequency synthesizer specifications, gives

$$(SN_\varphi R)_{\mathrm{SSB}} = 4e_{d,b}^2/\overline{e_d^2(t)} \tag{9.48}$$

or

$$(SN_\varphi R)_{\mathrm{SSB}} = 20 \log \frac{e_{d,b}}{e_{d,\mathrm{rms}}} + 6 \quad \mathrm{dB} \tag{9.49}$$

The noise power in a narrow bandwidth is proportional to the bandwidth; with white noise this is strictly true, but with other types of noise, such as the $1/f$ distribution that we observe for synthesizers, it is very close to the truth provided the bandwidth is small compared with the offset frequency. Thus using the wave analyser with the appropriate bandwidths the 'SSB signal to 1 Hz bandwidth ratio' is easily calculated using (9.48) or (9.49). Assuming that both synthesizers in Fig. 9.6 contribute equally to the phase-noise power, a

3-dB enhancement must be introduced. The frequency multipliers, shown dashed, serve to increase the sensitivity of the measuring set-up; hence

$$(SN_{\varphi} R)_{\mathrm{SSB}/1\ \mathrm{Hz}} = 20 \log \frac{e_{d,b}}{e_{d,\mathrm{rms}}} + 20 \log N + 10 \log f_{an} + 9 \ \ \mathrm{dB} \quad (9.50)$$

where N is the multiplication factor and f_{an} the wave analyser bandwidth in Hz. Typical results attained by this method are shown in Fig. 9.7; note the independence of the synthesizer output phase noise from the output frequency.

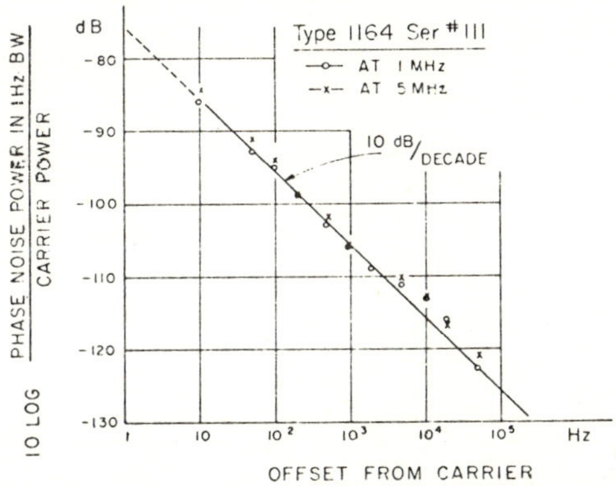

Fig. 9.7 Spectral densities $S_{\varphi}(f)$ of the output phase noise for two different settings of the frequency synthesizer type 1164 evaluated by the measuring set-up in Fig. 9.6. (Courtesy of the General Radio Company.)

(C) Measurement of Discrete Spurious Components

Discrete (single-frequency) spurious components are most easily measured by means of a sharply selective voltmeter-receiver, which may be of conventional type. But in instances where narrow-band spectra with closely spaced lines have to be examined a special-purpose receiver-analyser of the type suggested in Section 8.1(C)(ii) must be used (see Fig. 9.8). There are no stability problems

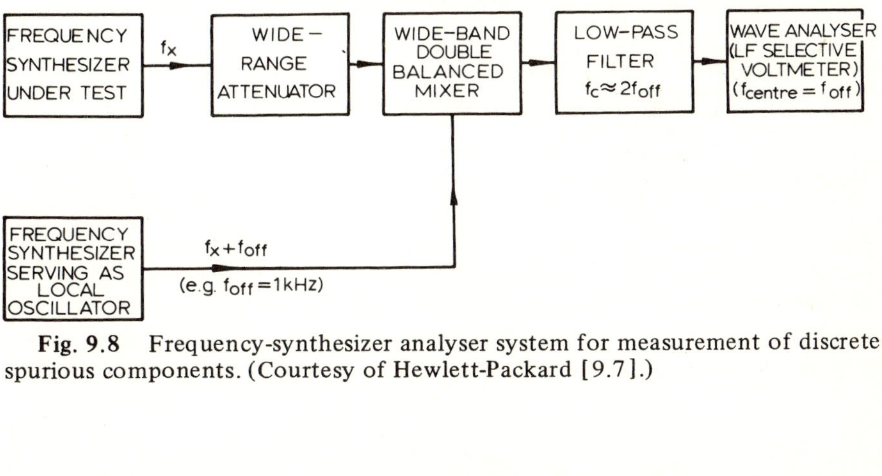

Fig. 9.8 Frequency-synthesizer analyser system for measurement of discrete spurious components. (Courtesy of Hewlett-Packard [9.7].)

because the local oscillator is another frequency synthesizer, and this is particularly true if its internal reference oscillator is phaselocked to the synthesizer being tested. In [9.7] an extremely sensitive measuring set-up of this type, and assembled from off-the-shelf instruments, is described; it is claimed that the system has a dynamic range of 120 dB and a sensitivity of 0·1 microvolt in 50 ohms.

9.5 CONCLUSIONS

Degradation of the output spectral purity of frequency synthesizers may come from several sources:

(i) in the ideal case the phase-noise power increases or decreases in proportion to $N^2 = (f_x/f_s)^2$;

(ii) in frequency synthesis systems in which a relatively high transporting carrier coherent with the reference frequency is used, the differential time delay τ will be responsible for an additional term proportional to $N_1 N_2 \omega^2 \tau^2$;

(iii) if there is a large contribution of additive noise due to the frequency synthesis circuits;

(iv) if there are discrete components due to intermodulation in nonlinear circuits, imperfect filtering, interference between individual stages of the synthesizer, or modulation by the power-line frequency, etc.

Experimental measurements of spectral densities $S_\varphi(f)$ of the output phase noise in different types of frequency synthesizer reveal the $1/f$ (10 dB/decade decline) characteristic at low offset-frequencies; whereas at higher offset-frequencies $S_\varphi(f)$ has approximately a constant value. This behaviour suggests that the 1 Hz intercept point and threshold noise levels together with the maximum discrete spurious components level might be sufficient to describe the output SNR or SSpR (signal-to-spurious-components ratio) of a frequency synthesizer. SSpR values below −70 dB require careful design. Beyond this point spurious suppression is also affected by layout, ground loops, and shielding. Levels as low as −90 to −100 dB have also been claimed [9.9, 9.10].

APPENDIX 9.1 SPECTRAL DENSITY OF SYNTHESIZER OUTPUT PHASE NOISE

Summarizing all the additive system noise components in (9.36) into $\varphi_s(t)$, the variance of the output phase fluctuations is

$$\begin{aligned}\overline{\varphi_{xn}^2(t)} = {} & N_1^2\,\overline{\varphi_{in}^2(t+\tau_1)} + N_2^2\,\overline{\varphi_{in}^2(t+\tau_2)} + \overline{\varphi_s^2(t)} \\ & - 2N_1 N_2\,\overline{\varphi_{in}(t+\tau_1)\,\varphi_{in}(t+\tau_2)} \\ & + 2N_1\,\overline{\varphi_{in}(t+\tau_1)\varphi_s(t)} - 2N_2\,\overline{\varphi_{in}(t+\tau_2)\,\varphi_s(t)} \qquad (9.51)\end{aligned}$$

Because $\varphi_{in}(t)$ and $\varphi_s(t)$ are generated in independent sources, both the last terms in (9.51) are zero. Introducing the autocorrelation according to (9.16) one may write

$$R_{\varphi_{xn}}(0) = [N_1^2 + N_2^2]\, R_{\varphi\,in}(0) - 2N_1\, N_2\, R_{\varphi\,in}(\tau_1 - \tau_2) + R_{\varphi_s}(0) \quad (9.52)$$

The output spectral density can be found, using equation (9.18):

$$S_{\varphi_{xn}}(\omega) = S_{\varphi\,in}(\omega)\, [N_1^2 + N_2^2 - 2N_1\, N_2 \cos\omega(\tau_1 - \tau_2)] + S_{\varphi_s}(\omega) \quad (9.53)$$

Evidently the output spectral density depends upon the differential time delay

$$\tau = |\tau_1 - \tau_2| \quad (9.54)$$

as well as on the offset frequency ω. For $\omega\tau \ll 1$ (9.53) simplifies to

$$S_{\varphi_{xn}}(\omega) \approx S_{\varphi_{in}}(\omega)\, [(N_1 - N_2)^2 + N_1\, N_2\, \omega^2\, \tau^2] + S_{\varphi_s}(\omega) \quad (9.55)$$

whereas for $\omega\tau \to \pi$, 3π, etc. the output spectral density may be enhanced even to

$$S_{\varphi_{xn}}(\omega) \to S_{\varphi_{in}}(\omega)\, (N_1 + N_2)^2 + S_{\varphi_s}(\omega) \quad (9.56)$$

REFERENCES

[9.1] CUCCIA, C. L. *Harmonics, Sidebands, and Transients in Communication Engineering.* New York: McGraw-Hill, 1952.

[9.2] CUTLER, L. S. and SEARLE, C. L. Some aspects of the theory and measurement of frequency fluctuations in frequency standards. *Proc. I.E.E.E.*, **54**, 136-54, 1956.

[9.3] BAGHDADY, E. J., LINCOLN, R. N. and NELIN, B. D. Short-term frequency stability: characterization, theory and measurement. *Proc. I.E.E.E.*, **53**, 704-22, 1965.

[9.4] DAVENPORT, W. B. and ROOT, W. L. *An Introduction to the Theory of Random Signals and Noise*, Ch. 6. New York: McGraw-Hill, 1956.

[9.5] LEESON, D. B. A simple model of feedback oscillator noise spectrum. *Proc. I.E.E.E.*, 54, 329-30, 1966.

[9.6] TYKULSKI, A. Phase noise in frequency synthesizers. *Hewlett-Packard Journal*, **19**, 14-6, Oct. 1967.

[9.7] HEWLETT-PACKARD. HP direct-type frequency synthesizers: theory, performance and use. *Application Note* 96, Palo Alto, 1968.

[9.8] NOYES, A., Jr. Phase noise in 1160-series frequency synthesizers. *General Radio Experimenter*, **41**, 11-15, Jan. 1967.

[9.9] MEYER, D. G. An ultra low noise direct frequency synthesizer. *24th Annual Frequency Control Symposium*, Fort Monmouth, N.J., 1970.

[9.10] MEYER, D. G. A test set for the accurate measurement of phase noise on high-quality signal sources, *I.E.E.E. Transactions on Instrumentation and Measurement*, **IM-19**, 215-27, Nov. 1970.

Name Index

Subject Index